Introduction to Materials Science *and* Engineering

A GUIDED INQUIRY

Introduction to Materials Science *and* Engineering

A GUIDED INQUIRY

Elliot P. Douglas
University of Florida

PEARSON

Upper Saddle River Boston Columbus San Francisco New York Indianapolis
London Toronto Sydney Singapore Tokyo Montreal Dubai Madrid Hong Kong
Mexico City Munich Paris Amsterdam Cape Town

For the students.

Vice President and Editorial Director, ECS: *Marcia J. Horton*
Executive Editor: *Holly Stark*
Editorial Assistant: *Carlin Heinle*
Executive Marketing Manager: *Tim Galligan*
Marketing Assistant: *Jon Bryant*
Senior Managing Editor: *Scott Disanno*
Production Project Manager: *Clare Romeo*
Operations Specialist: *Lisa McDowell*
Cover Designer: *Black Horse Designs*
Cover Photo: *Vincenzo Lombardo/Getty Images*
Manager, Rights and Permissions: *Mike Lackey*
Permissions Project Manager: *Karen Sanatar*
Composition: *GEX Publishing Services*
Full-Service Project Management: *GEX Publishing Services*
Printer/Binder: *Courier Kendallville, Inc.*
Cover Printer: *Lehigh-Phoenix Color*
Typeface: *10/12 Times*

Pearson Education Ltd., *London*
Pearson Education Singapore, Pte. Ltd
Pearson Education Canada, Inc.
Pearson Education—Japan
Pearson Education Australia PTY, Limited
Pearson Education North Asia, Ltd., *Hong Kong*
Pearson Educación de Mexico, S.A. de C.V.
Pearson Education Malaysia, Pte. Ltd.
Pearson Education, Inc., *Upper Saddle River, New Jersey*

Cataloging in Publications Data can be obtained from the Library of Congress.

V011
10 9 8 7 6 5 4 3 2

www.pearsonhighered.com

ISBN-13: 978-0-13-213642-6
ISBN-10: 0-13-213642-2

Contents

Preface to Instructors

This is not a typical textbook. Typical textbooks really serve as reference texts, a place where students can go to find information that they need. This text is designed with a very different goal in mind. It is a learning tool, a book that students use to help them learn. Thus, two very important philosophies underlie it:

1. *Students learn by being actively engaged.* The theory of constructivism states that learning occurs when learners "think about what the teacher tells them and interpret it in terms of their own experiences, beliefs, and knowledge."[1] One practical application of the constructivist approach is through the learning cycle model.[2–4] In this model there are three phases of learning. The first is *exploration*, in which the learner manipulates data or information. This leads to the second phase, *concept invention* or *term introduction*. In this phase the learner uses the data to develop general rules or concepts. Finally, in the *application* phase, the learner applies the concepts developed to new situations. This learning cycle models both the way scientific research is done and the way young children learn about their world. Traditional teaching skips the exploration phase and begins with concept invention. Studies have shown, however, that learning occurs better when the concept invention phase comes later in the sequence[3, 5, 6] and when the learners themselves invent the concepts (rather than being told about them). This approach to learning is the basis for constructivism. In a constructivist approach the roles of the instructor and students are quite different from those in a traditional class.[7] In the approach used in this text, students work together in teams to come to a common understanding of new concepts.
2. *Students do not need to be told every detail about every topic.* Rather, they need to learn the fundamental concepts. For example, this book does not have information about all 14 Bravais lattices. Rather, it focuses on the cubic crystal systems, with the idea that these systems serve as models from which students can learn the important concepts of crystals.

This book can be used in many ways, but was written based on the approach of Process-Oriented Guided Inquiry Learning (POGIL). POGIL was initially developed as a means of teaching general chemistry and has since spread to many other fields. POGIL is used in many different ways, but its basic approach shifts the primary responsibility for learning from the instructor to the student. There are many resources at the POGIL website (http://pogil.org/), including a detailed Instructor's Guide to POGIL (http://pogil.org/resources/implementation/instructors-guide). The preface you are reading now describes how I use it in my 100-student Introduction to Materials class. For more details and other tips on how to implement POGIL in your class, you should get the free Instructor's Guide from the POGIL website.

In my POGIL class I do not lecture. Rather, students work in teams, typically of four students, using the book as their guide. Each section of the text has three primary components: 1) data or information as background material; 2) guided inquiry questions, which are designed to lead the students to understanding the fundamental concepts represented by the data; and 3) application questions (end-of-chapter problems), which provide the students with practice in solving problems using the concepts they have derived. My role is to guide the students, walking around the room and probing them with questions to check their understanding. Farrell et al. have described the roles of students within the groups and the class procedures.[8, 9] Typical roles are Manager (responsible for ensuring that tasks are completed),

Recorder (records the group's answers), Presenter (presents group answers to the class), and Technician (the only person allowed to use a calculator). The typical class period proceeds as follows:

1. I give a brief recap of the previous day or an introduction to that day. Sometimes basic content is provided in a mini-lecture.
2. Students begin working on the day's section of the text.
3. I observe the groups and may interact with them in several ways. I may respond to questions from a particular group or may ask questions of particular members of a group. This latter technique is particularly useful if it appears that one member of a group is lagging behind the others in understanding.
4. If a particular question is causing difficulty for several groups, I may choose to interrupt all groups and have the Presenters from each group discuss their group's answer. In this way, different approaches can be compared and a consensus answer obtained.
5. Concept Checks are provided throughout the class period using student response systems ("clickers"). These multiple-choice questions are an important way to give students feedback on their understanding of the concepts. Without this feedback students often feel "lost" and may think they have not learned anything. If most members of the class get the question correct, we move on to the next section of the text. If a substantial number of students get it wrong, we discuss the possible answers and then try the question again.
6. With about 5 to 10 minutes left in class I stop the activity. I may summarize the day's activities myself, or ask the Presenters to present some aspect of their group's work as a means of providing a summary. Reviewing the learning objectives for that day's section of the text provides a useful means of summarizing the class period and pointing out to the students what they have learned that day.
7. Students are then given a brief period of time to answer a series of questions, such as: *What was the most important thing you learned today? What questions do you still have about the material? What was a strength of your group's performance? What could be improved about your group's performance?* The answers to these questions serve as the basis for the review and introduction for the next day.

This approach has been used successfully in a wide variety of classrooms, from large introductory classes of 300 students to smaller classes of 20. It has many variations, and each instructor must decide what works best for their classroom. Ultimately, however, the focus should be on enabling students to actively engage in the content and discover those concepts for themselves.

Resources for Instructors

MasteringEngineering™. www.masteringengineering.com. The Mastering™ platform is the most effective and widely used online tutorial, homework, and assessment system for the sciences and engineering. Now including Materials Science and Engineering, this online tutorial homework program provides instructors customizable, easy-to-assign, and automatically graded assessments, plus a powerful gradebook for tracking student and class performance.

Pearson eText. The integration of Pearson eText within MasteringEngineering gives students with eTexts easy access to the electronic text when they are logged into MasteringEngineering. Pearson eText pages look exactly like the printed text, offering additional functionality for students and instructors including highlighting, bookmarking, and multiple view formats.

Instructor's Manual. The instructor's manual includes solutions to all of the problems from the book as well as to the Guided Inquiry Activities. The manual also includes suggested guidelines and best practices for implementing Guided Inquiry in class.

Presentation Resources. Photographs, figures, tables, and charts from the book are available as PowerPoint slides. A set of lecture slides written by the author will also be available.

All requests for instructor resources are verified against our customer database and/or through contacting the requestor's institution. Contact your local Pearson Education representative for additional information.

1. Jonassen, D. H. *Computers as Mindtools for Schools, Engaging Critical Thinking,* 2nd ed. Upper Saddle River, NJ: Prentice-Hall, 1996.
2. Lawson, A. E. *Science Teaching and the Development of Thinking*. Belmont, CA: Wadsworth Publishing Company, 1995.
3. Abraham, M. R., and Renner, J. W. "The sequence of learning cycle activities in high school chemistry." *J Res Sci Teach* (1986), 23(2): 121–43.
4. Renner, J.W. et al. "The importance of the form of student acquisition of data in physics learning cycles." *J Res Sci Teach* (1985), 22(4): 303–25.
5. Hall, D. A., and McCurdy, D. W. "A comparison of a biological sciences curriculum study (BSCS) laboratory and a traditional laboratory on student achievement at two private liberal arts colleges." *J Res Sci Teach* (1990), 27(7): 625–36.
6. Renner, J. W., and Paske, W. C. "Comparing two forms of instruction in college physics." *Amer J Phys* (1977), 45(9): 8519.
7. Spencer, J. N. "New directions in teaching chemistry: A philosophical and pedagogical basis." *J Chem Educ* (1999), 76(4): 566–9.
8. Farrell, J. J., Moog, R. S., and Spencer, J. N. "A guided inquiry general chemistry course." *J Chem Educ* (1999), 76(4): 570–4.
9. Hanson, D., and Wolfskill, T. "Process workshops–A new model for instruction." *J Chem Educ* (2000), 77(1): 120–30.

your work...

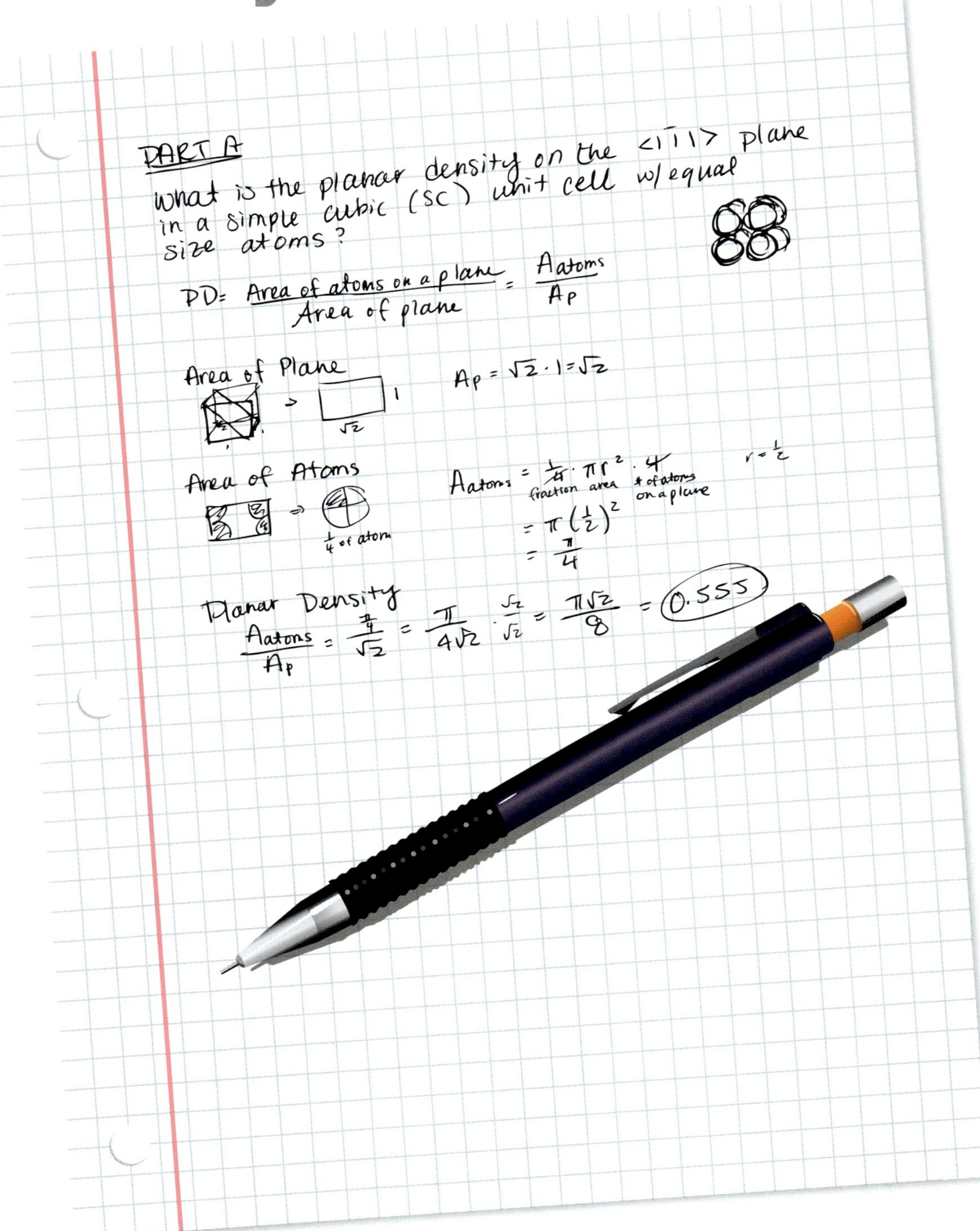
PART A
What is the planar density on the <11̄1> plane
in a simple cubic (SC) unit cell w/ equal
size atoms?
PD= Area of atoms on a plane / Area of plane = Aatoms / Ap
Area of Plane
√2
1
Ap = √2 · 1 = √2
Area of Atoms
1/4 of atom
Aatoms = 1/4 · πr² · 4
fraction area
# of atoms on a plane
r = 1/2
= π(1/2)²
= π/4
Planar Density
Aatoms / Ap = (π/4) / √2 = π / 4√2 · √2/√2 = π√2 / 8 = 0.555

your answer specific feedback

Part A

What is the planar density on the $\langle 1\bar{1}1 \rangle$ plane in a simple cubic (SC) unit cell with equal size atoms?

(Figure 1)

Express your answer numerically.

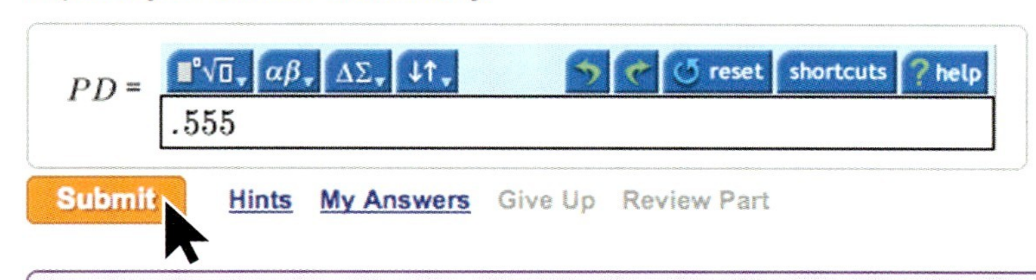

Try Again; 5 attempts remaining

It appears that you calculated the planar density on the wrong plane. Recall that each index that defines the plane is the reciprocal of the intercept.

www.MasteringEngineering®.com

Preface to Students

Teaching and learning are correlative or corresponding processes, as much so as selling and buying. One might as well say he has sold when no one has bought, as to say that he has taught when no one has learned.

John Dewey (1910). *How We Think.* Boston: D.C. Heath and Company, p. 29.

Botany is the study of plants, not the study of books.

Charles E. Bessey (1889). *The Essentials of Botany.* New York: Henry Holt, p. ix.

Take a moment and flip through this book. It looks a lot different from most other textbooks, doesn't it? That is because this is not a typical textbook. This book is based on the concept of *guided inquiry*—an approach to learning based on research on how people think and learn. Cognitive scientists (people who study how we think) have developed models to describe how we learn. A common feature of these models is that we don't learn by having information deposited in our brain. Instead, we learn when we try to figure out how something works, process information as we do that, and compare it to information we already have. A simple example would be how a toddler learns to eat with a spoon. We can't just tell the toddler what to do—the toddler has to play with the spoon and figure out how to make it work. Now think about how scientific discoveries occur—they occur when someone gathers a lot of data and then plays around with it trying to make sense out of it.

The same approach applies to learning in classrooms. It turns out that lectures are not a very good way to learn material. This may seem wrong to you. Because we've gotten used to lectures, many of us believe we learn best from lectures. If you believe this to be true about yourself, think back carefully to how you learned in a lecture class. Did you really understand the material at the moment the professor said it? Or did you need to play around with the material—rewriting your notes, solving the homework, discussing it with your friends—in order to understand it? Research clearly shows that learning improves in a classroom that is active, one in which the students are engaged by doing things. If you want to read a good summary of this research, find the article "Inductive teaching and learning methods: Definitions, comparisons, and research bases" by Michael Prince and Richard Felder (*J Eng Educ* (2006), 95(2): 123–138).

There are many different types of active learning. This text is based on one approach, called *guided inquiry*. Guided inquiry is in many ways like scientific research. You are asked to sort through information to come up with the general concepts. However, it is guided in the sense that you are not left completely on your own. To make sure that you get to the important concepts and don't go off in the wrong direction, you are asked to answer specific questions that have been written to help you sort through the information and get to the right place.

The Guided Inquiry Activities are comprised of two types of guided inquiry questions:

Exploration questions ask you to find information that is already presented in the text, or to give answers from common knowledge. You should not have to go searching other texts or websites to find the answers.

Concept invention questions ask you to use the answers from the exploration questions to figure out a general concept or approach to solving a problem. Again, you should not have to go searching other books or websites to find the answers. Use the answers to the exploration questions to figure these out. Concept invention question are the key questions of each chapter, because this is where the discovery occurs. In a typical lecture class you would be told the answers to these questions. However, if you work to figure them out yourself, you will understand the material better.

There is also a third type of guided inquiry question that is not part of the activities in the middle of the chapters. These are application questions, which give you practice using the concepts. The application questions are the end-of-chapter problems. There are two types of problems: *skill problems* and *conceptual problems*. Skill problems are very similar to the example problems in the chapter, and applying the concepts to them should be fairly straightforward. Conceptual problems ask you to think a bit beyond the example problems. They are still based on the same concepts, but they apply them in different ways from what you have seen in the chapters.

In addition to the guided inquiry questions, this book has several other features that will help you:

Concept Checks: These are questions that test your understanding of key points. Answers to the Concept Checks are in the back of the book, but before looking at the answers you should make sure to try to solve the questions yourself. It's much easier to see how an answer was obtained than to try to figure it out yourself.

Example Problems: Once you go through the questions and figure out the main concepts, you'll see example problems with solutions that can help you understand how to apply the concepts. Sometimes the example problems include hints that will help you when you have to solve similar problems. As with the Concept Checks, you should try to solve problems yourself before you look at the Example Problems for hints.

Application Spotlights: These are brief descriptions of how some of the concepts and ideas are actually used in engineering, with an emphasis on issues related to the environment.

Using this book may take some getting used to. You are probably being asked to think about new material in a way that you haven't been before. Give it a chance—if you embrace the approach and make an effort, you will find in the end that you will have a deeper understanding and that the extra work was worth it.

Acknowledgements

While this textbook was largely written over a three-year period, its origins go back much further. I took the first step on my path to this book when I was hired at the University of Florida in 1996. In my very first semester I gave some guest lectures in the Introduction to Polymers class and asked the students to give me end-of-semester evaluations. The response to my teaching was not positive, with comments such as, "Keep Douglas out of the classroom." I taught the Introduction to Materials course for the first time the following semester, spring of 1997, with results that were not much better. After struggling for a few semesters I was fortunate to be accepted to the Teaching Teachers to Teach Engineering (T4E) program at West Point in 1998, now known as the ExCEEd Teaching Workshop run by the American Society of Civil Engineers. At this program I learned how to be an effective lecturer, including my first introduction to some basic active learning techniques. I was fortunate to be invited back as a facilitator to help other faculty for the next eight years. Over that time many people gave me advice, too many to mention them all, but I would specifically like to thank Col. Stephen Ressler and Ltc. Jim O'Brien (ret) for their guidance and mentorship.

By the early 2000s I was growing dissatisfied with lectures and was looking for ways to more fully involve my students in their learning. At the American Chemical Society National Meeting in 2003 I happened onto a symposium on POGIL and was intrigued. I investigated it further and since I was teaching a freshman materials chemistry class the following spring, decided to use the existing POGIL materials for general chemistry in that class. I also requested that an experienced POGIL practitioner visit my classroom and host a workshop, mainly so I could learn more about POGIL. After that one-semester experiment that particular class was dropped from our curriculum (not because I had used POGIL!). The only choice remaining to me was to begin to write my own activities for my Introduction to Materials class. The first one I wrote was on mixtures, which eventually became Section 6.1 of this book.

A number of people associated with the POGIL Project supported and encouraged me from those first days. Again, there are too many to mention, but I would specifically like to thank Rick Moog of Franklin & Marshall College, Director of POGIL, and Frank Creegan, retired from Washington College. Also, critical support was provided by the National Science Foundation, who provided two grants for the initial development and testing of the materials that became this textbook.

This book would not have been possible without the vision of the staff at Pearson. Holly Stark, Executive Editor, had the vision to see the potential of moving beyond the traditional textbook to a new kind of learning. Without her support, both to me and within Pearson, this book would not have been possible. Clare Romeo, my production project manager, turned the manuscript into something that looks like a real textbook and that reflects the guided inquiry philosophy. Many others at Pearson whom I don't know had an integral hand in creating the book you are holding, and my appreciation goes out to those unsung and unknown contributors.

I used various versions of this manuscript over the course of seven years. The students in those classes suffered through mistakes in the manuscript, hand-drawn figures, and my fumbling attempts at using guided inquiry. To them I apologize and assure them that the input they provided, no matter how harsh it seemed to me at the time, only served to make this text better and to improve the experience of future students.

I also want to thank the reviewers who found mistakes and provided suggestions on how to improve the book: Richard Hennig, Cornell University; Patrick Ferro, Gonzaga University; David Bahr, Washington State University; Debbie Chachra, Olin College; Amy Moll, Boise State; Satya Shivkumar, Worcester Polytechnic Institute; Cindy Waters, NCA&T State University; Mark Weaver, University of Alabama; Lia Stanciu, Purdue University; Trevor Harding, California Polytechnic State University; Doug Irving, North Carolina State University; Stephen Krause, Arizona State; Jerry Floro, University of Virginia; Eunice Yang, University of Pittsburgh.

Finally, I want to thank my wife Heidi. While many authors thank their wives for support, Heidi played an active role in making this text what it is. She actively pushed me to write and publish this book, even before I started working with Pearson. During the writing production she made specific suggestions to improve it, and if you look carefully you will see she is credited with providing two of the photos (which are pictures of our two children, Spencer and Gracie). This text would not be what it is without her.

Elliot Douglas

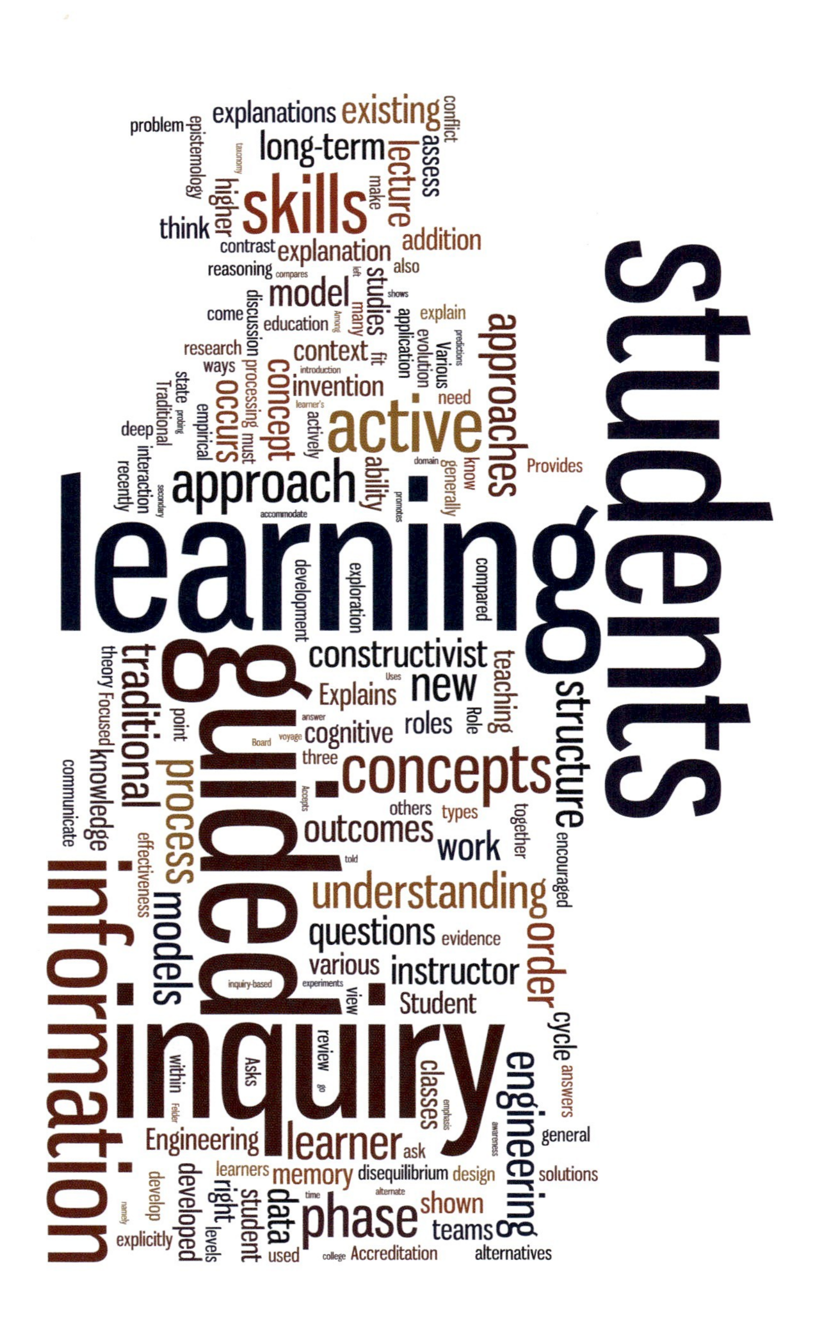
students
learning
guided
inquiry
information
active
skills
approach
approaches
concepts
understanding
constructivist
traditional
process
models
questions
instructor
engineering
learner
phase
outcomes
order
structure
new
explanations
existing
long-term
lecture
assess
explanation
addition
model
context
invention
concept
occurs
cognitive
roles
work
Explains
teaching
various
Student
Engineering
memory
data
developed
student
disequilibrium
design
shown
teams
solutions
general
alternatives
Accreditation
explicitly
develop
learners
right
levels
used
classes
Asks
review
within
cycle
answers
evidence
view
types
others
together
encouraged
three
point
knowledge
communicate
theory
Focused
effectiveness
compared
exploration
development
ability
generally
know
Provides
need
Various
evolution
application
explain
fit
many
studies
education
discussion
come
research
ways
processing
state
Traditional
empirical
deep
interaction
recently
must
actively
think
higher
epistemology
problem
contrast
reasoning
also
make
conflict
ask

What Is Guided Inquiry?

Take a moment and flip through the pages in this book. You should notice that this text is very different from texts you have used in other classes. What you probably see is that there are a lot of questions for you to answer throughout the chapters, not just problems to solve at the end of each chapter. As you use this text you will notice some other important differences. One big difference is that not all the information you need to learn is given to you. The questions, which are called Guided Inquiry (GI) questions, are used to help you discover the concepts and ideas for yourself, instead of just reading about them. This approach, in which you figure out the information you need to know, is called active learning. Although active learning may not be something you have experienced before, a lot of research has shown that it works better than just being lectured to.

Since this text is based on active learning, you are not going to be told how it works or why it might be better than lectures. Instead, you will go through an active learning exercise to figure this out for yourself. We will begin with an activity on the First Law of Thermodynamics. This topic should already be somewhat familiar to you, so it provides a good way to compare how you will learn in this class compared to a typical lecture class. By the end of this chapter you will:

- Understand the First Law of Thermodynamics.
- Understand what is meant by guided inquiry.
- Understand the differences between a traditional lecture class and a guided inquiry class.

1.1 First Law of Thermodynamics

LEARN TO: Define heat and work.
Explain the First Law of Thermodynamics.

Thermodynamics is the study of changes in energy and flow of heat. Although thermodynamics is heavily mathematical, it also has a lot of practical applications. We can use it to understand the behavior of engines, power plants, and materials. In reality, much of science and engineering would not be possible without an understanding of thermodynamics.

Much of thermodynamics can be explained by three basic laws. We can't actually prove that any of these laws are true, but we have never found anything that contradicts them. We will begin our examination of the first law by looking at how temperature changes when two gases at different temperatures are put into contact with each other. **You should answer the first set of questions based on Figure 1.1.1.**

Figure **1.1.1**

Containers of two gases at different temperatures in contact with each other. The initial and final temperatures of the gases are shown.

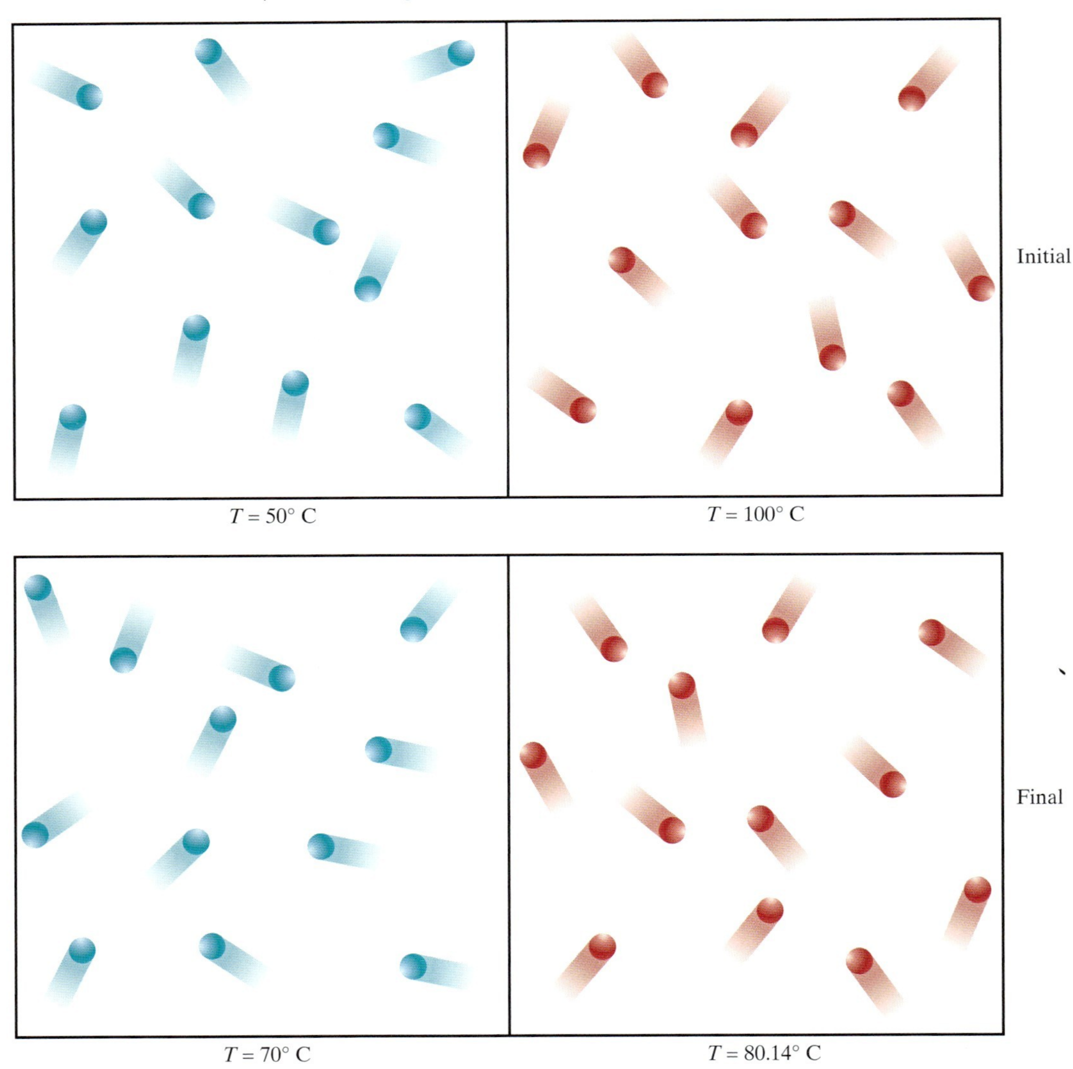

Guided Inquiry: Heat

1.1.1 What is the initial temperature of the cold gas? What is the initial temperature of the hot gas?

Work as a group to answer questions 1.1.1–1.1.9 and make sure everyone agrees on each answer before moving on to the next question.

1.1.2 What is the final temperature of the cold gas? What is the final temperature of the hot gas?

1.1.3 In which direction does energy flow in this system?

1.1.4 What is the initial energy of the cold gas? What is the final energy of the cold gas?

The energy of a molecule of a gas can be calculated by $E = {}^{3}\!/_{2}\,kT$*, where* k *is Boltzmann's constant and* T *is the temperature in Kelvin.*

1.1.5 What is the initial energy of the hot gas? What is the final energy of the hot gas?

1.1.6 What is the energy change of the hot gas? What is the energy change of the cold gas?

1.1.7 How much energy was transferred between the hot gas and the cold gas?

1.1.8 What is the total energy change of the system?

1.1.9 Write an equation relating energy loss and gain for the gases.

This is a concept invention question, where you have discovered something about thermodynamics.

Concept Check **1.1.1**

- A hot piece of copper is placed in contact with a cold brick. If the copper loses 5 calories of energy, how much energy does the brick gain?

The first set of questions only considered thermal energy. When there is a temperature difference, the energy that is transferred is called "heat." However, other kinds of energy can be transferred. When energy is transferred without a difference in temperature, we call that energy transfer "work." **In the next set of questions you will examine the relationship between heat and work.**

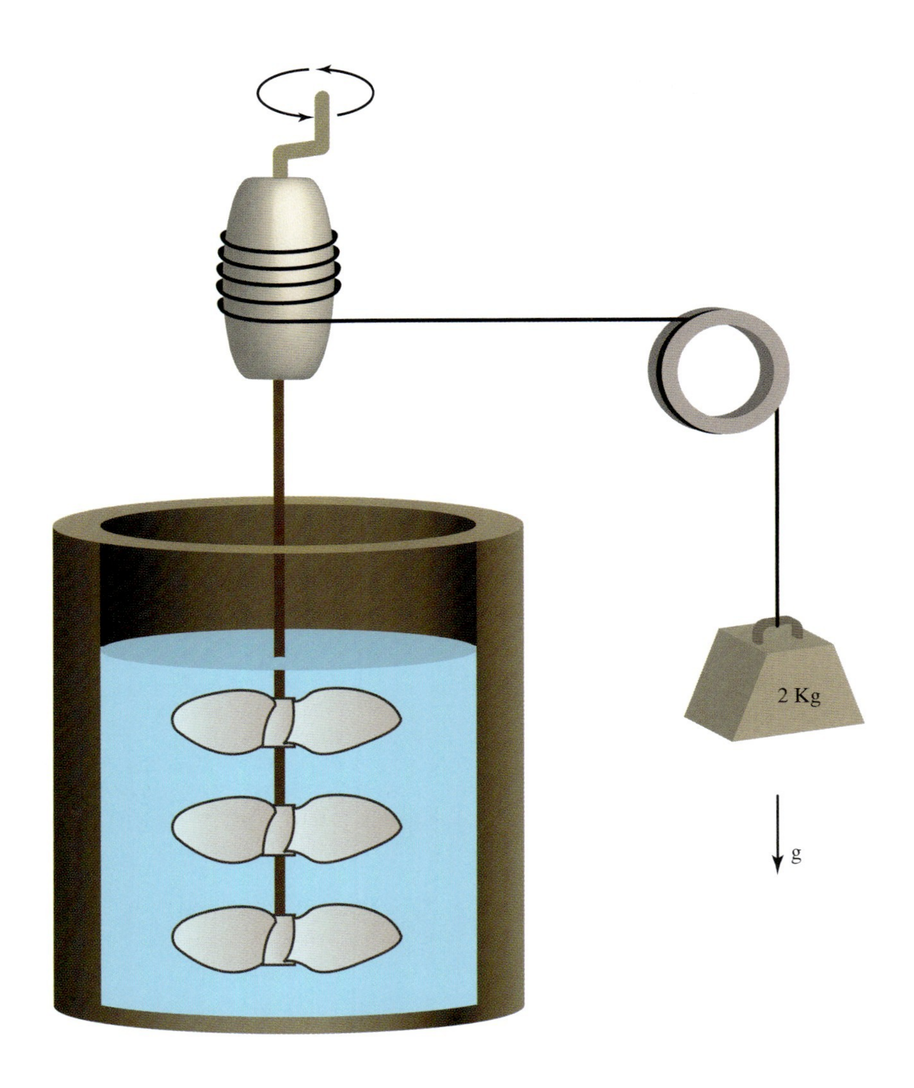

Figure 1.1.2

Illustration of James Joule's experiment on heat and work.

Guided Inquiry: Work

1.1.10 What is the mass of the weight hanging off the pulley in Figure 1.1.2?

1.1.11 If the mass drops, what happens to the paddles in the water?

1.1.12 The mass drops 5 cm. How much does its potential energy change?
Remember, the potential energy of an object is mgh, *where* m *is the mass,* g *is* 9.8 m/s², *and* h *is the height. When* m *is in* kg *and* h *is in meters, the potential energy is in joules.*

1.1.13 When the mass drops 5 cm, how much work is done by the paddles?

1.1.14 When the mass drops, the temperature of the water increases. Where did the energy come from to increase the temperature of the water?

1.1.15 How much energy was transferred to the water to raise its temperature?

1.1.16 Write an equation relating the work done by the mass and the energy change of the water.

1.1.17 The apparatus in Figure 1.1.2 is modified so that the container of water is placed on top of a hot brick. As the mass falls 5 cm, 1.2 J of heat are transferred from the brick to the water. What is the total change in energy of the water?

1.1.18 Write an equation that relates the work done by the mass, the heat transferred from the brick, and the energy change of the water.
The equation you have created is the First Law of Thermodynamics.

1.1.19 Describe the First Law of Thermodynamics in words.

Concept Check 1.1.2

- **According to the First Law of Thermodynamics, could heat be transferred from a cold object to a hot object, resulting in the hot object getting hotter and the cold object getting colder?**

1.2 Active Learning

LEARN TO: Describe the procedures of a guided inquiry class.
Compare the advantages and disadvantages of a traditional lecture class and a guided inquiry class.

Now that you have tried out an active learning exercise, let's think about how the active learning class works. Since you should have learned about the First Law of Thermodynamics in a previous chemistry or physics class, we're going to compare that class with the activity you just finished. Take a moment and remember what that class was like.

Figure 1.2.1 shows two ways of organizing a class. In Class 1, most or all of the time is spent by the instructor explaining concepts to the students and showing them practice problems. Students then practice using that information on homework problems. This type of class would be considered a lecture class. In Class 2, most or all of the time is spent by the students figuring out the concepts from information that is provided to them. There may be an occasional short mini-lecture or discussion. Class 2 is an active learning classroom, which means that the students are asked to be active during the class time. There are many different versions of active learning. And of course a class can be a mixture of a lecture class and an active learning class—you may have a lecture class in which the instructor occasionally has you work on something, or an instructor might mix some days of active learning and some days of lecture.

There is research that compares the effectiveness of lecture and active learning classrooms, but for most students what is more important is their own experience. **In the next set of questions you will reflect on your experience in the first part of this chapter and think about how this approach might affect your ability to learn.**

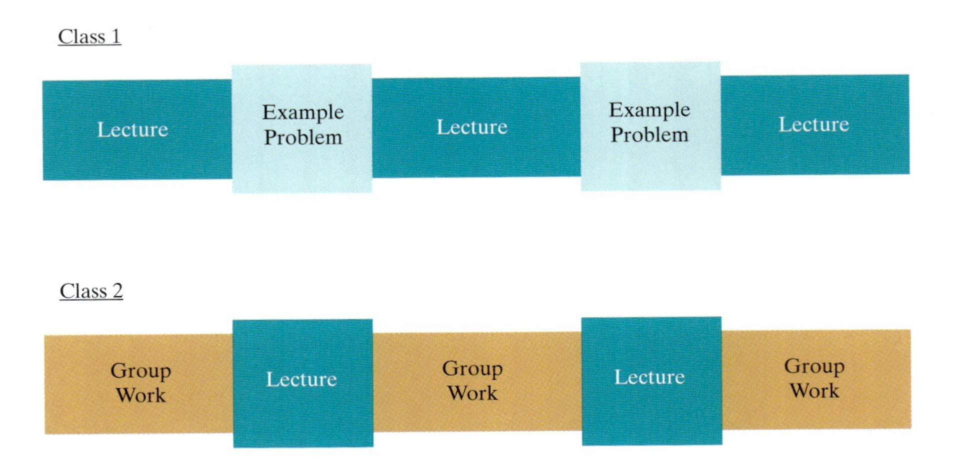

Figure 1.2.1 Schematic showing two ways of organizing a class.

Guided Inquiry: Active Learning

1.2.1 In Class 1, who is most responsible for explaining concepts? Who is responsible in Class 2?

1.2.2 Which type of class is more like scientific research?

1.2.3 Thinking back on your general chemistry or physics class where you first learned about thermodynamics, was that more like Class 1 or Class 2? Explain why.

1.2.4 Were questions 1.1.1–1.1.19 of this chapter more like Class 1 or Class 2? Explain why.

1.2.5 List three advantages and three disadvantages of Class 1.

1.2.6 List three advantages and three disadvantages of Class 2.

1.2.7 Which type of class do you think would be most effective for learning? Explain why.

Summary

While hopefully you learned something about the First Law of Thermodynamics, the primary goal of this chapter was for you to experience a guided inquiry lesson and compare it to a traditional lecture class. Neither approach is perfect, and you likely identified advantages and disadvantages for each. But on balance, active learning classes lead to better student learning and improvements in thinking skills. Be forewarned, however, that you may not be told all the answers by your instructor. The book does provide many opportunities for you to make sure you are on the right track: Concept Checks, Example Problems, and Mastering Engineering (the online homework system). Also, sometimes the way a Guided Inquiry question is asked will make you realize that you got a previous Guided Inquiry question wrong.

To be successful using this book you will need to accept a different way of learning. For example, in this chapter, you were never told what the First Law of Thermodynamics is, but you were able to figure it out. This is the way real-world engineering works, so think of this book as not just a way to learn materials science and engineering, but also as a way to learn how to be an engineer.

Smartphones rely on the use of all types of materials, from the ceramic glass screen, to the metal casing, to plastics used as parts of the various electronic components. (3Dstock/Shutterstock)

2 What Is Materials Science and Engineering?

Materials science and engineering (MSE) deals with the stuff things are made out of. Since everything is made out of something, materials engineering touches every aspect of our lives. The importance of materials can be seen by how we name prehistoric times in human history; Stone Age, Bronze Age, Iron Age. Development of new materials throughout human history has allowed civilization to advance because these new materials have allowed humans to do things they couldn't do before. For example, the development of iron led to more durable tools for farming and more effective weapons.

The goal of this chapter is to provide you with an introduction to the basic idea of materials science and engineering: structure–property–processing relationships. Everything we do in materials science is related to how these elements interact with each other. By the end of this chapter you will:

Understand how we classify materials and how they differ from each other in their properties and uses.

Understand how the elements of structure, properties, and processing relate to each other.

2.1 Types of Materials

LEARN TO: Classify materials into different types.
Identify how different materials are used.

There are many different ways to classify materials and many different names people use; nanomaterials, biomaterials, functional materials, self-assembled materials, biomimetic materials, etc. But fundamentally we can say that, based on structure (that is, the type of bonding present), there are three types of materials: metals, polymers (which include plastics and rubbers), and ceramics. **The next set of questions will get you to think about what distinguishes each of those types of materials and how they are used.**

Guided Inquiry: Types of Materials

2.1.1 List three examples each of a metal, a polymer, and a ceramic.
Make sure to discuss these questions together in your group.

2.1.2 List three unique properties each for metals, polymers, and ceramics.

2.1.3 List three applications that are uniquely suited for each type of material.

Concept Check **2.1.1**

- **What type of material would be best suited for the thermal insulation of a high-temperature furnace?**
- **What type of material would be best suited for lightweight air freight containers?**

Another way to classify materials is based on function—that is, some unique properties it has or how it is used. **In the next set of questions, you will think about the different properties needed for various applications.**

Guided Inquiry: Applications of Materials

2.1.4 List three properties that are important for a material that will be used for a bridge.

2.1.5 List three properties that are important for a material that will be used as an internal fixation plate for a broken bone (see Figure 2.1.1).

2.1.6 List three properties that are important for a material that will be used as a computer chip.

Concept Check **2.1.2**

- **What properties are important for a material that will be used for a trash can?**

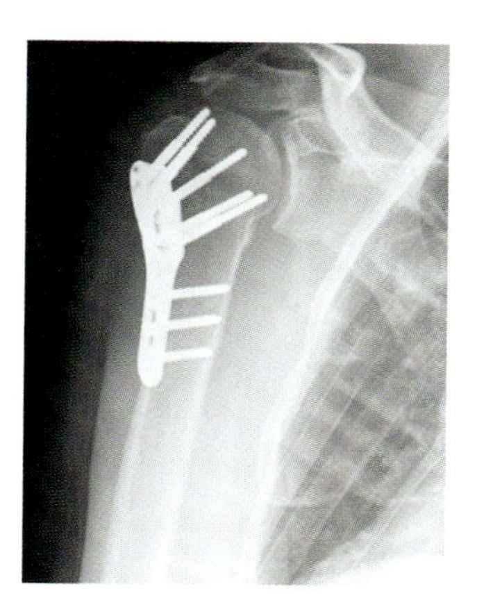

Figure 2.1.1

X-ray image of an internal fixation plate that was implanted in the author when he broke his arm in a cycling accident.

Materials whose primary function is to provide mechanical strength or stiffness in an engineered structure are called *structural materials*. Materials used in the body are called *biomaterials*. Materials used specifically for their electronic properties are called *electronic materials*. There are many other possible functional categories, and you can probably think of several. For example, there are magnetic materials which are used for their magnetic properties; high temperature materials which are important for applications such as jet engines; nuclear materials, which are used in nuclear reactors; and many others. Obviously one material can be in more than one of these functional categories. For example, stainless steel is used both as a structural material in applications such as pressure vessels and as a biomaterial for metal plates and screws for bone repair. The important point to keep in mind is that one material can have many different uses. We select materials for these uses based on the particular advantages and disadvantages they have in those particular applications. In the next set of questions, you will have a chance to decide for yourself how to classify different material, using the examples shown in Figure 2.1.2. **For these questions just consider the three categories of structural materials, biomaterials, and electronic materials.**

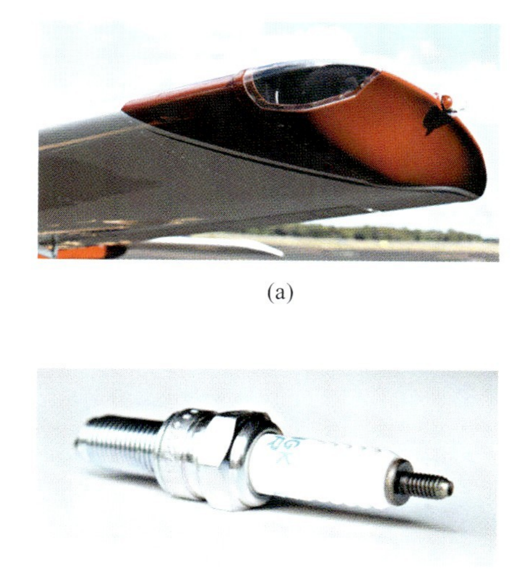

(a)

(c)

(b)

Figure 2.1.2

Examples of materials applications: (a) airplane wing; (b) soda bottle; (c) spark plug; the white piece is the insulating component.

(Photo courtesy of Lifeprints Photography)

Guided Inquiry: Classification of Materials

2.1.7 For each item in Figure 2.1.2, identify the type of material based on its structure.

2.1.8 For each item in Figure 2.1.2, identify the type of material based on its function.

2.1.9 For each item in Figure 2.1.2, identify what you think are the unique properties of the material that make it especially suitable for that application.

2.1.10 Many of the ages of human civilization are named after the primary materials that were used (Stone Age, Bronze Age, etc.). What materials age do you think we are we in right now?
There is no single correct answer to this question. Think about the different ways materials are used.

Concept Check **2.1.3**

- **What types of materials can be used as electronic materials?**
- **What types of functions can a metal be used for?**

Application Spotlight **Nanotechnology**

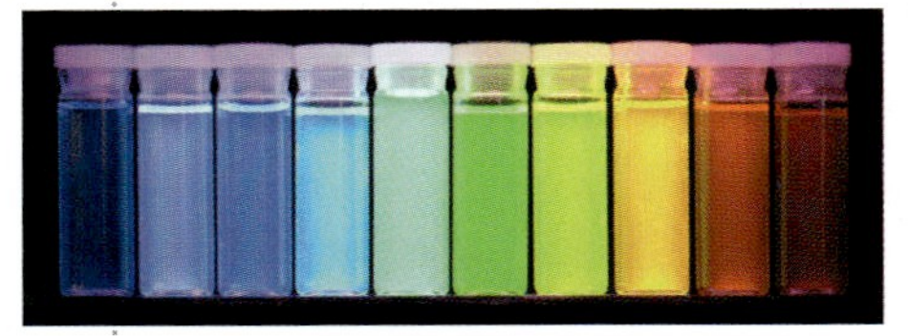

(Shuming Nie/National Institute of General Medical Sciences)

We often hear about nanotechnology, but what is it? Although many people think of miniature devices when they hear the term "nano," nanotechnology actually refers to the materials that are used. We can define nanotechnology as the use of materials that have some dimension on the order of nanometers and that have unique properties because of that small size. The figure shows one example of a size-dependent property. "Quantum dots" of semiconductors that are less than 100 nm in diameter will fluoresce at different wavelengths depending on the size of the nanoparticles. Larger particles emit light of longer wavelengths, resulting in red emission. As the particles get smaller, the wavelength of the emitted light gets shorter. These types of particles are being used in photovoltaics for solar energy applications. This is just one example of how materials are used in nanotechnology. Other applications for nanomaterials include new methods of delivering drugs, ultrasensitive pollution detectors, and novel types of electronic devices for computers.

2.2 The MSE Triangle

LEARN TO: Describe the relationship among structure, properties, and processing.

In the previous section you looked at different kinds of materials and how they are used. But materials science and engineering is more than that. Materials scientists and engineers have a particular way of looking at the world. If you want to understand how to think like a materials scientist, you need to understand the relationship among structure, properties, and processing. In fact, that is what this whole text is about: giving you examples of how these elements are related to each other. But first, what do we mean by these three terms? Here are some definitions:

Structure: By structure, we roughly mean the arrangement of atoms in the material. This arrangement can be considered at many different levels. We could look at how atoms are connected together (Chapter 3), how the atoms are lined up (Chapter 4), or how they come together to form microscopic particles inside a material (Chapter 6).

Properties: Many different kinds of properties are of interest. The first ones we usually think of are mechanical properties, such as stiffness, strength, and toughness (Chapter 9). But also important are how materials interact with the environment (Chapter 10), electrical properties (Chapter 11), and thermal properties (Chapter 12).

Processing: Processing is what we do to a material as we make it. This usually involves heat (Chapters 7 and 8), but could also include mechanical deformation, such as stretching.

In order to understand how these three things are related, here are some results of experiments on a bobby pin, which is made out of plain carbon steel:

1. When you buy a bobby pin, you can bend it and it will not break.
2. If you take a bobby pin and heat it in a propane torch then dunk it into water, once it cools to room temperature, it will break easily if you bend it.
3. If you heat and then dunk the bobby pin as in experiment 2, but then heat it in a candle flame before you bend it, it will once again no longer break when you bend it.

Here is some additional data that will be helpful to you:

Composition of plain carbon steel: Approximately 99 wt% Fe and 1 wt% C.

Melting temperature of plain carbon steel: Approximately 1450° C.

Temperature in the flame of a propane torch: Approximately 1300° C.

Temperature in the flame of a candle: Approximately 1000° C.

Guided Inquiry: Structure-Properties-Processing

2.2.1 List the processing steps for each of the experiments. Provide specific temperatures.
Processing can mean heating and cooling the material, or doing something to change its shape.

2.2.2 Describe the properties of the bobby pin for each experiment before and after processing.

2.2.3 What is the composition of the steel before processing?

2.2.4 Has the composition of the steel changed as a result of the processing? Explain why or why not.

2.2.5 What do you think has changed in the steel to cause the change in the properties? How was that change created?
Did any atoms get boiled away to change the composition? If the composition didn't change, what else might have happened?

Concept Check 2.2.1

- **After experiment 3, you again heat the bobby pin in the propane torch and dunk it into water. What will the properties be now?**

As you go through the text you will learn the information needed to understand specifically what has happened to the structure of the steel and how that affects the properties. For now, what is important is the relationship among structure, properties, and processing as demonstrated in this example. Figure 2.2.1 demonstrates this relationship. The MSE triangle shows that structure, properties, and processing all affect each other. For example, if we want certain properties, we can understand what structure we need to get those properties, and what processing technique will provide that structure. Always keep this triangle in mind because this is what defines materials science and engineering.

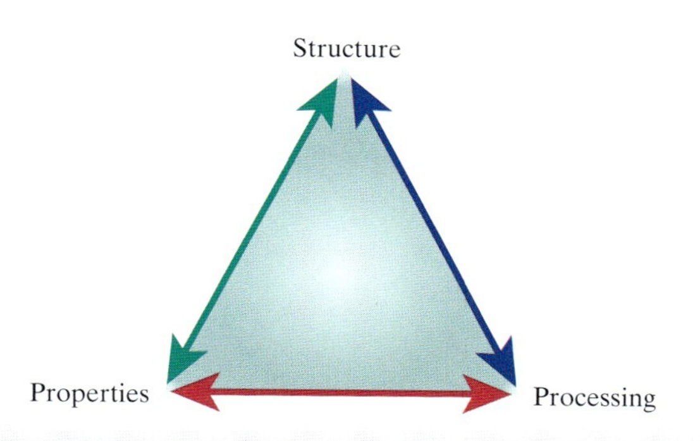

Figure 2.2.1
The MSE triangle.

Summary

The goal of this chapter was to introduce you to the different ways we think of materials. As you have seen, we can classify materials in two ways: based on their structure (ceramics, polymers, metals) or based on their function (structural materials, biomaterials, etc.). The same type of material can be used in many different ways, so it is important to understand the potential advantages and disadvantages of each material when choosing it for an application.

One major difference between the field of materials science and engineering and other engineering fields that use materials is the MSE triangle. A civil engineer designing a bridge needs to select a material that will have the needed strength. To do that, she will calculate the strength and then look up materials properties to find one that works. In contrast, a materials engineer doesn't just accept properties as given. As shown with the bobby pin example, a materials engineer understands how to control the properties of materials by using processing to change the structure. This chapter, and specifically Figure 2.2.1, has provided you with the overall theme of materials science and engineering that will continue throughout the book.

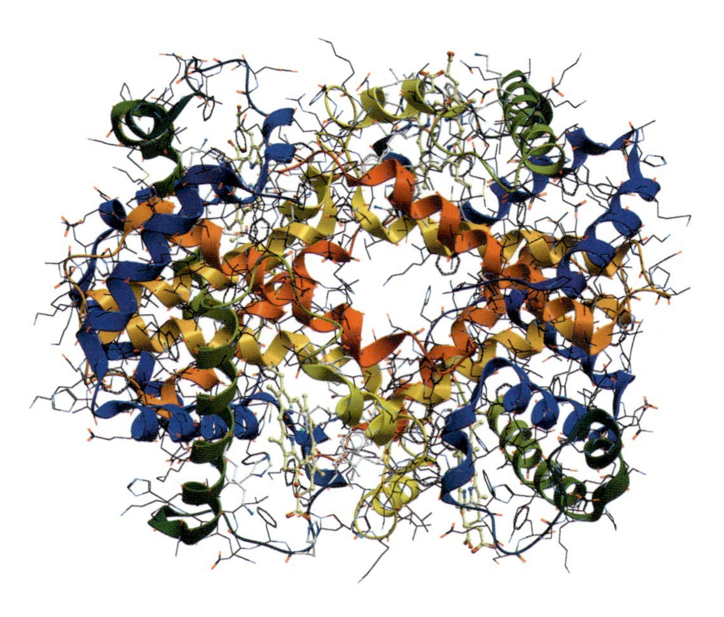

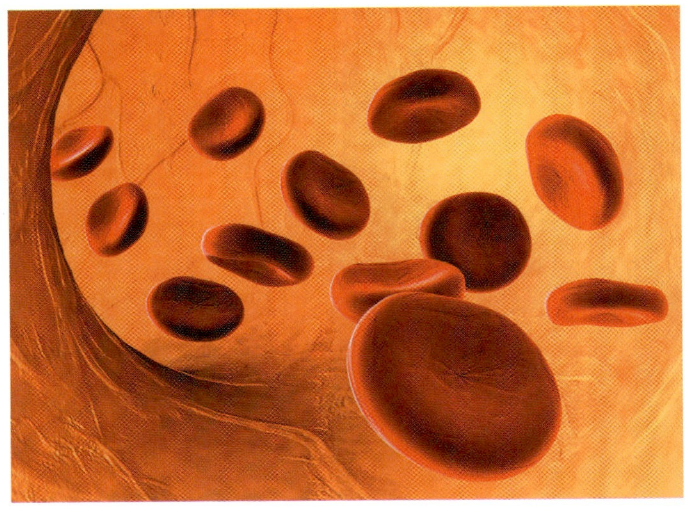

Top: Hemoglobin relies on nonbonding interactions to bind oxygen in red blood cells (bottom). If those nonbonding interactions are disrupted hemoglobin can no longer function properly. [Top (Leonid Andronov/Shutterstock); (bottom) (Shilova Ekaterina/Shutterstock)]

Bonding

This chapter begins our exploration of the structure of materials, the first element in the MSE triangle. The first level of structure we need to consider is the types of bonds that form between atoms. You probably learned about bonds in chemistry. For materials scientists, the importance of bonds is the way in which they affect the properties of materials. Some of how bonding affects properties will be evident in this chapter, but we will also come back to it in later chapters. By the end of this chapter you will:

Understand the different types of bonds and nonbonding interactions present in materials.

Be able to predict the types of bonds and nonbonding interactions present in a material.

Be able to predict properties of materials based on the types of bonds and nonbonding interactions that are present.

3.1 Electronegativity

LEARN TO: Define electronegativity.
Predict the distribution of electrons in a bond.

Electronegativity is an atomic property that describes how tightly an atom holds onto its electrons. An atom with a higher electronegativity has a stronger pull on electrons than an atom with a lower electronegativity. Many different reasons have been given for this difference between atoms. Possible explanations have included that it is related to the ionization energy (how strongly an atom holds on to its own electrons), the charge at the "surface" of an atom, and the average energy of the valence electrons. Unfortunately it is not possible to measure electronegativity directly, so to get electronegativity values researchers need to measure something else and come up with a way to relate that measurement to electronegativity. This means that many different electronegativity scales are used. The most common one is the first one that was created, the Pauling scale, developed by Linus Pauling in 1932. He used values of the energy required to break bonds to derive a set of electronegativities. His approach requires that a reference value be defined to set the values of all the other elements. Pauling assigned hydrogen an electronegativity of 2.1, and values of all the other elements are derived from that.

Although electronegativity values might seem arbitrary, they are very useful to describe how electrons are distributed between two atoms in a chemical bond. If two atoms come together to form a bond, the atom with higher electronegativity will attract the electrons away from the atom with lower electronegativity. The degree to which the electrons are attracted to the more electronegative atoms depends on the difference in electronegativities. Figure 3.1.1 is a periodic table showing the Pauling electronegativities of the elements. You will now use this table to answer questions about the electronegativity of atoms; you will learn how electronegativity affects the way two atoms interact.

Figure 3.1.1

Periodic table showing the Pauling electronegativities.

IA	IIA	IIIB	IVB	VB	VIB	VIIB	VIII	VIII	VIII	IB	IIB	IIIA	IVA	VA	VIA	VIIA	0
1 H 2.1																	2 He –
3 Li 1.0	4 Be 1.5											5 B 2.0	6 C 2.5	7 N 3.0	8 O 3.5	9 F 4.0	10 Ne –
11 Na 0.9	12 Mg 1.2											13 Al 1.5	14 Si 1.8	15 P 2.1	16 S 2.5	17 Cl 3.0	18 Ar –
19 K 0.8	20 Ca 1.0	21 Sc 1.3	22 Ti 1.5	23 V 1.6	24 Cr 1.6	25 Mn 1.5	26 Fe 1.8	27 Co 1.8	28 Ni 1.8	29 Cu 1.9	30 Zn 1.6	31 Ga 1.6	32 Ge 1.8	33 As 2.0	34 Se 2.4	35 Br 2.8	36 Kr –
37 Rb 0.8	38 Sr 1.0	39 Y 1.2	40 Zr 1.4	41 Nb 1.6	42 Mo 1.8	43 Tc 1.9	44 Ru 2.2	45 Rh 2.2	46 Pd 2.2	47 Ag 1.9	48 Cd 1.7	49 In 1.7	50 Sn 1.8	51 Sb 1.9	52 Te 2.1	53 I 2.5	54 Xe –
55 Cs 0.7	56 Ba 0.9	57–71 La–Lu 1.1–1.7	72 Hf 1.3	73 Ta 1.5	74 W 1.7	75 Re 1.9	76 Os 2.2	77 Ir 2.2	78 Pt 2.2	79 Au 2.4	80 Hg 1.9	81 Tl 1.8	82 Pb 1.8	83 Bi 1.9	84 Po 2.0	85 At 2.2	86 Rn –
87 Fr 0.7	88 Ra 0.9	89–102 Ac–No 1.1–1.7															

Source: Data from Pauling, L., *The Nature of the Chemical Bond*, 3rd ed., Cornell University Press, 1960, p. 93.

Guided Inquiry: Electronegativity

3.1.1 What is the electronegativity of carbon?
Even though some of these questions may seem easy, make sure everyone in your group has the same answer before moving on.

3.1.2 What is the electronegativity of sodium?

3.1.3 What element has the highest electronegativity?

3.1.4 What element has the lowest electronegativity?

3.1.5 Describe the trends in electronegativity on the periodic table.

3.1.6 If a bond is formed between sodium and chlorine, which atom will tend to attract the electrons more?

3.1.7 If a bond is formed between carbon and chlorine, which atom will tend to attract the electrons more?

3.1.8 Is the distribution of electrons in the sodium–chlorine bond and the carbon–chlorine bond the same? For each case, describe how the electrons that make up the bond are distributed between the two atoms.
Remember, the degree to which the electrons are attracted to the more electronegative atom depends on the difference in electronegativities.

3.1.9 How are the electrons distributed in a bond between two carbon atoms?

Concept Check 3.1.1

- In which of the following bonds are the electrons most strongly pulled to one of the atoms?
  - Fe–Br
  - Br–Br
  - K–Br
  - C–Br
- In which of the following bonds are the electrons distributed evenly between the two atoms?
  - Fe–Br
  - Br–Br
  - K–Br
  - C–Br
  - None

3.2 Primary Bonds

LEARN TO: Predict the type of bonds formed for different atoms.
Predict properties from knowledge of bonding.

Materials are not made of individual atoms, they are made of groups of atoms. Sometimes these are groups of all the same kind of atom (like a piece of iron), and sometimes these are groups of different kinds of atoms (like polyethylene, which is made of carbon and hydrogen). While this is probably obvious, it is important because the bonds that form between the atoms have a significant influence on the material's properties. For example, melting point, stiffness, and thermal expansion are all directly related to the strength of the bonds. This relationship between structure and properties is one of the major themes of materials science and engineering, so understanding the types of bonds that are present in different materials is very important.

One way to define chemical compounds is that they are groups of atoms held together by *primary bonds*. Primary bonds involve the sharing of electrons between different atoms. Electrons are shared to satisfy the octet rule. This rule states that atoms are most stable when they have a filled valence shell, which for many atoms means 8 electrons. (One important exception is hydrogen, which has 2 electrons in its filled valence shell.) There are three basic types of primary bonds, each of them formed by different ways of sharing the electrons to satisfy the octet rule. These three types are ionic, covalent, and metallic bonds.

Schematic illustrations of each of these bonds are shown in Figures 3.2.1–3.2.3, and the strengths of the different types of primary bonds are shown in Table 3.2.1. Electrons are shared between atoms in each of these bonds as follows:

Ionic bond: Bond formed when an electron is transferred from one atom to another to satisfy the octet rule for each of them, resulting in positive and negative ions. These ions are then attracted to each other through electrostatic interactions. As shown in Table 3.2.1, ionic bonds are the strongest bonds, and the melting temperatures of ionic compounds are the highest of any material.

Covalent bond: Bond formed when two atoms share electrons equally. These atoms are then bound together because the octet rule is satisfied only while the shared electrons are considered to be in the valence shell of both atoms. Covalent bonds are considered strong, but are generally weaker than ionic bonds.

Metallic bond: Bond formed when a group of atoms contribute their valence electrons to form a "sea of electrons" around the atoms. The octet rule is satisfied on average for all the atoms. Metallic bonds are considered to be the weakest of the primary bonds, although some metals can form bonds that are stronger than covalent bonds.

You should be aware that these descriptions of the bonds are only general ones that do not capture all the intricacies. A complete description of bonding requires the use of quantum mechanics, which is at a level beyond what you need to know for this book.

Figure 3.2.1

Schematic diagram of ionic bonding.

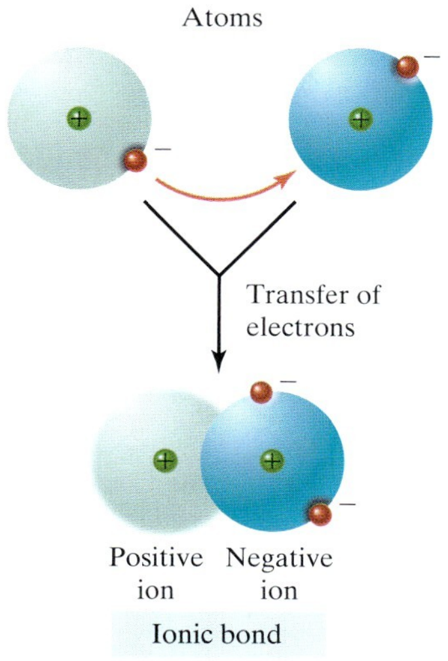

Atoms

Transfer of electrons

Positive ion

Negative ion

Ionic bond

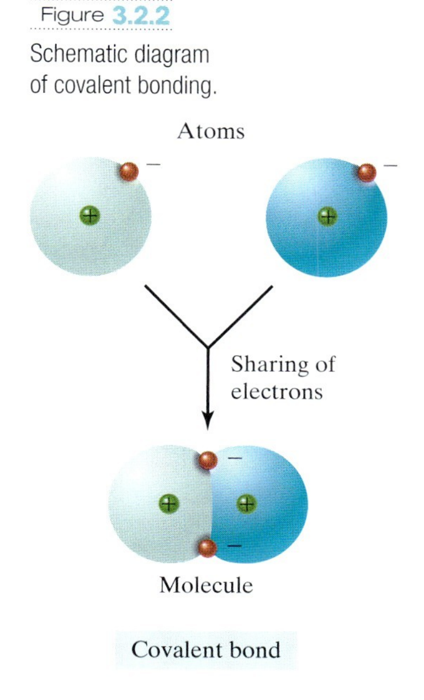

Figure 3.2.2

Schematic diagram of covalent bonding.

Figure 3.2.3

Schematic diagram of metallic bonding.

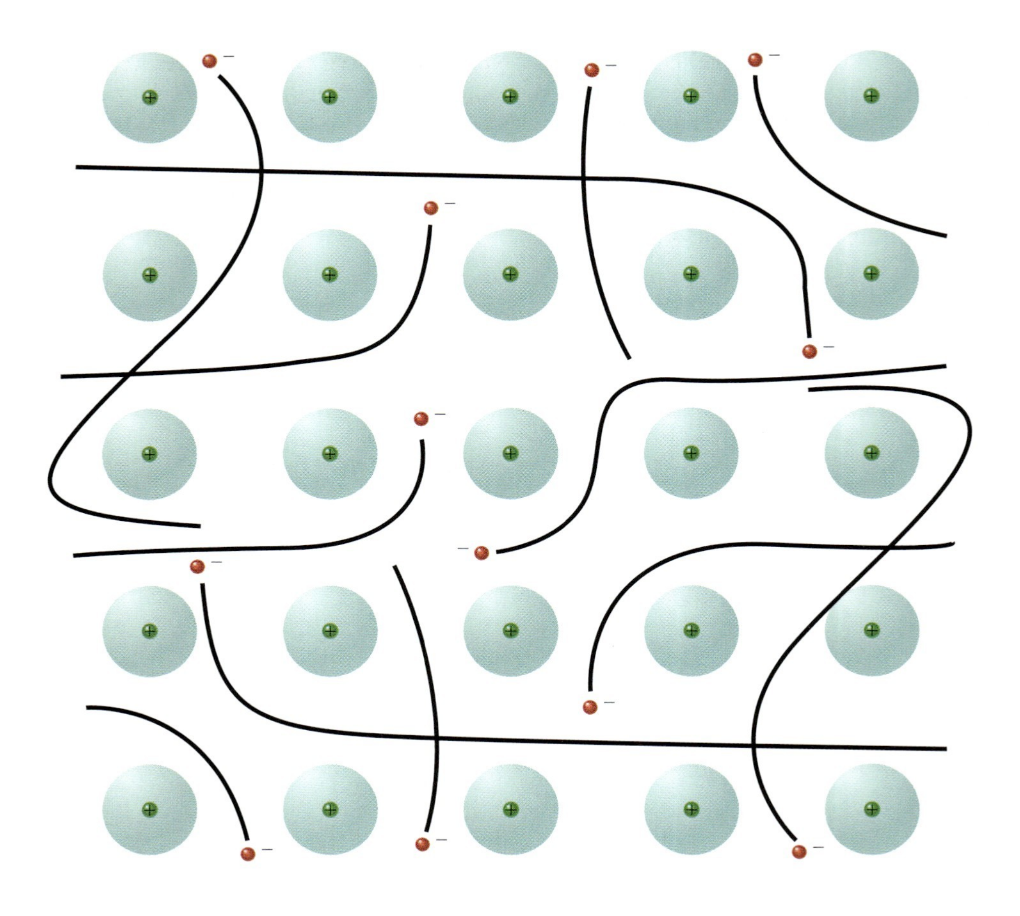

TABLE 3.2.1 Strength of primary bonds

Type of bond	Bond energy (kJ/mol)
Ionic	600–1500
Covalent	200–700
Metallic	70–850

TABLE 3.2.2 List of chemical compounds that are formed from different kinds of primary bonds

Compounds with ionic bonds	Compounds with covalent bonds	Compounds with metallic bonds
NaCl	CH_4	Fe
KBr	CH_3Cl	Na
Na_2S	H_2O	CaK
$CaCl_2$	NH_2	AlMg

In the next set of questions you will use your knowledge of electronegativity from the previous section to predict the types of bonds that will form between two atoms. Use Table 3.2.2 to help you answer these questions.

Guided Inquiry: Types of Bonds

3.2.1 What kind of a bond is present in NaCl?

3.2.2 What kind of atoms are in NaCl; atoms with high electronegativity or atoms with low electronegativity?

3.2.3 What kind of a bond is present in CH_4?

3.2.4 What kind of atoms are in CH_4; atoms with high electronegativity or atoms with low electronegativity?

3.2.5 What kind of bond is present in pure iron?

3.2.6 What kind of atoms are in Fe; atoms with high electronegativity or atoms with low electronegativity?

3.2.7 In general, what kinds of atoms form ionic bonds?
Use your answers from the previous questions to answer this. This is a concept invention question, in which you should be able to figure out the concept in your group rather than having the instructor tell you.

3.2.8 What kinds of atoms form covalent bonds?

3.2.9 What kinds of atoms form metallic bonds?

Concept Check **3.2.1**

- What kind of primary bonds are in the compound CaF_2?
- What kind of primary bonds are in the compound GaN?

As you tried to answer the Concept Check questions, you may have found yourself wondering, "How big of a difference in electronegativity is needed for a bond to be ionic?" In reality, electrons are not shared exactly equally between two atoms because one will have a higher electronegativity than the other (except for diatomic molecules with the same atoms, such as N_2). This means there is (almost) always some ionic character to a covalent bond. The percent ionic character of a bond can be calculated with the following equation:

$$\%\text{ionic character} = 100 \times (1 - \exp(-0.25(EN_A - EN_B)^2)) \qquad (3.2.1)$$

where EN_A and EN_B are the electronegativities of the two atoms. The one limitation of this equation is that it only compares the ionic and covalent characters of the bond and does not account for metallic bonding.

Another way to look at the type of bonding is through the bond-type triangle, shown in Figure 3.2.4. This triangle considers both the electronegativity difference between the two atoms and the average value of the electronegativities of the two atoms to determine the type of bonding. How do you use this triangle? **The next set of questions will show you.**

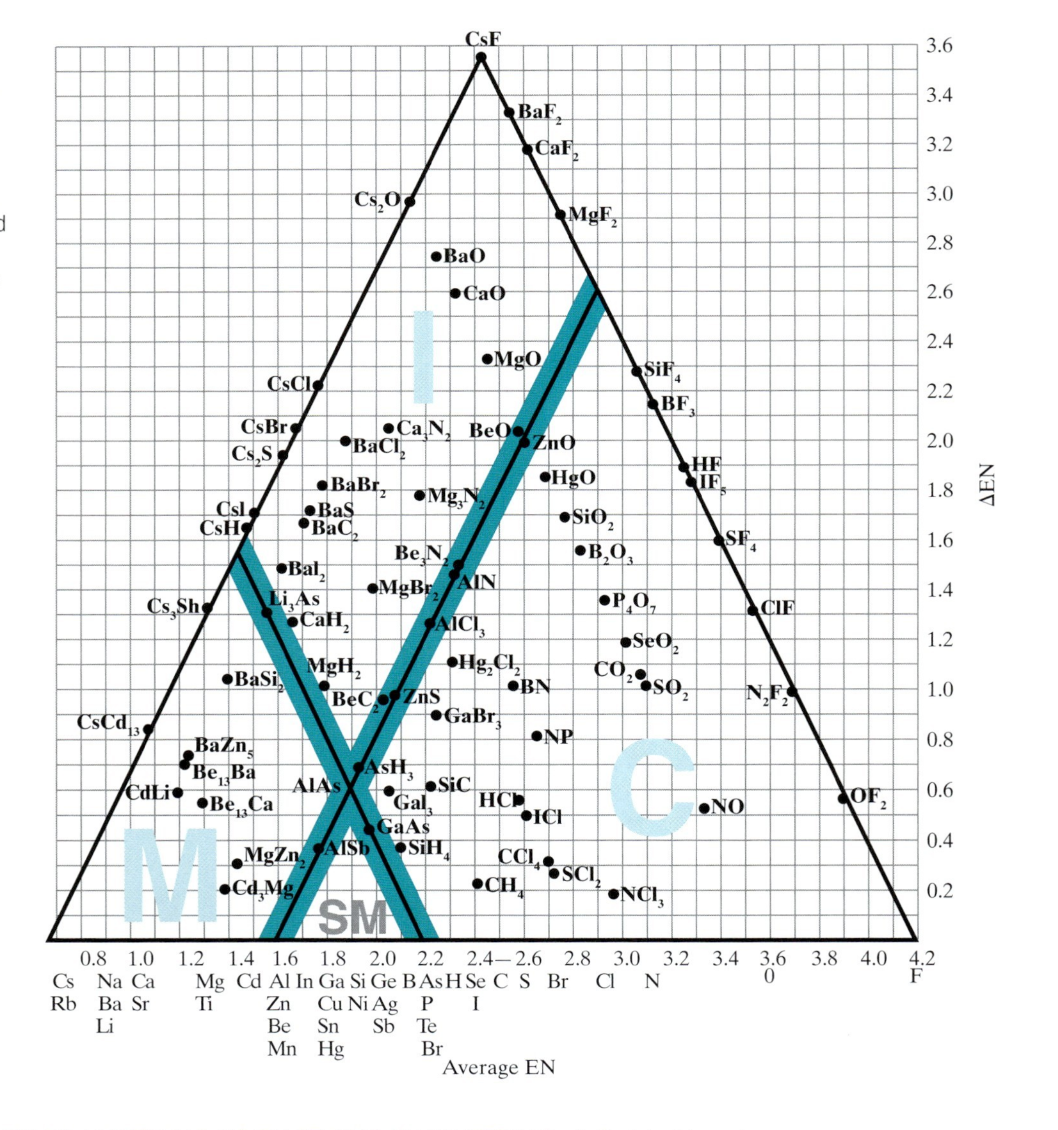

Figure 3.2.4

Bond-type triangle, showing the regions for metallic (M), semimetallic (SM), covalent (C), and ionic (I) bonding as a function of the average electronegativity and electronegativity difference for the two atoms that make up the bond.

Guided Inquiry: Bond Character

3.2.10 Do compounds with metallic bonds have low or high average electronegativity?

3.2.11 Do compounds with covalent bonds have low or high average electronegativity?

3.2.12 Do compounds with ionic bonds have a small or large difference in electronegativity?

3.2.13 Do compounds with covalent bonds have a small or large difference in electronegativity?

3.2.14 Identify three compounds that lie on the boundary between ionic and covalent bonding, and calculate the difference in electronegativity and %ionic character for each of those compounds.

3.2.15 Based on your answer to 3.2.14, is there a single value of electronegativity difference that defines the boundary between ionic and covalent bonding? Is there a single value of %ionic character that defines that boundary? If so, what is the value?

3.2.16 Based on your answer to 3.2.15, how would you determine if a bond is ionic or covalent?

What approach gives you a consistent answer? Is it sufficient to just consider the electronegativity difference or the %ionic character? Based on the questions above, is there something else you can use?

Concept Check 3.2.2

- What kind of bond is formed in the compound $FeBr_2$?

EXAMPLE PROBLEM 3.2.1

What type of bonding occurs in the compound NiO?

From the positions of Ni and O on the periodic table, you might expect the bonding to be ionic. Calculating the %ionic character, we find it is slightly more ionic than covalent. However, what we really need to do is use the bond-type triangle:

$$\text{EN(Ni)} = 1.8$$
$$\text{EN(O)} = 3.5$$
$$\text{EN (average)} = 2.65$$
$$\Delta\text{EN} = 1.7$$

This puts the compound NiO into the covalent region, so we would say it has covalent bonding. Keep in mind, however, that in calling it a covalent bond, we are just applying a convenient label. In reality, the distribution of electrons is between the atoms as in a covalent bond, but slightly more toward the oxygen atom, as in an ionic bond.

3.3 Nonbonding Interactions

LEARN TO: **Describe types of nonbonding interactions.**
Identify the nonbonding interactions present in a molecular material.
Predict properties of molecular materials based on the nonbonding interactions present.

Another important type of interaction between atoms are *nonbonding interactions*. Note that many textbooks call these interactions "secondary bonds," but that can be confusing because these are not actually bonds. We will stick to the term "nonbonding interactions." Unlike primary bonds, nonbonding interactions do not result from the sharing of electrons. They occur because of attraction between partial charges that are present in molecules. Nonbonding interactions are bonds that occur between molecules and are much weaker than primary bonds. Table 3.3.1 shows the strengths of the different types of nonbonding interactions. The types of nonbonding interactions are:

Hydrogen bond: Occurs between oxygen, nitrogen, or fluorine on one molecule, and hydrogen atoms on another molecule that are bound to oxygen, nitrogen, or fluorine. This is the strongest nonbonding interaction. A hydrogen bond is shown in Figure 3.3.1. Remember that even though this is commonly called a "bond," it is not a primary bond; it is a nonbonding interaction.

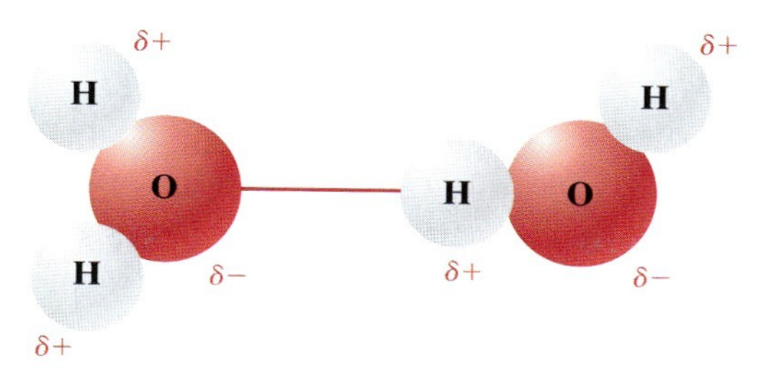

Figure 3.3.1
Schematic diagram of hydrogen bonds formed in water.

δ^+ is the symbol for a partial positive charge, δ^- is the symbol for a partial negative charge.

Permanent dipole: Occurs when a very electronegative atom forms a covalent bond with a less electronegative atom, resulting in partial positive and negative charges in the molecule. The partial charges on this molecule can then interact with the partial charges on another molecule, forming a permanent dipole interaction. A permanent dipole interaction is shown in Figure 3.3.2.

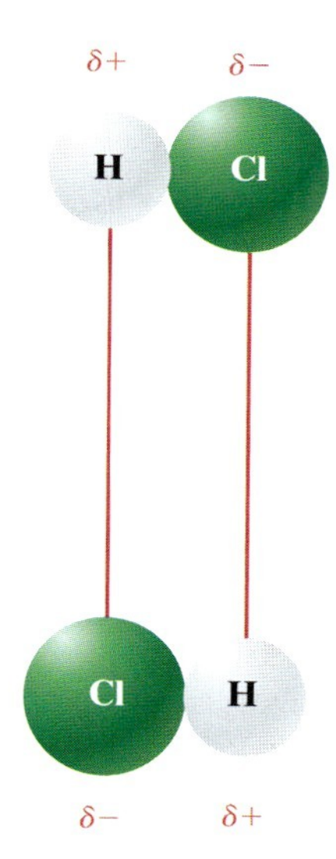

Figure 3.3.2
Schematic diagram of permanent dipole interactions formed in hydrogen chloride.

Induced dipole: This type of interaction is formed by random fluctuations in the electron distribution in atoms. As shown in Figure 3.3.3, a random fluctuation results in partial positive and negative regions on an atom. The charges that are randomly created on this atom can attract or repel electrons on a nearby atom, resulting in a weak attraction between the atoms. Induced dipole interactions are also called van der Waals bonds and London dispersion forces. (There are actually subtle differences among these three types of interactions, but these are not important for us.) Induced dipole interactions occur in all molecules, but are overwhelmed by hydrogen bonds and permanent dipole interactions if those are present. Normally if other types of nonbonding interactions are present, we ignore the induced dipole interactions and don't even list them as being present.

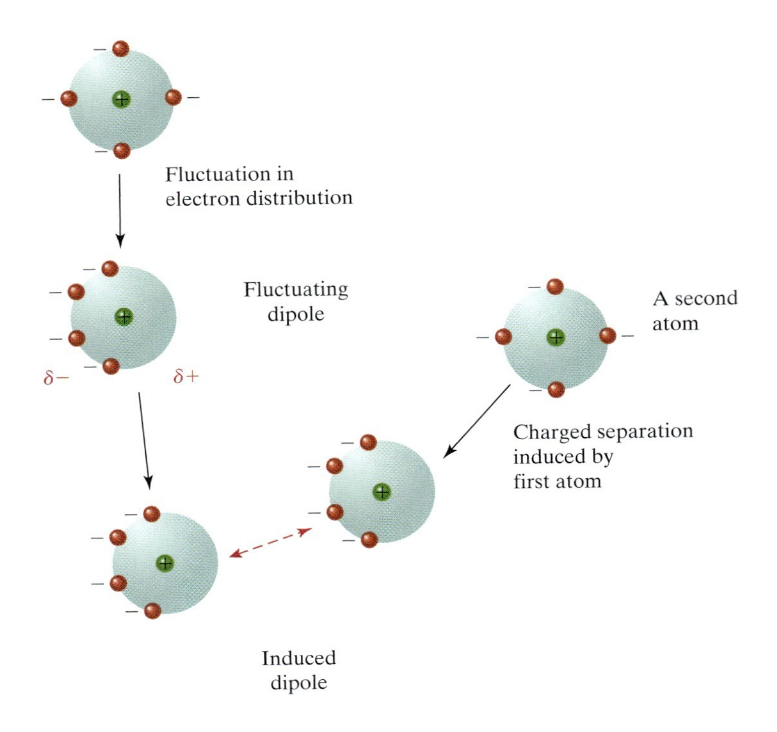

Figure 3.3.3 Schematic diagram of induced dipole interactions.

TABLE 3.3.1 Strengths of nonbonding interactions

Type of bond	Bond energy (kJ/mol)
Hydrogen bond	30–50
Permanent dipole	5–10
Induced dipole	< 5

Before we look at how nonbonding interactions affect the properties of materials, we need to make sure we understand the difference between primary bonds and nonbonding interactions. The next questions will help you to distinguish where these different types of interactions occur in a material.

Guided Inquiry: Bonds vs. Nonbonding Interactions

3.3.1 Circle one methane molecule in Figure 3.3.4.

If your instructor has given you a time limit for answering the questions, make sure you are staying on track.

3.3.2 Ignoring the other methane molecules in Figure 3.3.4, what types of bonds connect the atoms in the molecule you circled?

3.3.3 What kinds of bonds connect the atoms within the other methane molecules?

3.3.4 Is a carbon–hydrogen covalent bond considered polar or nonpolar?

A polar bond is a covalent bond in which the electrons are distributed very unequally between the two atoms. The difference in electronegativity should be relatively high, but there is no specific number to define it.

3.3.5 What kinds of interactions occur between the methane molecule you circled and the other methane molecules? Draw dashed lines in Figure 3.3.4 that show these interactions.

Concept Check 3.3.1

- **What kinds of primary bonds and nonbonding interactions are present in water?**

Figure 3.3.4

A group of methane molecules as they might be arranged in liquid methane.

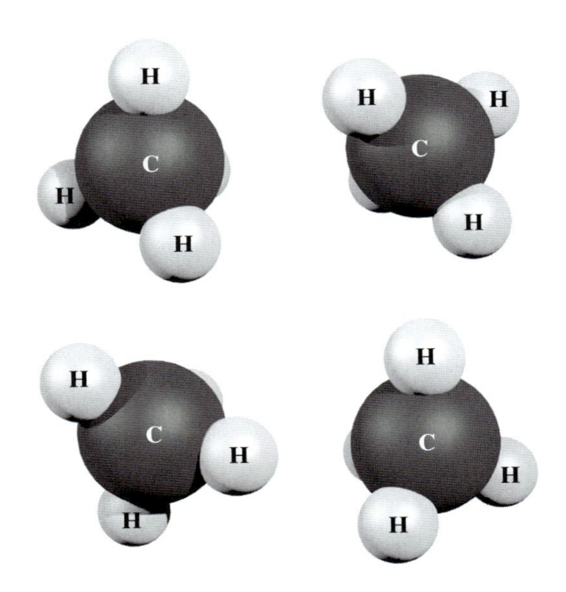

Application Spotlight **Recycling of Tires**

(Photo courtesy of Lifeprints Photography)

Waste from tires is a huge problem. Over 200 million tires per year are disposed of, most of them being heated under controlled conditions to decompose the tires into smaller organic molecules such as gas and oil, or being chopped for use in asphalt and cement, as garden mulch, and other filler applications. So why aren't they recycled? The answer comes from the strong covalent bonds that connect all the molecules. Vulcanization of rubber to make tires causes crosslinking, which is the formation of covalent bonds. The result is a network of covalent bonds that spans the entire tire, making it impossible to melt. Compare that to a plastic bottle made from polyethylene terephthalate (PET), which has nonbonding interactions between the individual molecules, making it possible to melt PET and form it into another shape. Unfortunately, the properties needed for a tire require this crosslinking process. Research is continuing on ways to more efficiently turn waste tires into value-added products.

An important distinction between primary bonds and nonbonding interactions is how they are used to define chemical compounds. As we have already said, primary bonds define compounds. For example, water (H_2O) is considered a single molecule because the three atoms that make up water are held together by primary bonds. In contrast, nonbonding interactions define the interactions between molecules. Depending on the type of compound, either primary bonds or nonbonding interactions are the deciding factor in determining properties. Compounds formed from ionic or metallic bonds have their properties determined by the primary bonds because in these compounds all the atoms are connected through primary bonds. For most molecular compounds that are formed from individual atoms connected by covalent bonds, the properties are determined by the nonbonding interactions. The reason is that the covalent bonds define distinct molecules that interact with other molecules only through nonbonding interactions. However, there are some important exceptions to this. For some materials with covalent bonding the properties are determined by the presence of those covalent bonds. For example, in diamond the carbon atoms are connected through a network of covalent bonds (see Figure 3.3.5). This means that any change to the diamond structure can only be accomplished by breaking the covalent bonds. A similar situation occurs in rubber. In order to understand the difference in how primary bonds and nonbonding interactions affect properties, you will next examine two similar molecules (see Figure 3.3.6) and determine how their structure affects their boiling points.

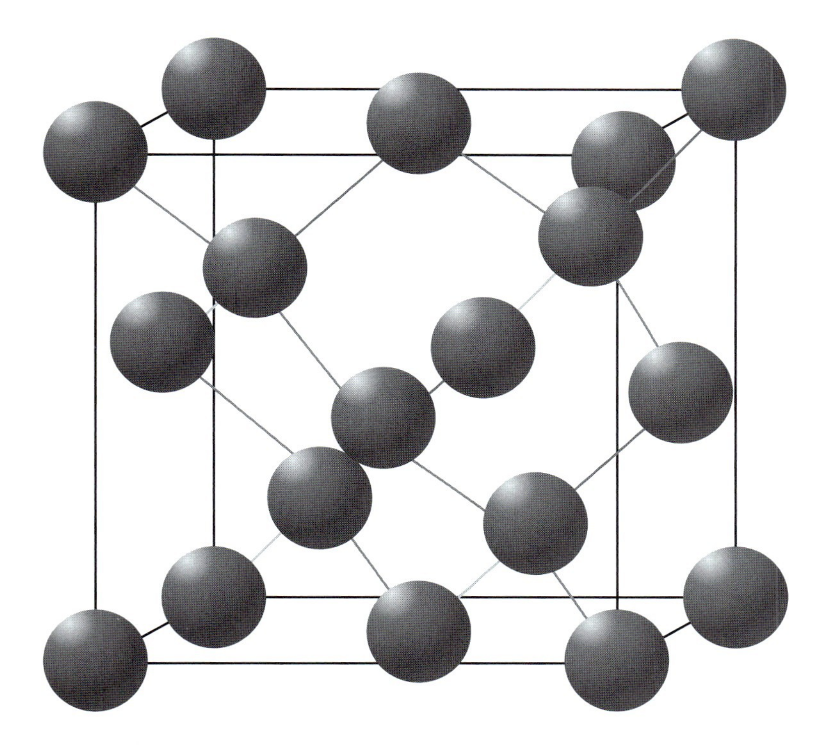

Figure 3.3.5
Structure of diamond, showing the network of covalent bonds that connects the carbon atoms.

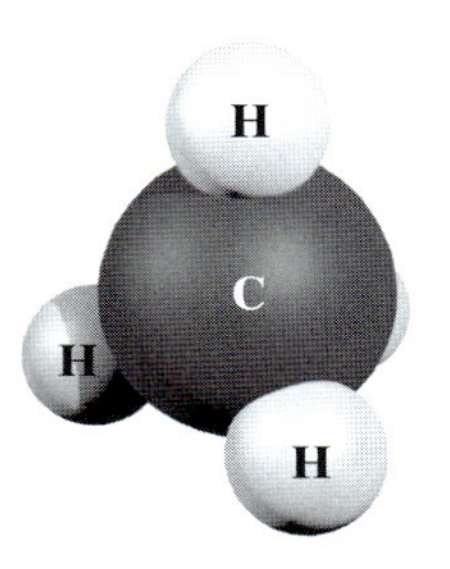

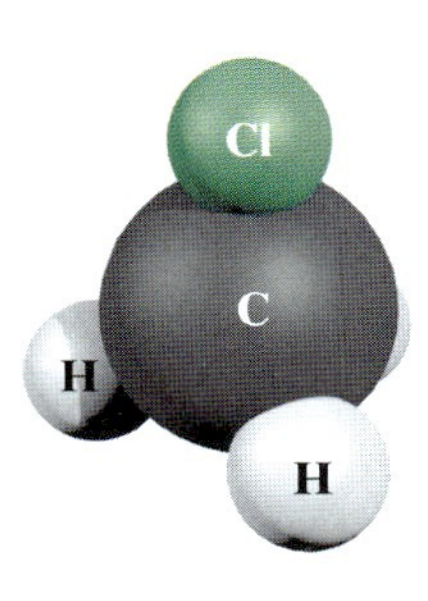

Figure 3.3.6
Molecular structures of methane (left) and chloromethane (right).

Guided Inquiry: Bonding and Properties

3.3.6 What kinds of primary bonds are present in chloromethane (see Figure 3.3.6)?

3.3.7 What kinds of nonbonding interactions are present in chloromethane?

3.3.8 Explain why chloromethane has a higher boiling point than methane.
Although the molecular weights are different, also look at nonbonding interactions.

3.3.9 How would you describe what would happen to methane if the primary bonds were to break?
There is no single correct answer to this question. Its purpose is to have you think about the difference between breaking primary bonds and breaking nonbonding interactions.

Concept Check 3.3.2

- **When water boils, what happens to the primary bonds? What happens to the nonbonding interactions?**

Summary

As you have seen in this chapter, the fundamental atomic property of electronegativity has an important influence on materials properties. The way in which atoms interact, as defined by electronegativity, affects both primary bonds and nonbonding interactions, which in turn affect the properties. In this chapter we considered only boiling point as a property. In later chapters we will see how bonding can also be used to explain engineering properties such as stiffness and thermal expansion. But the connection among electronegativity, bonding, and properties that we have seen illustrates an important aspect of materials science and engineering: connecting fundamental atomic behavior to engineering properties. Understanding this connection allows us to predict and control the properties of materials.

Key Terms

Covalent bond
Electronegativity
Hydrogen bond
Induced dipole
Ionic bond
Metallic bond
Nonbonding interaction
Permanent dipole
Primary bond

Problems

Skill Problems

3.1 For each of the following pairs of atoms, describe where the electrons would be in a bond between them.

a. Mg and Cl
b. Br and Cl
c. Na and K
d. Two S atoms

3.2 What kind of bond is formed in each of the following pure materials? Explain why.

a. Magnesium
b. Silicon
c. Copper
d Carbon

3.3 What kind of bond is formed in each of the following compounds? Explain why.

a. Na_2S
b. CH_4
c. AlMg
d. GaAs
e. CS_2

3.4 What kind of nonbonding interaction occurs in liquids of the following molecules? Draw a picture illustrating the interaction.

a. CH_4
b. CH_3Br
c. NH_3
d.

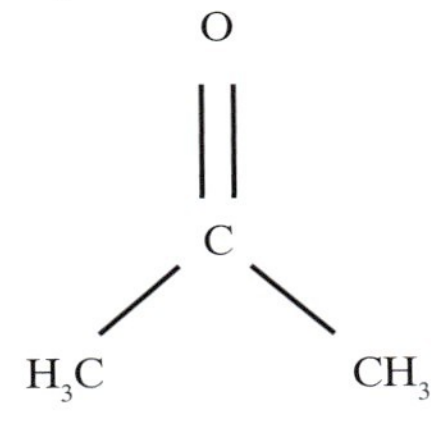

e.

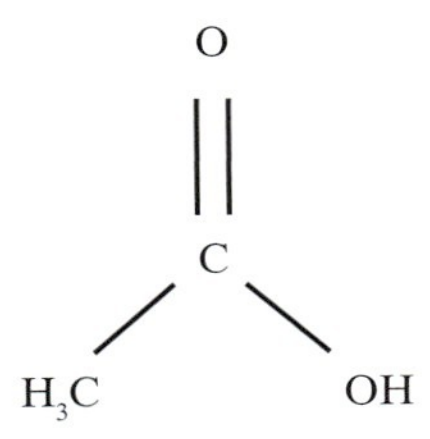

3.5 Draw a diagram of a mixture of H_2O and NH_3, and on this diagram illustrate all of the different types of bonds and nonbonding interactions that are present.

Conceptual Problems

3.6 CdLi and SiC have approximately the same ΔEN, but one is a good conductor of electricity and one is not. Identify which compound is a good conductor of electricity, and explain why ΔEN is not sufficient to distinguish between them.

3.7 Which compound do you expect to have the higher boiling point; $NH(CH_3)_2$ or $N(CH_3)_3$? Explain why.

3.8 For the two compounds below, which do you expect to have the higher boiling point? Explain why.

$$H_3C — CH_2 — OH \qquad\qquad H_3C — O — CH_3$$

3.9 Explain why hydrogen fluoride (HF) has a higher boiling point than hydrogen chloride (HCl) (19.4° versus –85° C), even though HF has a lower molecular weight.

3.10 The electrical conductivity of copper is approximately 10^{22} times greater than that of diamond. Explain this difference on the basis of the type of bonding present in the two materials.

3.11 In their native state, proteins have what are called secondary and tertiary structures. For example, one type of secondary structure is the helix, in which the protein chain winds around itself. This helical structure is stabilized by various interactions, such as hydrogen bonds, covalent bonds, and dipole interactions. When a protein is denatured, it loses this secondary and tertiary structure. Explain why adding urea (structure shown below) to a protein solution will cause the protein to become denatured.

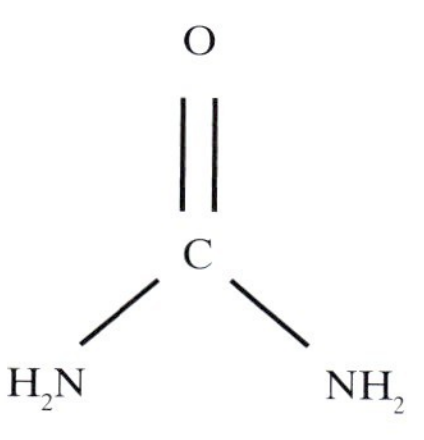

3.12 Is it possible for a pure element to exhibit ionic bonding? Explain why or why not.

Gemstones are single crystals, with facets that correspond to specific crystallographic planes. (Olga Miltsova/Shutterstock)

Atomic Arrangements in Solids

4

For a chemist, the most important things that define the structure of a material are the chemical bonds and the nonbonding interactions. For a materials engineer, although bonds are very important, we go beyond the bonds to examine how atoms are arranged within the material. There are many different levels at which we examine that arrangement. In this chapter we begin by looking at the repeating arrangement of atoms that defines a crystal, as well as defects that occur in those crystals. Both the specific arrangements and the defects have important effects on properties such as strength, as we will see in Chapter 9. For now, we will just be looking at the structure of crystals. By the end of this chapter you will:

- Know the atomic arrangements that define crystals.
- Be able to conduct calculations of various crystal properties.
- Know the defects that occur in crystals.
- Know how to determine the crystal structure of a material.

4.1 Crystals and Glasses

LEARN TO: Compare the structure and properties of crystals and glasses.

Before we begin to look at crystals in detail we need to know what a *crystal* is. Figure 4.1.1 shows the atomic arrangement for two basic types of materials; crystals and *glasses* (also called amorphous materials). At the microscopic level a diamond is crystalline, while a window is amorphous. You will use this figure to identify the difference in structure and properties between a crystal and a glass.

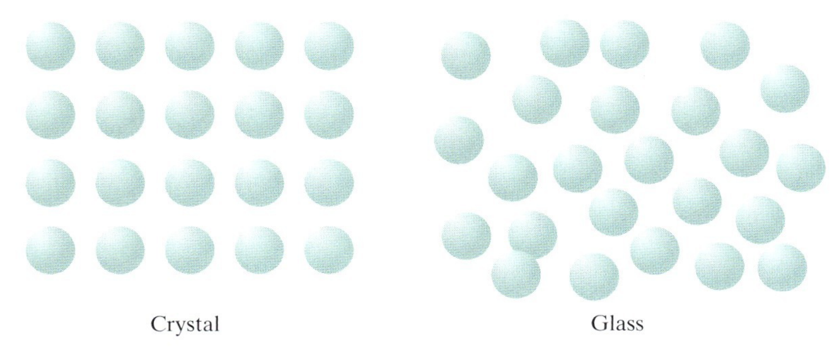

Figure 4.1.1 Atomic arrangements in a crystal and a glass.

Guided Inquiry: Crystals and Glasses

4.1.1 Which has the more ordered arrangement of atoms; a crystal or a glass?

4.1.2 Use a ruler to estimate the average distance between atoms in the crystal parallel to the bottom of the previous page and along a diagonal. Are these average distances the same or different?

4.1.3 Use a ruler to estimate the average distance between atoms in the glass parallel to the bottom of the previous page and along a diagonal. Are these average distances the same or different?

4.1.4 Based on your answer to question 4.1.2, would you expect a crystal's properties to be isotropic or anisotropic?

Isotropic means the properties will be the same in all directions. Anisotropic means the properties will be different in different directions.

4.1.5 Based on your answer to question 4.1.3, would you expect a glass's properties to be isotropic or anisotropic?

Concept Check **4.1.1**

- **A glass rod has a stiffness of 70 GPa when pulled along its length. How would you expect the stiffness to change if you pulled it perpendicular to its length?**

4.2 Unit Cells

LEARN TO: **Draw and identify common unit cells.**
Calculate quantities associated with unit cells.

In a crystal, the atoms form a repeating, periodic array. What is meant by a periodic array? We will start by looking at some art. M. C. Escher was a Dutch artist who lived from 1898 to 1972. He is famous mostly for his drawings that depict physically impossible situations. Figure 4.2.1 shows an example of the kinds of things he drew.

Escher is also famous for his tessellations, which are images that show a regular repeating pattern. Figure 4.2.2 shows an example of a tessellation. One way to create a tessellation is to make a stamp of the smallest unique piece of the pattern and then use that stamp to make the whole figure. The red box in Figure 4.2.2 shows one possible stamp, although there are many others. The green box shows how the stamp can be used to create the whole image. You should notice that what is in the two boxes is identical. Also notice how some

Figure 4.2.1

A paradox drawn in the style of Escher. Follow the path of the water to see why this image depicts a physically impossible situation. (Paul Fleet/Shutterstock)

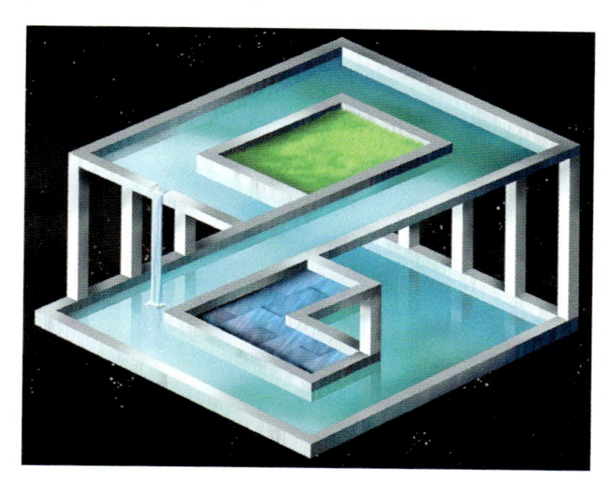

Figure 4.2.2

A tessellation in the style of Escher. Notice how the entire figure can be created from the section outlined in red. (Wendy Middleditch/Shutterstock)

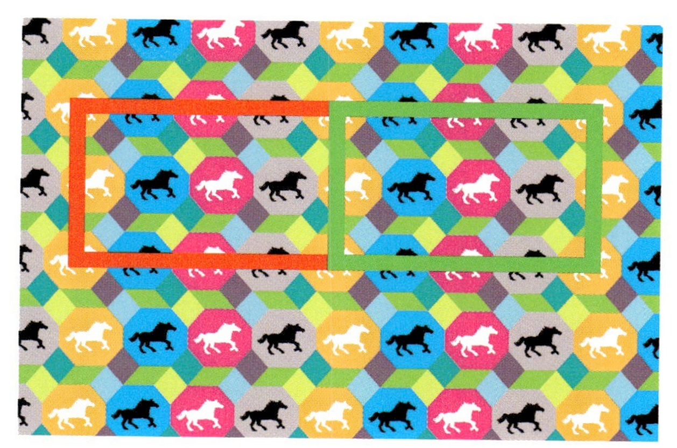

of the horses are not in a single box. Instead they are shared by several boxes. A periodic array is just an arrangement of objects that repeats in this way. You can create the entire group of objects by creating a stamp of the smallest unique piece of the group and then use that stamp to create the entire group.

At the atomic level a crystal is just like Escher's tessellations, except that it is in three dimensions and the objects are atoms. The *unit cell* of a crystal is a piece of the crystal that repeats throughout the whole material. We can think of the unit cell as a three-dimensional stamp that can be used to create the entire crystal, just like the two-dimensional stamp of the horses can be used to create the entire tessellation. One trick in working with crystals is making sure you are visualizing the arrangement in three dimensions and not just two.

There are 14 possible unit cells for crystals, called Bravais lattices. We will consider just three of them. Examples of these three unit cells are shown in Figure 4.2.3. One important property of unit cells that we will use throughout the book is the number of atoms present in a unit cell. By looking at a unit cell as a "stamp" you will now figure out how many atoms are present in different unit cells.

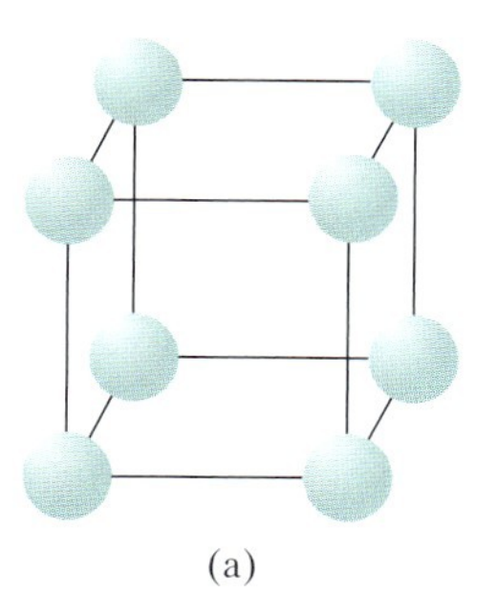

(a)

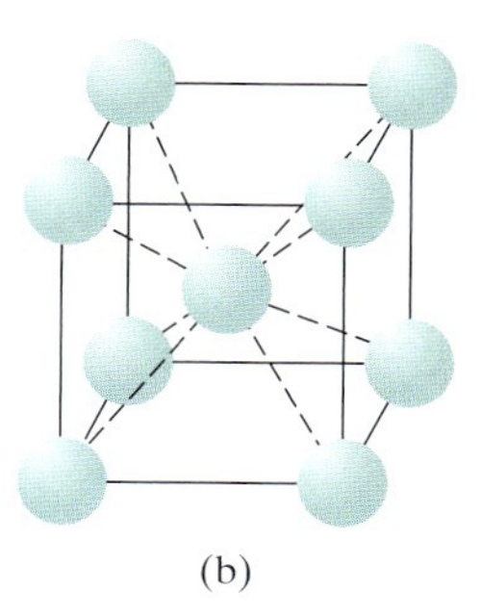

(b)

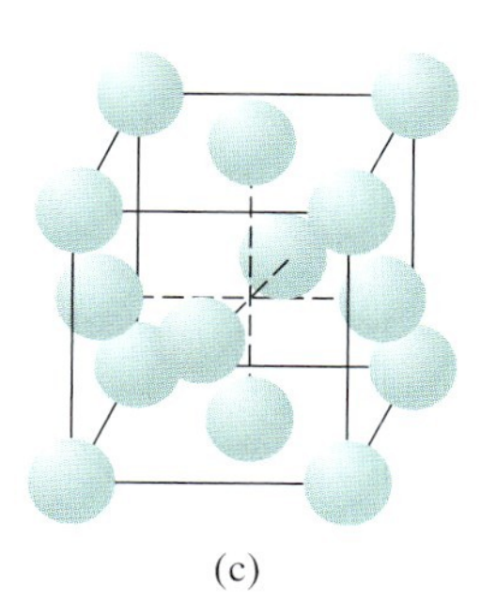

(c)

Figure 4.2.3

The three cubic crystal structures: (a) *simple cubic* (SC); (b) *body-centered cubic* (BCC); (c) *face-centered cubic* (FCC).

Guided Inquiry: Unit Cells

4.2.1 How many atoms do you see in the figure of the simple cubic (SC) unit cell?

4.2.2 How many atoms are within one SC unit cell?
Consider how you would use this unit cell as a stamp to create a three dimensional arrangement of atoms and how the atoms are shared with other unit cells. Also note that the answer is different from the answer to question 4.2.1.

4.2.3 How many atoms do you see in the figure of the BCC unit cell?

4.2.4 How many atoms are within one BCC unit cell?

4.2.5 How many atoms do you see in the figure of the FCC unit cell?

4.2.6 How many atoms are within one FCC unit cell?

4.2.7 Develop a general approach for determining how many atoms are in a unit cell.

Concept Check **4.2.1**

- **For the crystal structure shown to the right, how many blue atoms are there per unit cell?**
- **How many yellow atoms are there per unit cell?**

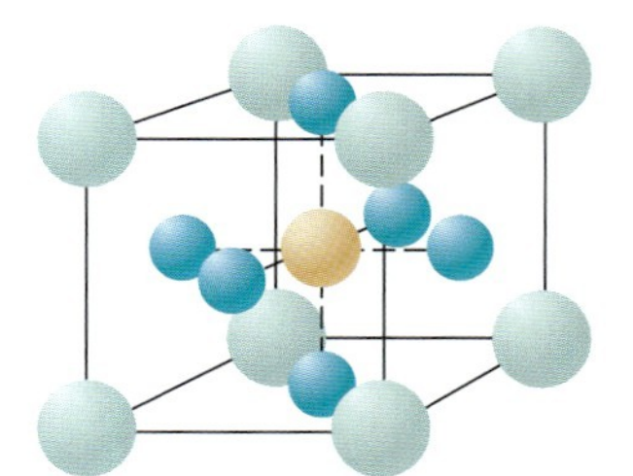

The unit cells shown in Figure 4.2.3 are not completely realistic because they do not show the atoms at the correct size relative to the unit cell. In Figure 4.2.3 the atoms are smaller than they should be so that the overall arrangement is easier to see. Figure 4.2.4 shows the atoms as spheres that are touching, which is a better representation of their size.

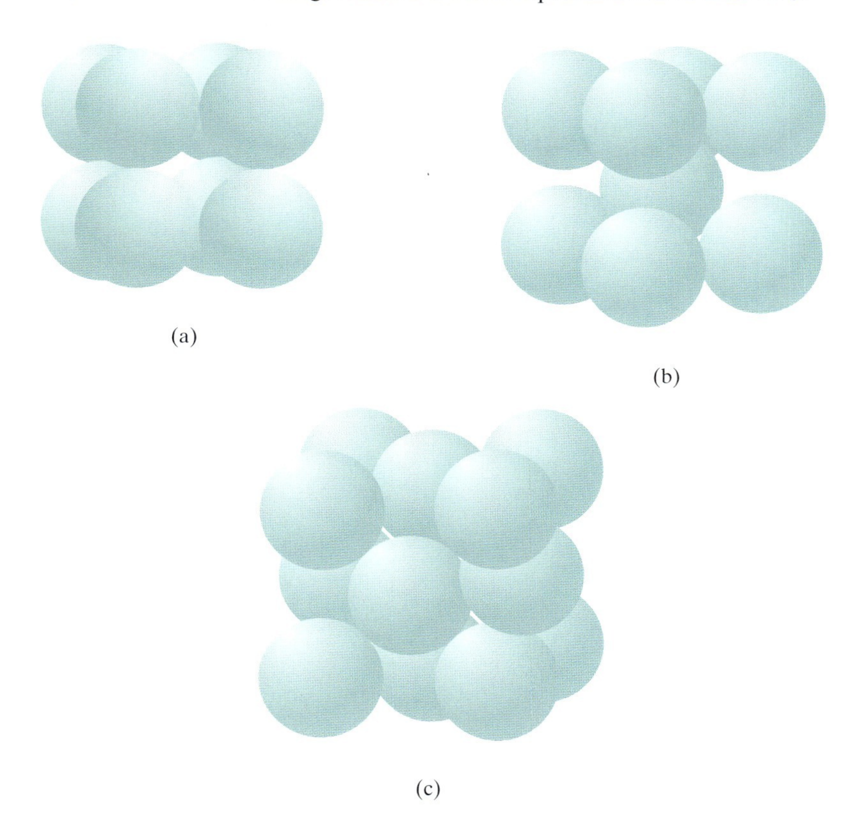

Figure 4.2.4

The three cubic crystal structures shown as hard-sphere models with full-size atoms: (a) simple cubic (SC); (b) body-centered cubic (BCC); (c) face-centered cubic (FCC).

Now we need a few definitions. The length of the unit cell edge is called the *lattice parameter*, and is given the symbol a. The atomic radius is given the symbol R. The *atomic packing factor* (APF) is the fraction of space occupied by atoms in a crystal. It can be calculated using the following equation:

$$APF = \frac{\text{volume of atoms in unit cell}}{\text{volume of unit cell}} \tag{4.2.1}$$

You will now see how to use this equation.

Guided Inquiry: Atomic Packing Factor

4.2.8 Look at the figure of the SC unit cell in Figure 4.2.4. Along what direction do the atoms touch?

Don't spend too much time on this question. It's just to get you to look at the unit cell in the way needed to answer the next questions.

4.2.9 Using your answer from question 4.2.8 develop an equation that expresses the lattice parameter, a, as a function of the atomic radius, R, for the SC unit cell.

You might want to draw a figure and label the lattice parameter and radius to help you see the relationship.

4.2.10 Based on your answer to question 4.2.2, what is the volume of atoms per unit cell in a SC unit cell?

4.2.11 What is the volume of a SC unit cell? For this question, ignore the atoms.

4.2.12 What is the APF for a SC unit cell?

You should get a number. There should be no variables in your answer.

4.2.13 What does your answer to question 4.2.12 mean in terms of how much empty space there is in a SC unit cell? Does this result surprise you?

Concept Check 4.2.2

- **What is the APF for a SC unit cell?**

The density of a crystal is the mass of atoms in a unit cell divided by the volume of the unit cell. The density will depend on the type of atom and the crystal structure (unit cell type) of the crystal. Data for many elements is given in Table 4.2.1. As you will now see, calculating the density is similar to calculating the APF.

TABLE 4.2.1 Properties of elements

Element	Density (g/cm³) at 20° C	Crystal structure	Atomic radius (nm)	Ion	Ionic radius (nm)
Aluminum	2.70	FCC	0.143	Al^{3+}	0.057
Barium	3.51	BCC	0.222	Ba^{2+}	0.134
Calcium	1.55	FCC	0.197	Ca^{2+}	0.099
Carbon	2.27 (graphite) 3.52 (diamond)	Hexagonal (graphite) Diamond cubic (diamond)	0.077	C^{4+} C^{4-}	0.016 0.260
Cesium	1.93	BCC	0.265	Cs^{+}	0.167
Chlorine	—	—	0.102	Cl^{-}	0.181
Chromium	7.19	BCC	0.128	Cr^{3+} Cr^{6+}	0.063 0.052
Copper	8.94	FCC	0.128	Cu^{+} Cu^{2+}	0.096 0.072
Fluorine	—	—	0.064	F^{-}	0.133
Gold	19.30	FCC	0.144	Au^{+} Au^{3+}	0.137 0.085
Hydrogen	—	—	0.031	H^{-}	0.154
Iodine	4.93	Orthorhombic	0.140	I^{-}	0.220
Iron	7.87	BCC	0.126	Fe^{2+} Fe^{3+}	0.074 0.064
Lead	11.34	FCC	0.175	Pb^{2+} Pb^{4+}	0.120 0.084
Lithium	0.534	BCC	0.152	Li^{+}	0.078
Magnesium	1.74	Hexagonal	0.160	Mg^{2+}	0.066
Molybdenum	10.28	BCC	0.139	Mo^{4+} Mo^{6+}	0.070 0.062
Nickel	8.91	FCC	0.124	Ni^{2+}	0.069
Oxygen	—	—	0.066	O^{2-}	0.132
Platinum	21.45	FCC	0.139	Pt^{2+} Pt^{4+}	0.080 0.065
Potassium	0.862	BCC	0.227	K^{+}	0.133
Silver	10.49	FCC	0.144	Ag^{+} Ag^{2+}	0.126 0.089
Sodium	0.968	BCC	0.186	Na^{+}	0.097
Sulfur	2.07	Orthorhombic	0.105	S^{2-}	0.184
Tin	7.37 (white) 5.77 (gray)	Tetragonal (white) Diamond cubic (gray)	0.140	Sn^{2+} Sn^{4+}	0.093 0.071
Tungsten	19.25	BCC	0.139	W^{4+} W^{6+}	0.070 0.062
Vanadium	6.0	BCC	0.134	V^{3+} V^{5+}	0.074 0.059
Zinc	7.14	Hexagonal	0.134	Zn^{2+}	0.074

Guided Inquiry: Density

4.2.14 What crystal structure does aluminum have?

4.2.15 How many aluminum atoms are present in one unit cell?

4.2.16 What is the atomic weight of aluminum?

4.2.17 What is the mass of one aluminum atom in grams?
Remember, if you have one mole of atoms you have Avogadro's number of atoms.

4.2.18 What is the mass of one aluminum unit cell in grams?

4.2.19 Look at the figure for the unit cell that is the crystal structure of aluminum. Along what direction do the atoms touch?

4.2.20 Using your answer from question 4.2.19, develop an equation that expresses the lattice parameter, a, as a function of the atomic radius, R, for this unit cell.

4.2.21 What is the volume of the unit cell in terms of the radius R?

4.2.22 What is the volume of one aluminum unit cell in cm^3?

4.2.23 Use your answers from questions 4.2.18 and 4.2.22 to calculate the density of aluminum. How does this compare to the value in Table 4.2.1?

Concept Check 4.2.3

- **What is the expression relating the lattice parameter and the atomic radius for a FCC unit cell?**
- **What is the expression relating the lattice parameter and the atomic radius for a BCC unit cell?**

EXAMPLE PROBLEM 4.2.1

Calculate the density of copper.

Copper is a FCC unit cell, so there are four atoms per unit cell. We calculate the mass of the copper atoms in a unit cell as follows:

Mass of one atom = 63.54 g/mol / 6.023×10^{23} mol^{-1} = 1.055×10^{-22} g/atom

Mass of one unit cell = 1.055×10^{-22} g/atom $\times$ 4 atoms/unit cell = 4.22×10^{-22} g/unit cell

Now we need the volume of the unit cell. The volume is a^3 and for a FCC unit cell $a = 2R\sqrt{2}$, so

volume/unit cell = $(2R\sqrt{2})^3 = (2(0.128 \times 10^{-7}\text{ cm})\sqrt{2})^3 = 4.745 \times 10^{-23}\text{ cm}^3$

We can now use these values to calculate the density:

$\rho = 4.22 \times 10^{-22}\text{ g}/4.745 \times 10^{-23}\text{ cm}^3 = 8.89\text{ g/cm}^3$

4.3 Miller Indices

LEARN TO: **Calculate Miller indices for directions and planes.**
Given Miller indices, draw directions and planes.

We are going to ignore the atoms in the unit cells for a little while to develop a concept called *Miller indices*. Miller indices are sets of coordinates for directions and planes in crystals. Miller indices were developed by several British mineralogists, and the idea was published by William Hallowes Miller in 1839. They were created to provide a simpler method for calculating the positions and angles between the faces of crystals. As you saw in Section 4.1, crystals have anisotropic properties. This anisotropy can be very important in applications where crystals are used. One example is the case of computer chips. The conductivity of the silicon depends on the crystallographic direction, and so controlling the orientation of the crystals during production is very important to getting a reproducible product. Another example is the emerging field of optical computing, in which photons take the place of electrons. In order to create the switches and other devices needed to control the flow of photons, materials with different optical properties in different crystallographic directions must be used. In order to talk about anisotropy in crystals, we need a way to categorize the different directions and planes. Miller indices provide the means to do that.

The first step in determining Miller indices is to define atomic positions in a unit cell. Figure 4.3.1 shows a unit cell with some axes. A point within this unit cell is defined by a coordinate system placed on the edges of the unit cell. There are three important rules to be followed when assigning the coordinates of a point:

1. The coordinate system must be right-handed.
2. The unit length is the lattice parameter, a. If the unit cell has unequal sides, the unit length for each direction is the length of the cell edge in that direction. This means that the unit lengths may be different in different directions.
3. When listing the coordinates of the point, no parentheses are used, and the unit vector is not included. So for example, point A in Figure 4.3.1 has the coordinates 1, ½, 0.

The next step is to learn how to define Miller indices for directions. We will do this by determining them for the direction shown in Figure 4.3.2, using the following rules:

1. Draw the axes. Make sure that they define a right-hand coordinate system. You can put them anywhere, but the convention is to put them in the bottom left, back corner, as shown.
2. Determine the coordinate for the head and tail positions of the vector that defines the direction. For the direction shown, the head is at 0, 0, ½ and the tail is at 0, ½, 1.
3. Determine the projection of the vector on each of the axes by subtracting the tail coordinates from the head coordinates. So for our example we have 0–0 (for the x-coordinates), 0–½ (for the y-coordinates), and ½–1 (for the z-coordinates), resulting in 0, −½, −½.
4. Clear fractions and reduce the answer to lowest integers. In our example we can do this by multiplying through by 2, resulting in 0, −1, −1.
5. The answer from step 4 is the set of Miller indices for that direction, but we have to use the correct notation. In order to write the indices for a direction correctly we must do the following:

   a. Put the indices in square brackets.
   b. Not use commas.
   c. Indicate negatives with a bar over the number.

Following all these rules gives us our final answer, $[0\bar{1}\bar{1}]$. Now you will have a chance to apply this procedure.

Figure 4.3.1

Defining atomic positions in a unit cell. Point A has the coordinates 1, ½, 0.

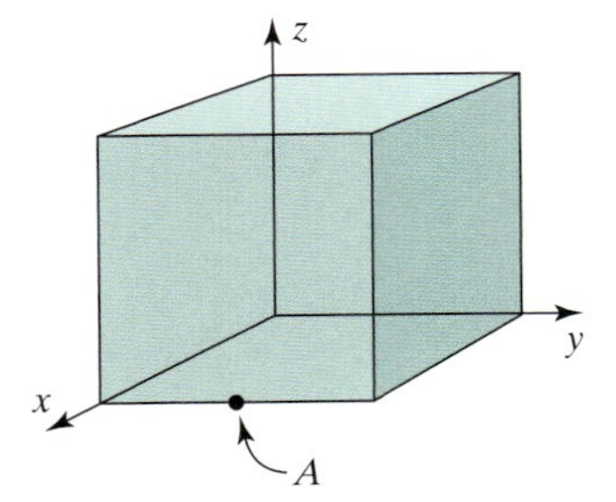

Figure 4.3.2

Example used in the text for how to determine Miller indices of directions.

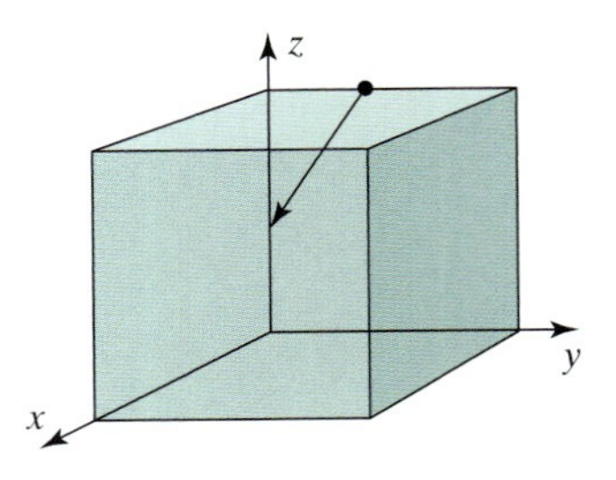

Guided Inquiry: Directions

4.3.1 Determine the Miller indices for the directions in Figure 4.3.3.
Try this in your group without looking at the example problem.

4.3.2 Draw [110].
Look at the procedure for determining Miller indices of directions. How would you use this procedure backward?

4.3.3 Draw [112].

4.3.4 Describe how you would draw a direction if you were given its Miller indices.

Concept Check **4.3.1**

- **What are the correct Miller indices for the direction shown on the right side of Figure 4.3.3?**
- **Draw [112].**

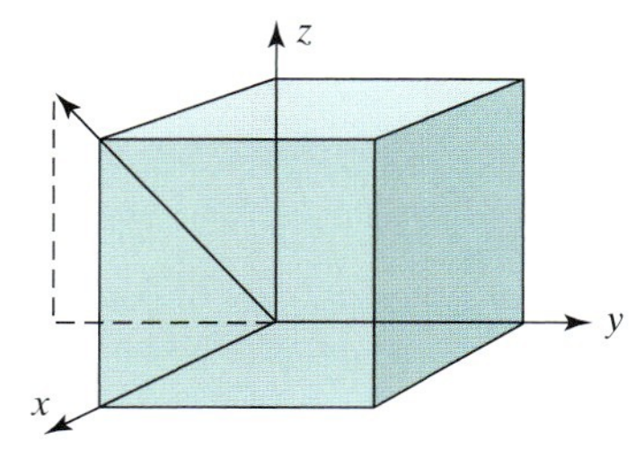

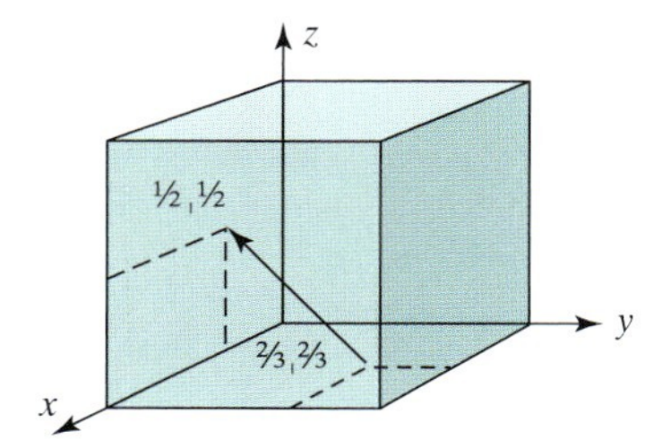

Figure **4.3.3**
Directions for use with question 4.3.1.

EXAMPLE PROBLEM 4.3.1

Determine the Miller indices for the direction shown in the figure below.

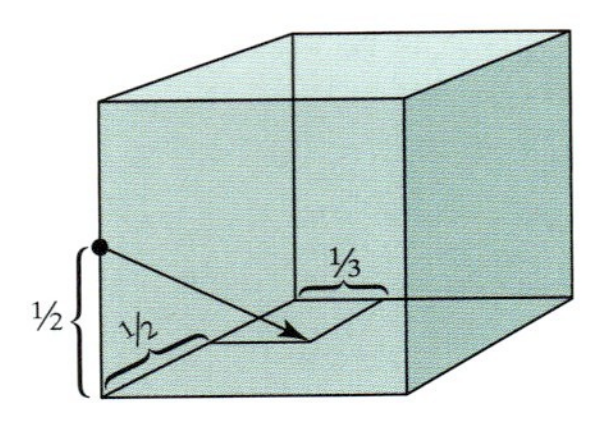

First we determine the coordinates of the head and the tail:

head = $1/2$, $1/3$, 0

tail = 1, 0, $1/2$

Now subtract head – tail:

$1/2, 1/3, 0 - 1, 0, 1/2 = -1/2, 1/3, -1/2$

Multiply through by a common factor to eliminate fractions. The lowest common factor is 6:

$6 \times -1/2, 1/3, -1/2 = -3, 2, -3$

These are already the lowest integers, so we do not have to divide by a common factor. Just place these into the correct notation:

$[\bar{3}2\bar{3}]$.

EXAMPLE PROBLEM 4.3.2

Draw the direction [012].

If we think about how we determine Miller indices for directions, it is by subtracting head – tail. If we put the tail at 0, 0, 0 the Miller indices define the coordinates for the head. This would put the head at 0, 1, 2. However, that extends beyond one unit cell. To keep it inside a single unit cell no coordinate should be greater than 1. We can easily take care of this by dividing the head coordinates by a common factor. In this case we divide by 2, so the head is at 0, $1/2$, 1. This direction is drawn in the figure shown here.

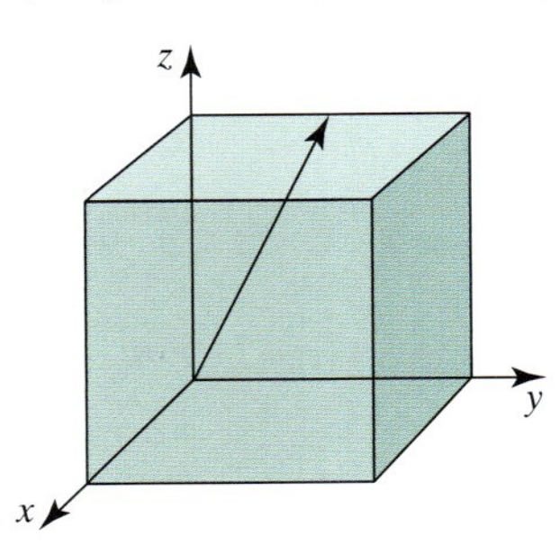

The next step is to define Miller indices for planes. We'll use the plane shown in Figure 4.3.4 as our example as we go through the steps:

1. Choose the origin for your coordinate system. This step is not as easy as it is for directions because you need to keep a few things in mind. First, the plane cannot go through the origin; in other words, the point 0, 0, 0 cannot lie on the plane. Second, you should choose an origin that makes it easy to figure out the indices. That second rule is vague and may not make much sense. How do you know if an origin is in a good spot or not? We'll get a partial answer to that as we move along, but it also just takes some experience. For our example the axes shown in Figure 4.3.4 are a good choice.

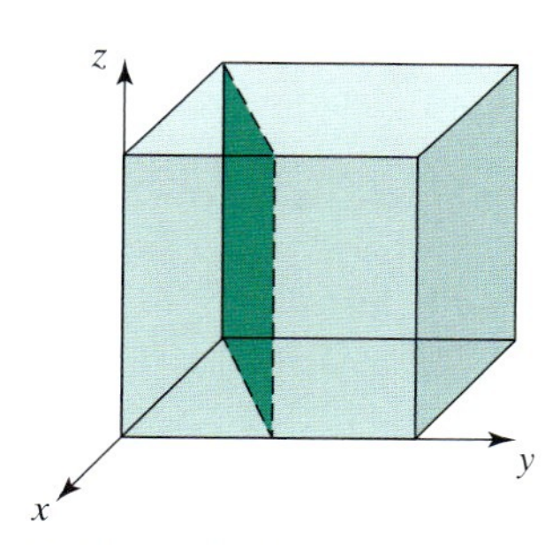

Figure **4.3.4**

Example used in the text for how to determine Miller indices of planes.

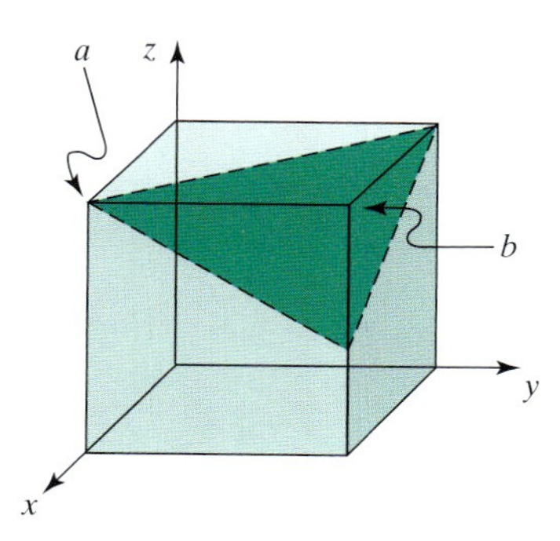

Figure **4.3.5**

Many students will make the error of saying this plane intercepts the *x*-axis at point *a*. This is incorrect, as the *x*-axis is along the bottom edge. It is actually very difficult to determine the intercepts with the axes shown. In this case it would be much easier to move the origin to point *b*. If you do this, the intercepts are easily determined to be $-1, -1, -\frac{1}{2}$.

2. Identify the points where the plane intercepts each of the axes. If the plane is parallel to the axis the intercept is ∞. Some students get confused by this—Figure 4.3.5 shows the error they make. As shown in this figure, sometimes you need to move the axes to find the intercepts. You may also need to extrapolate the plane to outside of a single unit cell. For the example in Figure 4.3.4, the intercepts are $-1, \frac{1}{2}, \infty$.
3. Take the reciprocal of the intercepts. For our example this gives $-1, 2, 0$.
4. Clear any fractions, but do not reduce to lowest integers. This means, for example, that if our answer to this point is 2, 0, 0, we will leave it like that. Note that this is different from what we do with directions. You will learn the reason later in this section and in the next section. For our example we have no fractions, so it just stays as $-1, 2, 0$.
5. The answer from step 4 is the set of Miller indices for that plane, but again we have to use the correct notation. In order to write the indices for a plane correctly we must do the following:

   a. Put the indices in parentheses.

   b. Not use commas.

   c. Indicate negatives with a bar over the number.

Following all these rules gives us our final answer, $(\bar{1}20)$. Note that because of the notation we don't need to specify if we are talking about a direction or a plane, we can tell by the type of brackets that are used. Now you will have a chance to try this out.

Guided Inquiry: Planes

4.3.5 Determine the Miller indices for the planes in Figure 4.3.6.

4.3.6 Draw (110).

Look at the procedure for determining Miller indices of planes. How would you use this procedure backward?

4.3.7 Draw (112).

4.3.8 Describe how you would draw a plane if you were given its Miller indices.

Concept Check 4.3.2

- **What are the correct Miller indices for the plane shown on the left side of Figure 4.3.6?**
- **Draw (110).**

Figure **4.3.6**

Planes for question 4.3.5.

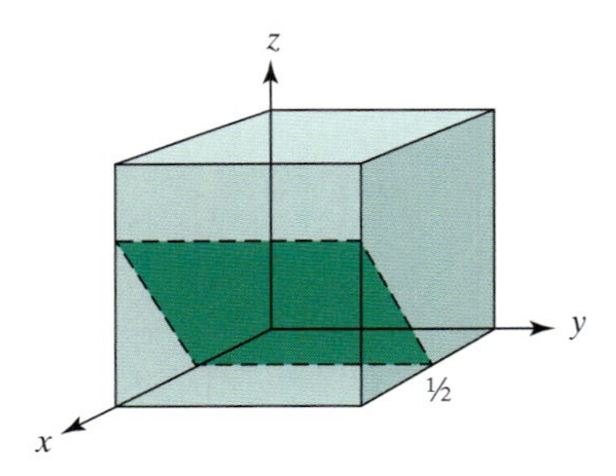

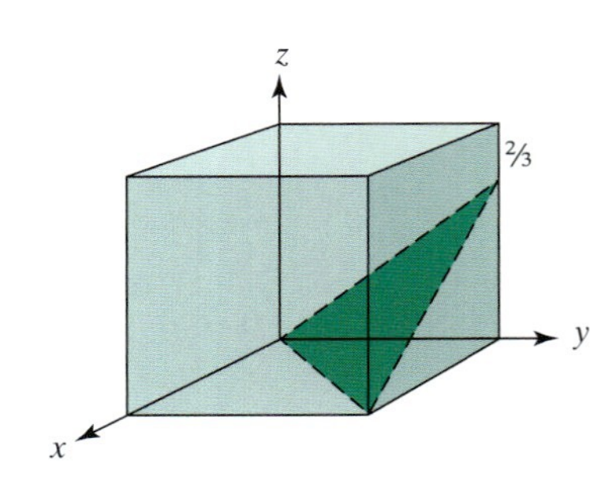

EXAMPLE PROBLEM 4.3.3

Determine the Miller indices for the plane shown in the figure below.

The first step is to decide where to put the axes. The best choice is shown in the figure. Next we find the intercepts with each of the axes. Note that this plane is parallel to the z-axis. The points where this plane intercepts the axes are $-\frac{1}{2}$, $\frac{1}{2}$, ∞. The next step is to take the reciprocals, which results in -2, 2, 0. Remember that for planes we do not reduce to lowest integers, so these numbers are the Miller indices. All we need to do is put them in the proper format: $(\bar{2}20)$.

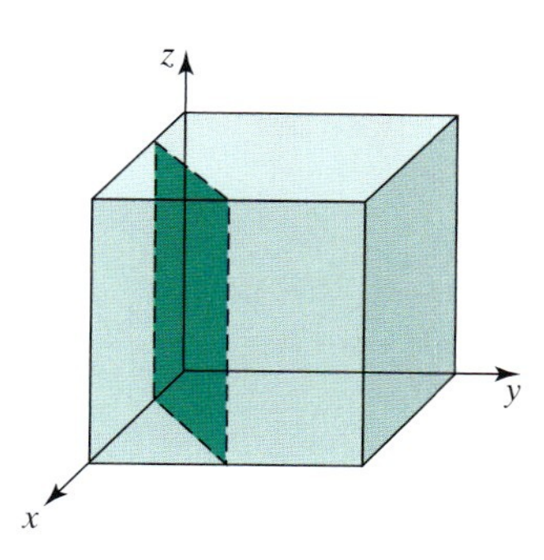

EXAMPLE PROBLEM 4.3.4

Draw the plane $(10\bar{2})$.

In order to do this, think about how the Miller indices were created: by finding the intercepts and then taking reciprocals. So to draw the plane we do the opposite: take reciprocals and then use those as the intercepts. The reciprocals are 1, ∞, $-\frac{1}{2}$. We now take these values and place those points on the axes as shown in the figure below. Since the y-intercept is ∞, we need to draw the plane parallel to the y-axis. The answer is shown in the figure below.

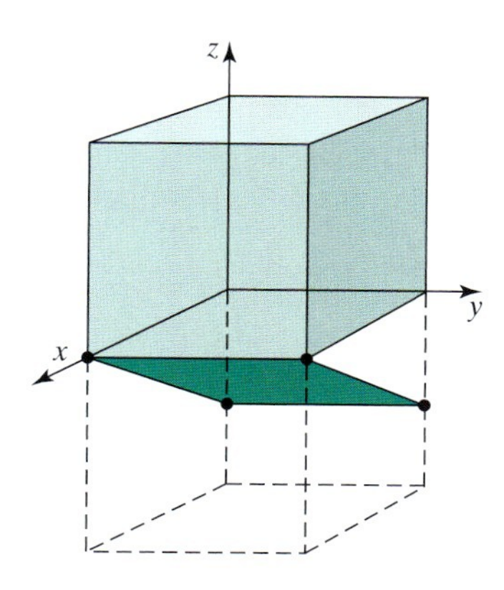

Now we need to look at the relationship between certain planes and directions. This will be important when we get to Chapter 9, but for now just answer the following questions based on the geometry.

Guided Inquiry: Planes and Directions

4.3.9 Compare your answers for questions 4.3.2 and 4.3.6. What do you notice about their geometrical relationship in space? Are they parallel, perpendicular, or something else?

4.3.10 Compare your answers for questions 4.3.3 and 4.3.7. What do you notice about their geometrical relationship?

4.3.11 What is a general rule for the relationship between [*hkl*] and (*hkl*)?

This rule is true only for cubic unit cells!

4.3.12 For unit cells, directions have no magnitude. Only the direction of the vector matters in determining what that direction is. Given this, compare [100] and [200]. How are they related?

4.3.13 Compare (100) and (200). How are they related?

Concept Check 4.3.3

- **What is the relationship between [*hkl*] and (*hkl*)?**

4.4 Planes and Directions in Crystals

LEARN TO: **Calculate linear density.**
Calculate planar density.
List members of families of directions and planes.
Identify close-packed directions, close-packed planes.

Many properties of materials depend on the packing of the atoms. Bulk properties, like density, depend on the packing in three dimensions, as we saw in Section 4.2. However, other properties depend on the packing in a particular plane or along a particular direction. For example, the stiffness of a perfect single crystal of copper depends on which direction the modulus is measured in: it varies from 67 GPa in the [100] direction, to 130 GPa in the [110] direction, to 191 GPa in the [111] direction—the stiffness triples just by changing the way we pull on it! The packing of atoms in a direction is called the *linear density*; the packing of atoms on a plane is the *planar density*. **The next sets of questions will guide you on how to calculate these quantities.**

Guided Inquiry: Linear Density

4.4.1 How long is the [110] direction shown in Figure 4.4.1(a) in terms of the lattice parameter?

4.4.2 How long is the [110] direction shown in Figure 4.4.1(b) in terms of the atomic radius?

4.4.3 How many atomic radii lie along the [110] direction shown in Figure 4.4.1? *Use either part (a) or (b) of this figure to help you answer.*

4.4.4 The linear density is defined as

$$LD = \frac{\text{length of atoms in terms of } R}{\text{Length of line}}$$

What is the LD for the [110] direction in a SC unit cell? *Your answer should be a number.*

4.4.5 Which direction in SC has the highest linear density? What is the value of the linear density for that direction?

Concept Check 4.4.1

- **What is the LD for [110] in a SC unit cell?**
- **What is the LD for [110] in a FCC unit cell?**

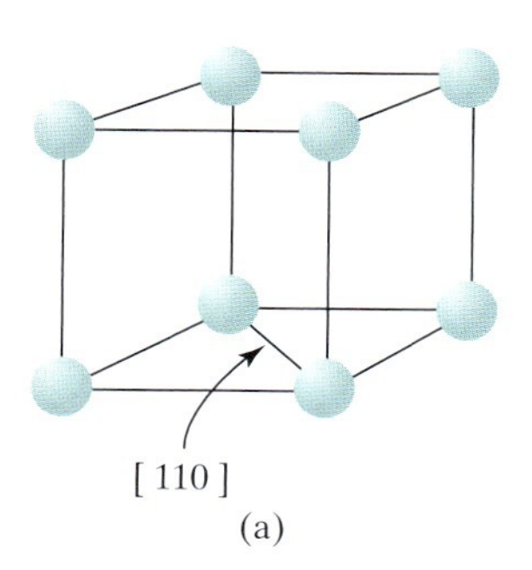

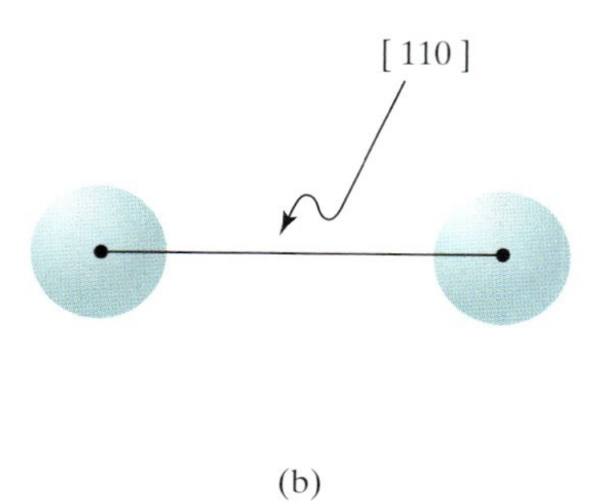

Figure **4.4.1**
Example for questions 4.4.1–4.4.5.

A direction with a linear density of 1.0 is called a *close-packed direction*. All crystal structures have a close-packed direction. You should be able to determine which directions are close-packed for SC, FCC, and BCC by looking at the hard-sphere models in Figure 4.2.4.

EXAMPLE PROBLEM 4.4.1

Calculate the linear density of the [110] direction of a BCC unit cell.

We begin by drawing the unit cell, and from that pulling out the direction with the atoms on it:

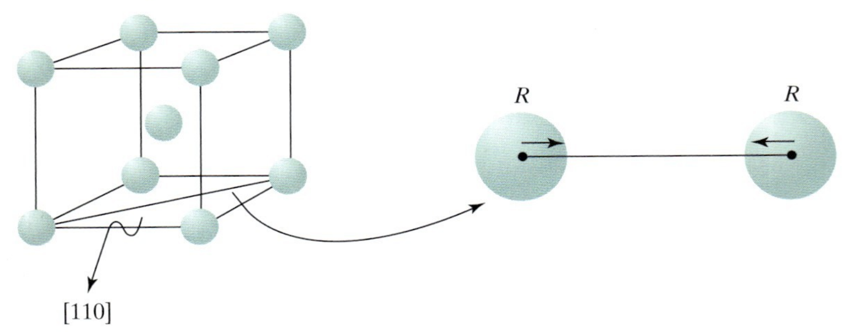

To get the length of the line in terms of the lattice parameter, use the Pythagorean theorem:

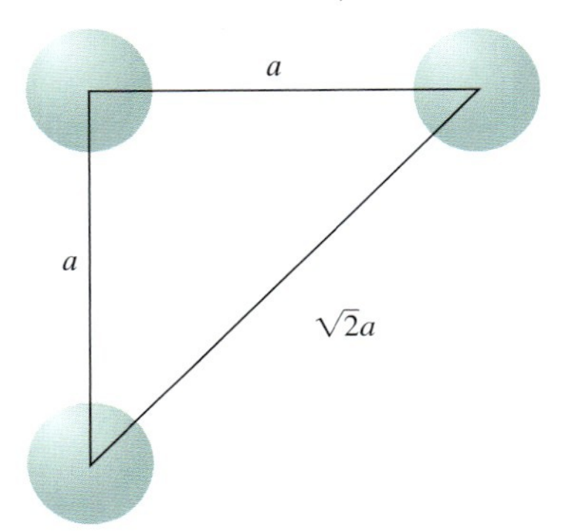

$$a^2 + a^2 = x^2$$

$$x = \sqrt{2}a$$

Since this is a BCC unit cell,

$$a = \frac{4R}{\sqrt{3}}$$

$$x = \sqrt{2}\frac{4R}{\sqrt{3}}$$

Now that we know the length of the line, we need the length of the atoms on the line. As shown above, there are 2 radii that fit on the line, so the length of atoms is $2R$. Now we can calculate the linear density:

$$LD = \frac{2R}{4R\sqrt{2}/\sqrt{3}} = \frac{2\sqrt{3}}{4\sqrt{2}} = 0.612$$

Guided Inquiry: Planar Density

4.4.6 What is the area of the (110) plane in Figure 4.4.2(a) in terms of the lattice parameter?

4.4.7 What is the area of the (110) plane in Figure 4.4.2(b) in terms of the atomic radius?

4.4.8. What is the total area of atoms that lie on the (110) plane in Figure 4.4.2? *Use either part (a) or (b) of this figure to help you answer.*

4.4.9 The planar density is defined as

$$PD = \frac{\text{Area of atoms on a plane}}{\text{Area of plane}}$$

What is the PD for the (110) plane in a SC unit cell? *Your answer should be a number.*

Concept Check 4.4.2

- **What is the PD for (110) in a SC unit cell?**

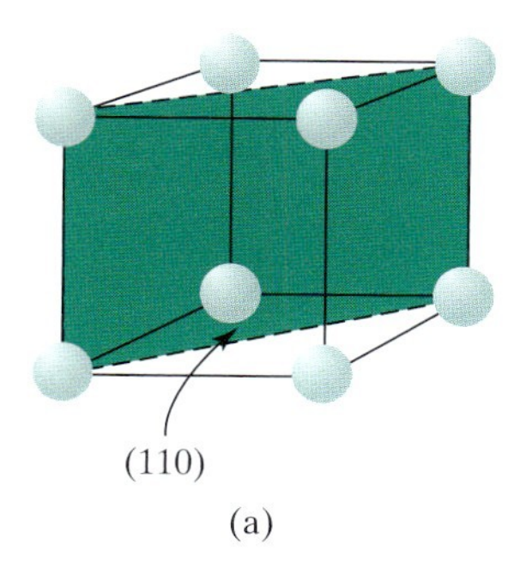

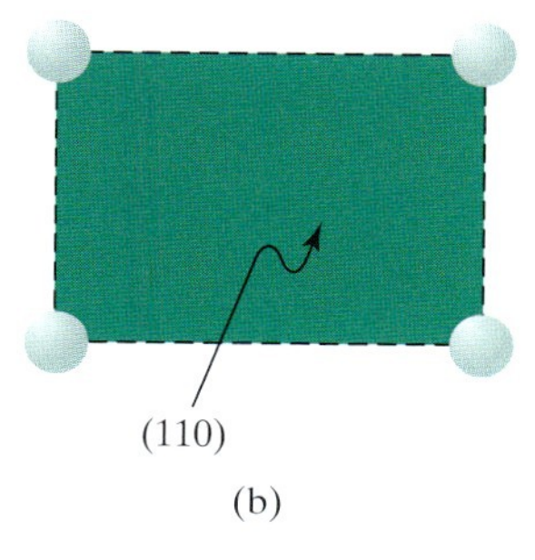

Figure 4.4.2 Example for questions 4.4.6–4.4.9.

EXAMPLE PROBLEM 4.4.2

Calculate the planar density of the (111) plane in FCC.

We begin by drawing the unit cell, and from that pulling out the plane with the atoms on it:

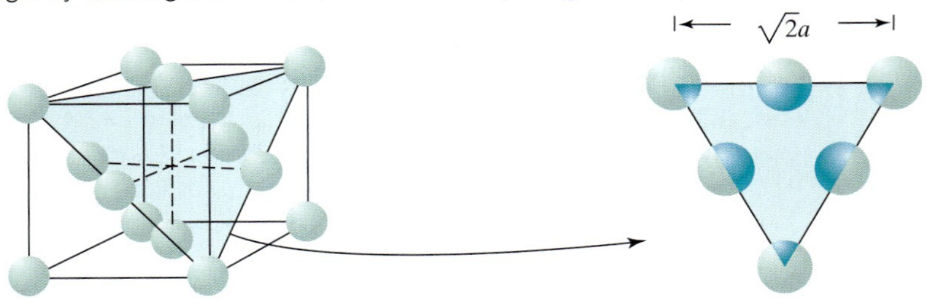

Since this is a triangle, the area of the plane is ½ × base × height. The base is the face diagonal, which has a length of $\sqrt{2}$a. (See Example Problem 4.4.1 if you are not sure how to do that.) We get the height from the Pythagorean theorem as follows:

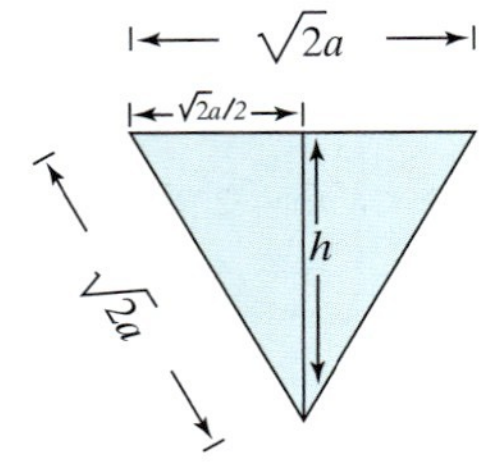

$$h^2 + \left(\frac{\sqrt{2}a}{2}\right)^2 = \left(\sqrt{2}a)^2\right)$$

$$h^2 = \left(\sqrt{2}a\right)^2 - \left(\frac{\sqrt{2}a}{2}\right)^2 = a^2\left(\left(\sqrt{2}\right)^2 - \left(\frac{\sqrt{2}}{2}\right)^2\right)$$

$$= a^2\left(2 - \frac{2}{4}\right) = \frac{3a^2}{2}$$

$$h = \frac{\sqrt{3}a}{\sqrt{2}}$$

Now we can calculate the area of the plane in terms of the lattice parameter:

$$\textit{area of plane} = \frac{1}{2}bh = \frac{1}{2}\left(\sqrt{2}a\right)\left(\frac{\sqrt{3}a}{\sqrt{2}}\right) = \frac{\sqrt{3}a^2}{2}$$

Since this is a FCC unit cell:

$$a = 2R\sqrt{2}$$

$$\textit{area of plane} = \frac{\sqrt{3}(2R\sqrt{2})^2}{2} = 4R^2\sqrt{3}$$

Now we need to determine the area of atoms on the plane. The figure above shows how to divide up each atom.

There are 3 side atoms, each of which counts as ½, and 3 corner atoms, each of which counts as ⅙. From the area of a circle we can calculate the total area of atoms as follows:

$$\textit{area of atoms} = 3\left(\frac{1}{2}\pi R^2\right) + 3\left(\frac{1}{6}\pi R^2\right) = 2\pi R^2$$

We can now calculate the planar density:

$$PD = \frac{2\pi R^2}{4R^2\sqrt{3}} = \frac{2\pi}{4\sqrt{3}} = 0.91$$

To this point we have considered only individual directions and planes. But take a moment and look at the structure of the SC unit cell. All the edges have the same linear density, even though they have different Miller indices. Similarly, all the faces of the unit cell have the same planar density, even though they are planes with different Miller indices. Because these directions and planes have the same LD and PD, and thus the same properties, we group them together and say that they are all in the same family. More specifically, a *family of directions* contains all the directions that have the same linear density, and a *family of planes* contains all the planes that have the same planar density. We also change the notation to indicate that we are talking about a family, rather than an individual direction or plane. The notation for a family of directions is <*hkl*>. The notation for a family of planes is {*hkl*}. By using your understanding of linear and planar density you will now identify directions and planes that are in the same family.

Guided Inquiry: Families

4.4.10 List all the members of the <100> family in a SC unit cell.
There are six of them.

4.4.11 List all the members of the {100} family in a SC unit cell.
There are six of them.

4.4.12 Draw a set of equal sized circles that are packed together as close as possible.
A square lattice is not the closest possible packing.

4.4.13 Which plane in FCC has the arrangement of atoms that you drew in question 4.4.12?

Concept Check 4.4.3

- **List four planes that are in the {110} family in a BCC unit cell.**
- **Which planes in a BCC unit cell have the maximum possible packing?**

The plane with the maximum possible PD is called a *close-packed plane*. Among the cubic unit cells, the only close-packed planes are the {111} planes in the FCC structure. SC and BCC structures do not have any close-packed planes. The planes in the SC and BCC structures with the highest planar densities are called simply the *planes of highest density.*

4.5 Crystalline Defects

LEARN TO: List and draw crystal defects.
Calculate vacancy concentration.
Draw the Burgers vector and describe its characteristics.
Define "single crystal" and "polycrystal."

Up until now we have assumed that the atoms in crystals are arranged perfectly in the unit cells, and that this perfect arrangement extends everywhere in a crystal. In fact, this is not true. Crystals can have several types of defects. We can divide these up based on the dimension of the defect. There are point defects (zero dimensional), line defects (one dimensional), and planar defects (two dimensional). The possible point defects are shown in Figure 4.5.1. Before we define these point defects for you, you will decide for yourself what they are.

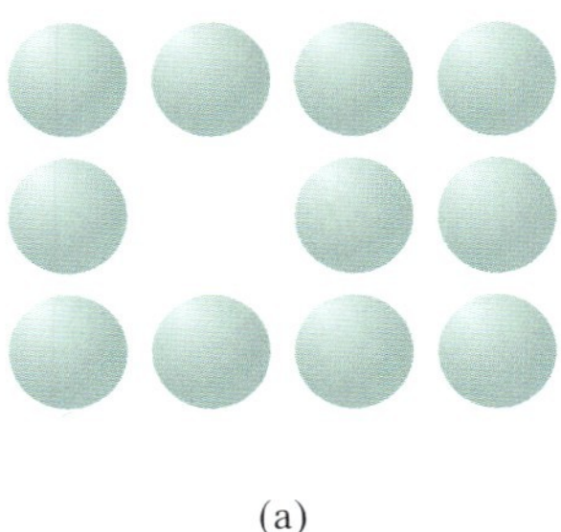

(a)

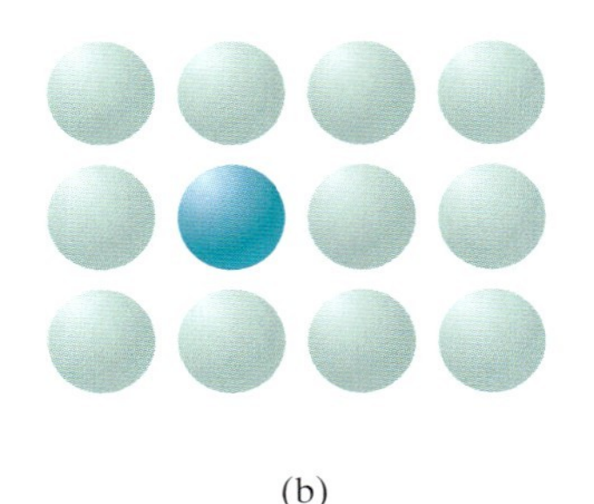

(b)

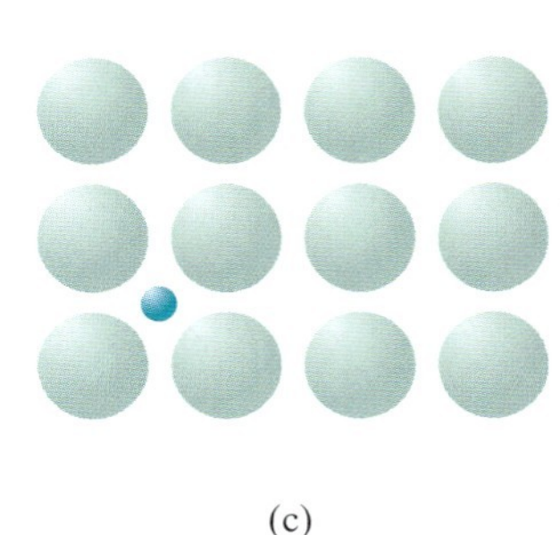

(c)

Figure **4.5.1**
Three types of point defects in crystals.

Guided Inquiry: Point Defects

4.5.1 List and describe in your own words three types of point defects in crystals.

There are specific names for each of these defects. Figure 4.5.1(a) shows a *vacancy*. Figure 4.5.1(b) shows a *substitutional impurity*; it is called substitutional because a different type of atom has been substituted for the normal atoms in that crystal. The spaces in between the normal atoms are called *interstitial sites*, and so Figure 4.5.1(c) shows an *interstitial impurity*. Impurities are very important in affecting the properties of materials. For example, 24-karat gold is pure gold and is extremely soft. In comparison 14-karat gold is only 58.33 wt% gold, with the rest being mostly copper and some silver. These additional elements make 14-karat gold much harder, but also cheaper. We will learn in Chapter 9 why those impurities increase the strength of gold.

Concept Check **4.5.1**

- The atomic radius of gold is 0.144 nm and the radius of copper is 0.128 nm. In 14-karat gold what kind of impurity is the copper?

Vacancies are required by thermodynamics; the presence of vacancies increases the entropy of the crystal. The number of vacancies in a crystal can be calculated using the Arrhenius equation:

$$N_v = Ne^{-E_v/kT} \tag{4.5.1}$$

where N is the total number of atoms, E_v is the activation energy for vacancy formation, k is Boltzmann's constant, and T is the temperature. You will now see how to use this equation.

Guided Inquiry: Vacancies

4.5.2 What are the units of k?

4.5.3 What are the required units for E_v?

4.5.4 What are the required units of T?

4.5.5 What is the mass of 1 m^3 of copper?

4.5.6 How many moles of atoms are in 1 m^3 of copper?

4.5.7 How many atoms are in 1 m^3 of copper?

4.5.8 The activation energy for vacancy formation in copper is 0.9 eV/atom. Room temperature is 25°C. How many vacancies are present in 1 m^3 of copper at room temperature?

4.5.9 Room temperature is 25°C. What percentage of the atoms in copper are vacancies at room temperature?

Concept Check 4.5.2

- **Under what conditions can a crystal have no vacancies?**

EXAMPLE PROBLEM 4.5.1

Determine the percentage of vacancies in copper at 800° C.

In Equation (4.5.1), N is the total number of atoms and N_v is the number of vacancies. That means the ratio N/N_v is the fraction of vacancies present, and $100*N/N_v$ is the percentage of vacancies. We can do the calculation from the following:

$$\%vacancies = \frac{N_v}{N} * 100 = 100 * e^{-E_v/k}$$

The variables are as follows:

E_V = 0.9 eV/atom (from question 4.5.8)

k = Boltzmann's constant = 8.62×10^{-5} eV/K

T = 800° C = 1073 K

Putting these into the equation

$$\%vacancies = 100 * \exp\left(-\frac{(0.9\text{ eV})}{\left(\frac{8.62 \times 10^{-5}\text{ eV}}{\text{K}}\right)(1073\text{ K})}\right) = 0.006\%$$

Besides point defects, crystals can have line defects. Line defects occur when a row of atoms does not line up correctly in the crystal. As with the point defects, before defining the types of line defects you will try to decide for yourself what these line defects are. Figures 4.5.2 and 4.5.3 show two types of line defects. Use these figures to answer the next question.

Guided Inquiry: Line Defects

4.5.10 Describe in your own words three types of line defects in crystals.

Line defects in crystals are called *dislocations*, and there are three kinds. An *edge dislocation* is shown in Figure 4.5.2. We can think of an edge dislocation as being formed by sticking an extra plane of atoms halfway through the crystal. It is important to note, however, that the dislocation itself is not this extra plane of atoms. The dislocation is actually the line that runs along the bottom of this extra plane of atoms (perpendicular to the plane of the paper in Figure 4.5.2). As you can see in the figure, the atoms around the dislocation do not have their normal bonding distances, and so the crystal experiences strain and is in a higher energy state around the dislocation.

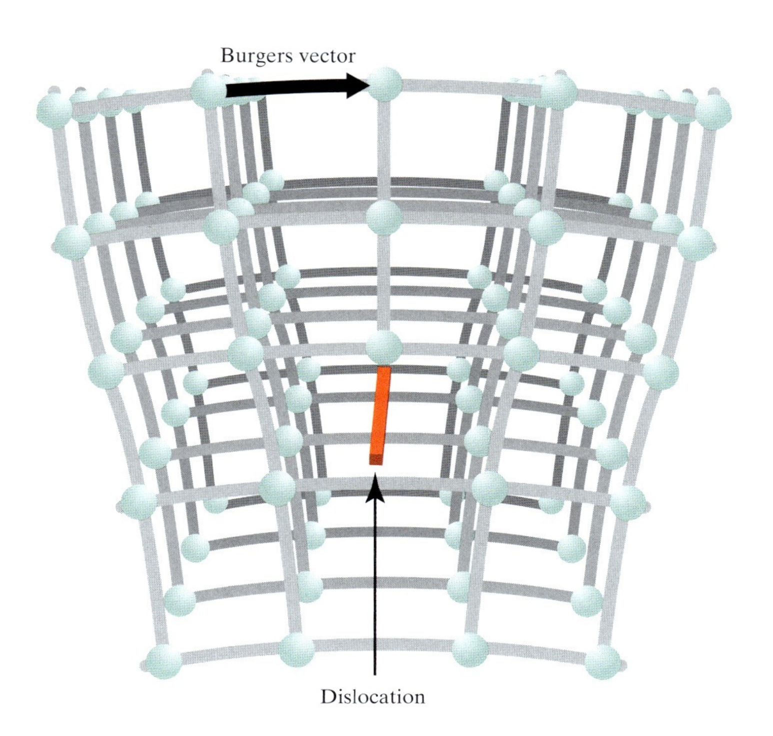

Figure 4.5.2
A type of line defect in a crystal.

Figure 4.5.3 shows a *screw dislocation*. If you could be inside a perfect crystal and walk around in a circle following the bonds, you would end up back where you started. However, if there is a screw dislocation you won't end up back where you started, you will end up one atom away from where you started. This is just like a parking garage; when you are in a parking garage and you drive around in a circle you end up one floor above or below where you started. The actual dislocation line is the axis around which you are traveling.

Dislocations can also be partially edge and partially screw. This type of dislocation starts off as screw on one side of the crystal and becomes edge on the other side. In between it is partially edge and partially screw, so we call it a *mixed dislocation*. A mixed dislocation

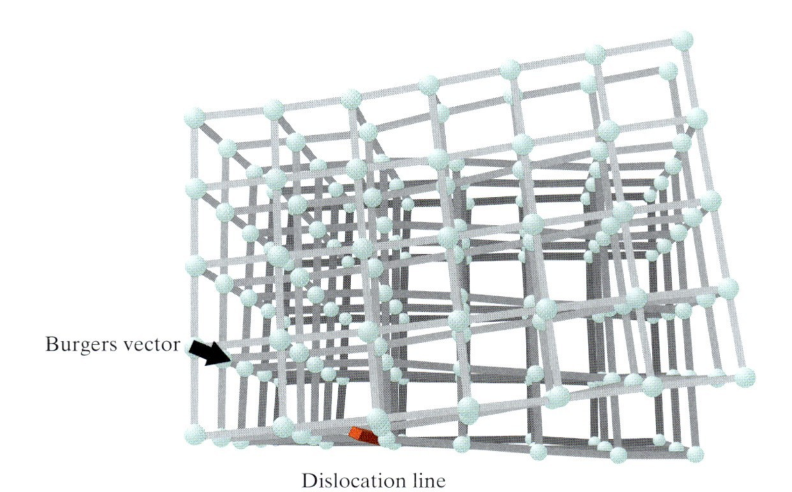

Figure 4.5.3
A type of line defect in a crystal.

is extremely hard to draw, so you don't need to know exactly what it looks like; just know that it does exist. Dislocations are important because of their influence on the properties of materials. We will see in Chapter 9 that dislocations control the strength of crystals.

An important characteristic of a dislocation is the *Burgers vector*. It was developed in 1939 by Johannes M. Burgers as part of a collaboration with his brother W. G. Burgers. W. G. Burgers was a crystallographer who was trying to understand how dislocations affect the strength of metals. Johannes was a physicist, and as part of their work together Johannes developed the Burgers vector as a way to describe the characteristics of the dislocation.

To understand how to determine the Burgers vector, we will use Figure 4.5.4, which shows a crystal containing an edge dislocation with the atoms approximately full size. To determine the Burgers vector, we want to travel around the dislocation by following these steps:

1. Pick an atom to start with. In Figure 4.5.4 our starting atom is the blue one.
2. Draw a vector from the center of that atom to the center of an adjacent atom. Continue to do this a specific number of times. The exact number doesn't matter. In Figure 4.5.4 the red vectors show that we did it 4 times.
3. Now do the same thing, turning 90° so you start to go around the dislocation clockwise. It is very important that you go the same number of steps as step 2. In Figure 4.5.4 this step is shown in green.
4. Keep doing this until you have gone in all 4 directions the same number of steps around the dislocation.
5. Now draw a vector from where you ended to where you started. This is the Burgers vector, which has the symbol $\vec{b}$. In Figure 4.5.4 the Burgers vector is shown in black.

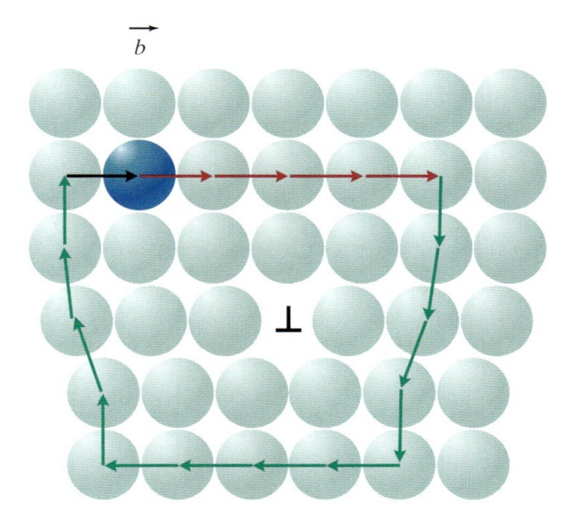

Figure 4.5.4 Illustration of how to determine the Burgers vector.

It doesn't matter where you start; you will always get the same vector in terms of magnitude and direction, although it might be translated to a different place in the crystal. Notice that if you follow these steps without going around a dislocation you will just make a square and end up exactly where you started. Figures 4.5.2 and 4.5.3 show the Burgers vectors for the different types of dislocations. We won't practice drawing the Burgers vector, but the next questions will ask you to look at how the Burgers vector relates to the dislocation.

Guided Inquiry: Burgers Vector

Try to draw a Burgers vector on your own. No matter where you start you should always get the same vector.

4.5.11 Look at the figure of an edge dislocation in Figure 4.5.2. What is the relationship between the direction of the Burgers vector and the direction of the dislocation?

4.5.12 Look at the figure of a screw dislocation in Figure 4.5.3. What is the relationship between the direction of the Burgers vector and the direction of the dislocation?

4.5.13 The Burgers vector usually lies along a close-packed direction. Given this fact, what is the magnitude of the Burgers vector in terms of the atomic radius R?

Concept Check **4.5.3**

- For iron (BCC), what is the direction of the Burgers vector?
- For iron, what is the magnitude of the Burgers vector?

Guided Inquiry: Planar Defects

4.5.14 List and describe in your own words three types of planar defects in crystals (see Figures 4.5.5–4.5.7).

The most important type of planar defect is shown in Figure 4.5.5 and is called a *grain boundary*. Up until now we have been talking about *single crystals*. A single crystal is a material in which the unit cells are all lined up with each other, even if there are some point or line defects. A diamond is a good example of a single crystal. Single crystals are transparent and they have facets. As we discussed in Section 4.4, the properties of single crystals can be anisotropic because of the different LDs of different directions. Single crystals are actually hard to make. Most crystalline materials exist as *polycrystals*, which is what is shown in Figure 4.5.5. You can think of a polycrystal as a bunch of tiny single crystals that have been stuck together (although that is not how they are formed). Each little single crystalline region is called a *grain* and the boundary between grains is the grain boundary. A grain is typically a few microns to several hundred microns

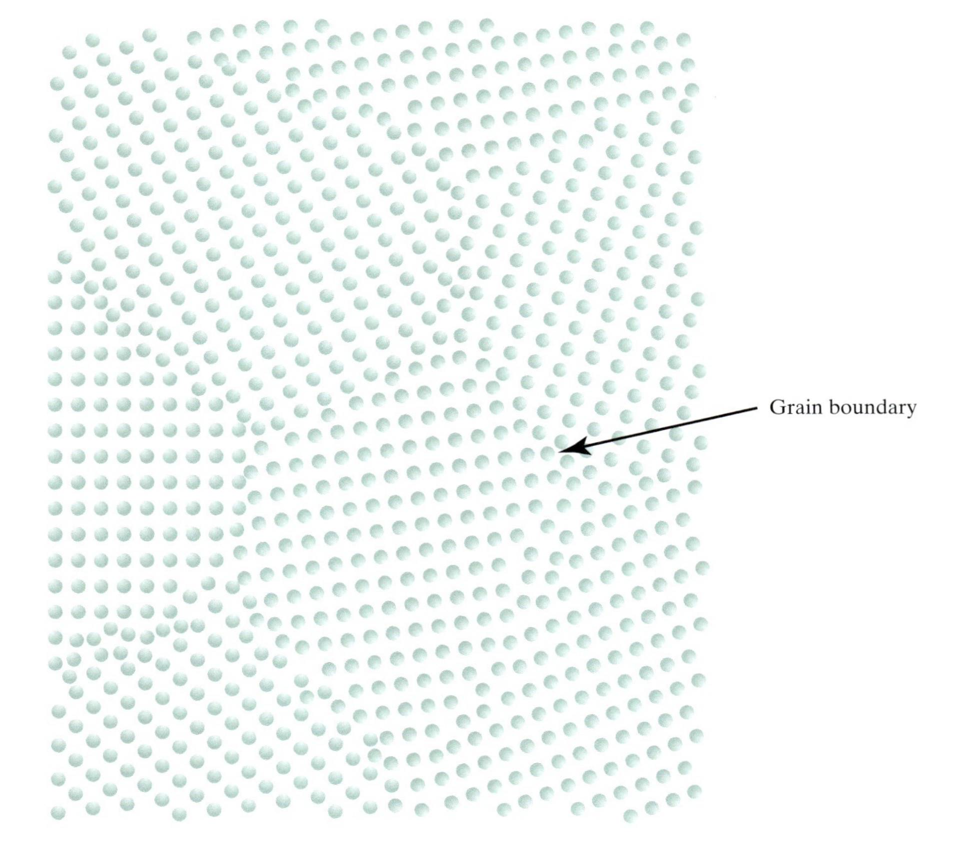

Figure **4.5.5**
A type of planar defect in a crystal.

in diameter. The grain boundaries scatter light, so polycrystals are opaque. If the grains are oriented randomly the properties are also isotropic; if you measure a property in one direction you are actually measuring it for millions of randomly oriented grains, so you get an average property corresponding to all possible crystal orientations. The presence of grains has a significant influence on the properties of materials, as we will see in Chapter 9.

Two other types of planar defects are the tilt boundary shown in Figure 4.5.6 and the twin boundary shown in Figure 4.5.7. These both are situations in which the alignment of the unit cells has been disturbed. The difference between a grain boundary and these other two is sometimes not clear. In general we can say that there is still some regular arrangement of atoms that defines the tilt boundary and twin, but the atoms are just randomly jumbled up at a grain boundary.

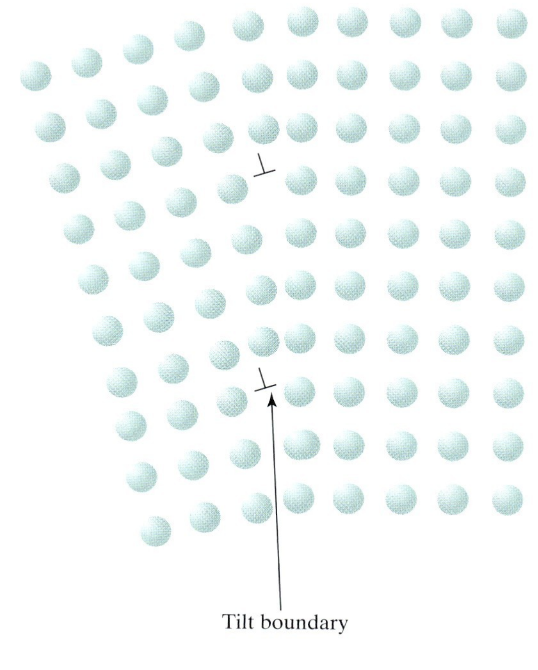

Figure 4.5.6
A type of planar defect in a crystal.

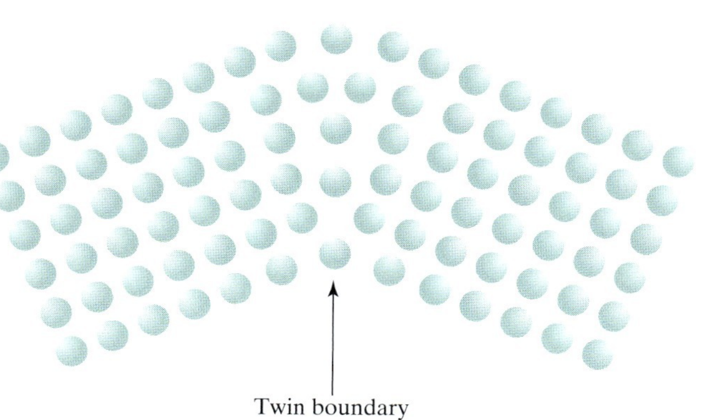

Figure 4.5.7
A type of planar defect in a crystal.

4.6 Ceramic Crystal Structures

LEARN TO: **Predict crystal structures of ceramic compounds.**

So far we have considered crystal structures of pure elements, such as copper or iron. But many crystals actually contain more than one type of atom. In some cases we can consider this extra atom to be an impurity, for example copper as a substitutional impurity in gold to make 14-karat gold, or carbon as an interstitial impurity in iron to make steel. In other cases, however, we need to consider both types of atoms together in determining the crystal structure. The crystal structures of ceramic compounds, such as NaCl, ZnS, CaF_2, and CsCl, can be determined by figuring out how the two different types of atoms pack. In order to understand how two different atoms can fit in a crystal, we will start by looking at the SC unit cell.

Guided Inquiry: Interstitial Sites

4.6.1 What percentage of a SC unit cell is empty space?
Remember the calculation of atomic packing factor.

4.6.2 Where could you fit another atom into the SC unit cell?

4.6.3 Will this atom be the same size as the other atoms, smaller than the other atoms, or larger than the other atoms?

4.6.4 How many of these extra atoms can you fit in the unit cell?

4.6.5 How many of the original SC atoms are surrounding the extra atom?

4.6.6 Based on the size of the atoms, if this were an ionic compound, which atoms would be the anions; the original atoms that make up the SC unit cell, or the atom you added? Explain your answer.

Concept Check 4.6.1

- **Sodium chloride is a FCC crystal structure with one of the types of atoms in the spaces between the other atoms. Which type of atoms (sodium or chlorine) are the original atoms that define the FCC lattice?**

An extra atom that fits in between the atoms of a unit cell is called an interstitial atom. The place where an interstitial atom can fit is called an *interstitial site*. Interstitial atoms can be present in all types of unit cells, but we will consider only SC and FCC unit cells. Figure 4.6.1 shows the interstitial site in a SC unit cell. This is called a cubic site because the 8 atoms that surround the interstitial site form a cube. The *coordination number* for the cubic interstitial site is 8, meaning that an atom in the interstitial site will touch 8 other atoms.

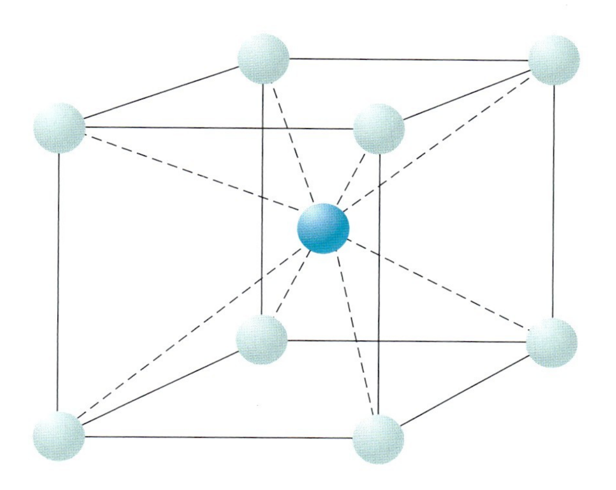

Figure **4.6.1**

Cubic interstitial site in a SC unit cell. The light green atoms form the SC lattice and the blue atom is in the interstitial site.

The FCC crystal structure has two types of interstitial sites. As shown in Figure 4.6.2, the octahedral sites are at the centers of each of the unit cell edges, and in the center of the unit cell. The octahedral site has a coordination number of 6, and those 6 atoms form an octahedron. This is most easily seen by looking at the octahedral site in the center of the unit cell. An atom in that site will touch the atoms on the center of each face.

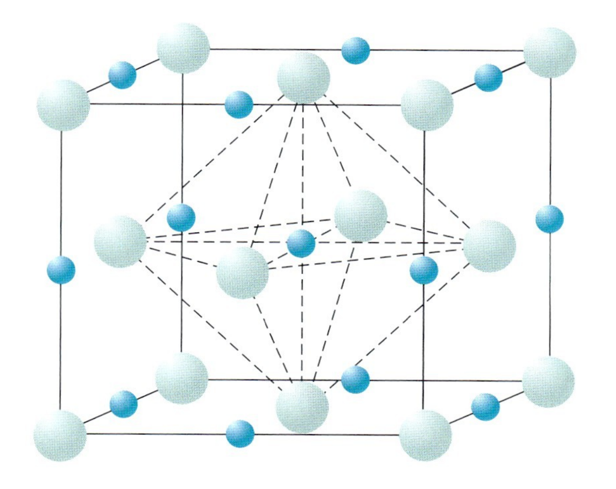

Figure **4.6.2**

Octahedral interstitial sites in a FCC unit cell. The light green atoms form the FCC lattice and the blue atoms are in the interstitial sites. The dotted black lines outline the octahedron formed from the nearest neighbors of the center interstitial atom.

FCC also has tetrahedral interstitial sites, shown in Figure 4.6.3. The tetrahedral sites are at the following atomic positions:

$\frac{1}{4}, \frac{1}{4}, \frac{1}{4}$
$\frac{3}{4}, \frac{1}{4}, \frac{1}{4}$
$\frac{1}{4}, \frac{3}{4}, \frac{1}{4}$
$\frac{3}{4}, \frac{3}{4}, \frac{1}{4}$
$\frac{1}{4}, \frac{1}{4}, \frac{3}{4}$
$\frac{3}{4}, \frac{1}{4}, \frac{3}{4}$
$\frac{1}{4}, \frac{3}{4}, \frac{3}{4}$
$\frac{3}{4}, \frac{3}{4}, \frac{3}{4}$

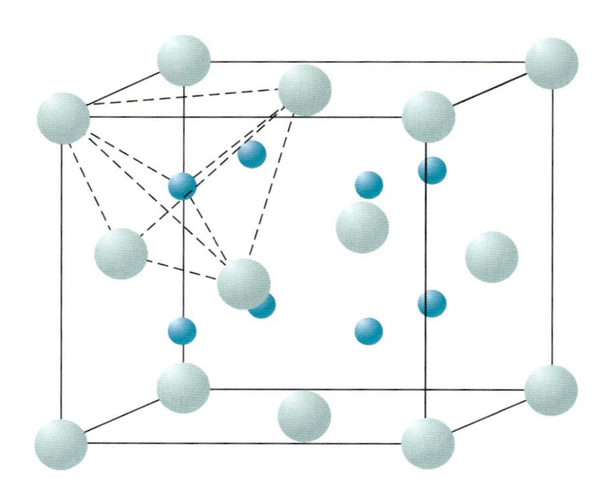

Figure 4.6.3
Tetrahedral sites in the FCC unit cell. The light green atoms form the FCC lattice and the blue atoms are in the interstitial sites. The dotted black lines show the tetrahedral formed by the nearest neighbors of one interstitial atom.

The tetrahedral site has a coordination number of 4, and those 4 atoms form a tetrahedron. Table 4.6.1 summarizes the characteristics of the different interstitial sites for all the crystal structures.

TABLE 4.6.1 Characteristics of the interstitial sites corresponding to close packing of atoms

Unit cell	Type of Interstitial site	Coordination Number	Number of Interstitial Atoms per Unit Cell
SC	Cubic	8	1
FCC	Octahedral	6	4
FCC	Tetrahedral	4	8

The type of interstitial site an atom can fit into depends on the ratio of the radii for the two different types of atoms. Because of bonding considerations, the anions and cations must touch, although the anions do not have to touch each other. Figure 4.6.4 illustrates the various possibilities for a planar triangle geometry (coordination number of 3). The largest interstitial site is the cubic site, so when the anion and cation are close to the same size the crystal structure will be a SC lattice of anions with the cations in the cubic interstitial sites. When the cation gets too small to fit into the cubic site, the geometry of the crystal will switch to they next largest interstitial site, which is the FCC octahedral site. By calculating the ratio of the radii for the cation and anion, we can determine which type of interstitial site the cation will be in, and therefore what the crystal structure will be. Table 4.6.2 shows the values of this ratio that correspond to each possible coordination number. You will now use the information in Tables 4.6.1 and 4.6.2 to predict the crystal structure for an ionic compound.

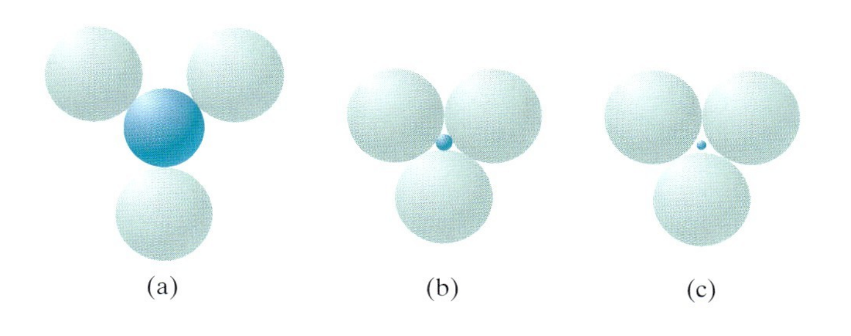

Figure 4.6.4

Illustration of the allowed geometries for nearest neighbor packing for anions (light green atoms) and cations (blue atoms): (a) geometry allowed because the anions and the cations are touching; (b) the smallest possible cation that would be allowed for this geometry; (c) geometry not allowed because the cation is not touching the anions.

TABLE 4.6.2 Coordination numbers and radius ratios corresponding to each of the possible interstitial site geometries

Coordination Number	Cation–Anion Radius Ratio	Coordination Geometry
2	< 0.155	
3	0.155–0.225	
4	0.225–0.414	
6	0.414–0.732	
8	0.732–1.0	

Guided Inquiry: Ionic Unit Cells

4.6.7 Table 4.2.1 shows two different radii for each atom. When calculating the radius ratio for ionic compounds, which radius should you use?

The following questions are all for CsCl.

4.6.8 What is the radius ratio for the two types of atoms?

4.6.9 What type of interstitial site is the cation in?

4.6.10 What type of unit cell do the anions form?

4.6.11 Describe the complete unit cell of CsCl that includes both the anions and cations.

4.6.12 Is your answer to question 4.6.11 the same as a BCC unit cell? Explain why or why not.

The following questions are all for ZnS.

4.6.13 Determine the complete unit cell for ZnS.

4.6.14 What is the total charge on the unit cell you determined in question 4.6.13?

4.6.15 What do you think the charge should be for an ionic unit cell?

4.6.16 If your answers for question 4.6.15 and 4.6.16 do not match; how can you adjust the unit cell to make them match?

You can't add new atoms to the unit cell, you can only take atoms out.

Concept Check 4.6.2

- **What is the crystal structure for ZnS?**

EXAMPLE PROBLEM 4.6.1

Determine the crystal structure of MgS.

We need to begin by determining the type of interstitial site from the radius ratio. The radii of the ions is as follows:

$$r(Mg^{2+}) = 0.066 \text{ nm}$$
$$r(S^{2-}) = 0.184 \text{ nm}$$
$$r(Mg^{2+}) / r(S^{2-}) = 0.066 \text{ nm} / 0.184 \text{ nm} = 0.36$$

This ratio falls in the range for tetrahedral coordination, so the base crystal structure is FCC with magnesium ions in the tetrahedral interstitial sites. However, we still need to check that the unit cell has the correct ratio of atoms. The formula MgS indicates there must be an equal number of magnesium and sulfur ions for charge neutrality. Now let's look at the actual number of each of these atoms in a FCC unit cell.

For sulfur, which forms the FCC lattice, there are four atoms per unit cell.

For magnesium, which is in the interstitial sites, there are eight atoms per unit cell.

There are too many magnesium atoms. We can fix this by filling only half the interstitial sites. So the crystal structure is a FCC lattice of sulfur ions with magnesium ions in half the tetrahedral interstitial sites. The structure is shown below.

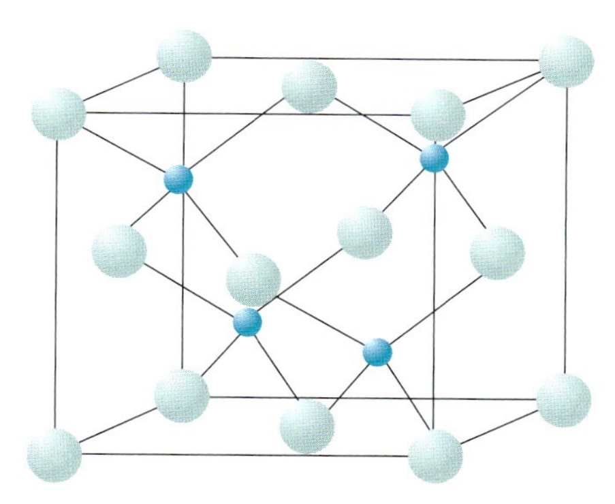

4.7 Defects in Ceramic Crystals

LEARN TO: **Determine types of defects in ceramic crystals.**

Just like metallic crystals, ceramic crystals can have defects. The difference is that when we identify the types of defects present in ceramic crystals we need to consider how they affect the overall charge on the crystal. Figure 4.7.1 shows a defect in an NaCl crystal. We will use this figure to start to understand the types of defects that can form in a ceramic crystal.

Figure 4.7.1

Schematic diagram of an NaCl crystal after the formation of an Na^+ vacancy.

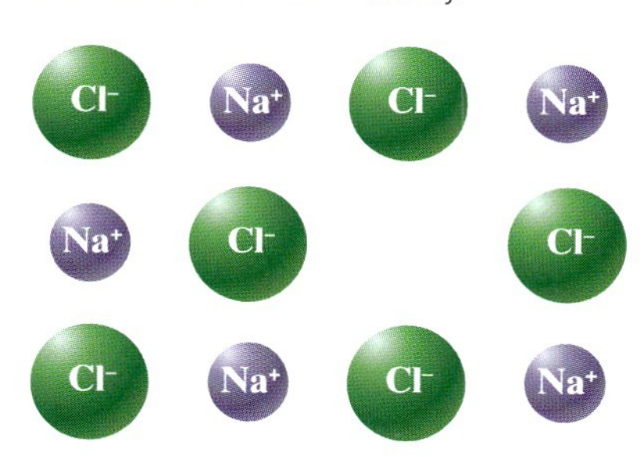

Guided Inquiry: Ionic Defects

4.7.1 What is the charge on the crystal before the vacancy is formed?

4.7.2 What is the charge on the crystal after the vacancy is formed?

4.7.3 What should the charge on an ionic crystal be?
What has the lowest energy; something that is positively charged, negatively charged, or neutral?

4.7.4 Describe two possible ways to change the crystal shown so that the charge on the crystal matches your answer to question 4.7.3.

Concept Check 4.7.1

- **How would your answer to question 4.7.4 change if the crystal were $CaCl_2$ (that is, all the Na^+ atoms in Figure 4.7.1 were changed to Ca^{2+})?**

The two types of defects you identified in question 4.7.4 have specific names. Figure 4.7.2 shows a cation vacancy paired with an anion vacancy, which is called a *Schottky defect* after Walter H. Schottky. Schottky is better known for his work on the behavior of electrons in solids, and several important aspects of semiconductor devices are named after him (Schottky barrier, Schottky diode, and others). The *Frenkel defect* is an ion vacancy and a corresponding ion interstitial impurity. The Frenkel defect was discovered in 1926 by Yakov Frenkel, a physicist.

Figure **4.7.2**

A Schottky defect, which is a cation–anion vacancy pair.

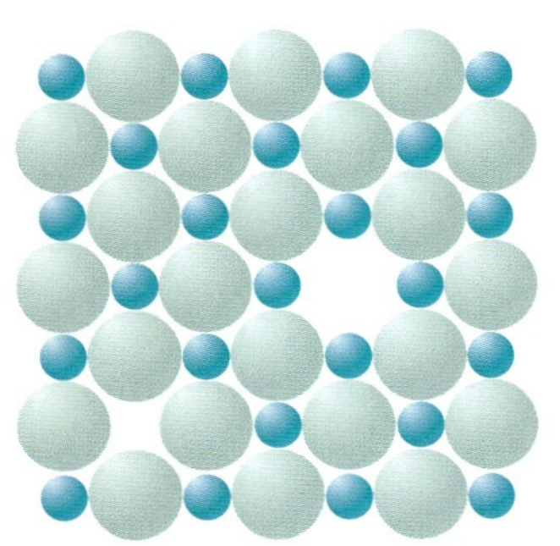

Figure **4.7.3**

A Frenkel defect, which is a vacancy and an interstitial impurity of the same ion.

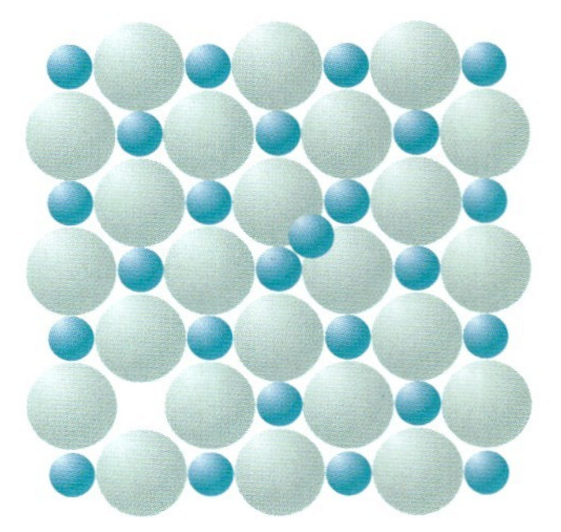

Other types of defects can exist, although they don't have specific names. One particular type involves multiple oxidation states. Transition metal atoms, such as iron, can have more than one oxidation state, or charge. For example, iron ions can exist as Fe^{+}, Fe^{2+}, or Fe^{3+}. If the oxidation state of an ion in a crystal changes, that is another type of defect. Figure 4.7.4 shows a crystal of FeO after one of these defects is formed. You will now use charge balance to determine all the defects that form when the oxidation state of the one atom is changed.

Figure 4.7.4

A crystal of FeO after one of the iron atoms has been oxidized to a charge of +3 instead of the normal +2.

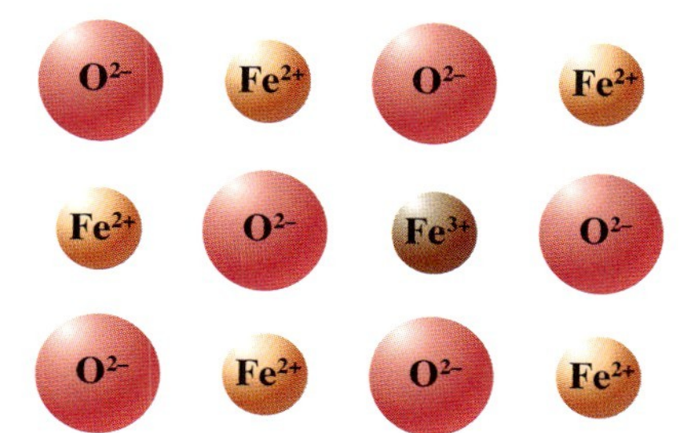

Guided Inquiry: Changing Oxidation State

4.7.5 What is the charge on the crystal before the defect is formed?

4.7.6 What is the charge on the crystal after the defect is formed?

4.7.7 What is one thing you could do to the crystal to try to make the charge zero? What is the charge after you do this?

The charge may not be zero at this point; you should just do one thing that starts to get you in the right direction.

4.7.8 What is the next thing you could do to the crystal to try to make the charge zero? What is the charge now?

4.7.9 At this point you should see a pattern that will eventually get you to zero charge. Describe all of the defects in the crystal once the charge is zero.

Concept Check 4.7.2

- In a crystal of FeO, one Fe^{2+} is oxidized to Fe^{3+}. What are all the defects that result from this?
- If a Ca^{2+} interstitial defect is added to a $CaCl_2$ crystal, what other defect will also form as a direct result?

EXAMPLE PROBLEM 4.7.1

In a CuO crystal, some Cu_2O_3 is added as an impurity so the Cu^{3+} substitutes for Cu^{2+}. What other type of defect is also formed, and how many of these defects are formed for each Cu^{3+} impurity atom?

The way to solve this problem is to just keep adjusting the defects to eventually balance the charge. First, notice that when one Cu^{3+} is added, the charge on the crystal becomes +1, so you need to create negative charge to balance this new positive charge. Some things you might consider are the following:

- Adding a Cu^{3+} interstitial. This won't work because it adds +3 of positive charge instead of the negative charge you want.
- Creating an O^{2-} vacancy. This won't work because you are taking away −2 of negative charge, or equivalently adding +2 of positive charge.
- Create an O^{2-} interstitial. Although this would work from a charge standpoint, you also need to think about atomic size. An oxygen anion is too big to fit into an interstitial site, so this won't work.
- Create a Cu^{2+} vacancy. When you do this you remove +2 of positive charge, or equivalently create −2 of negative charge. Combined with the Cu^{3+} substitutional impurity, the charge on the crystal is −1. Since we saw above that adding one Cu^{3+} impurity creates +1 of charge, adding a second Cu^{3+} impurity will balance out this negative charge.

So the final result is that we create one Cu^{2+} vacancy for every two Cu^{3+} impurity atoms.

Application Spotlight **Fuel Cells**

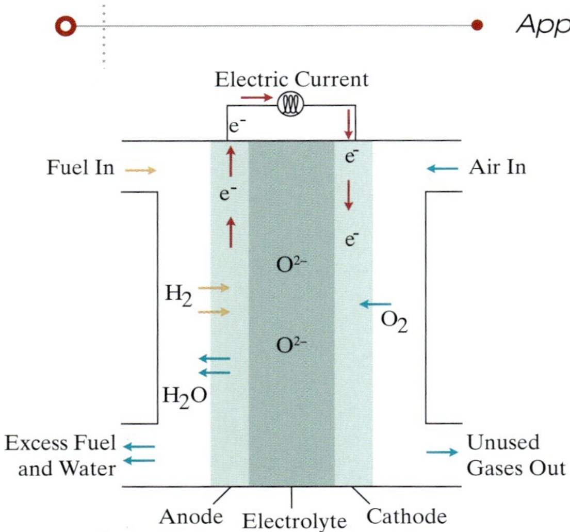

Fuel cells are devices for converting chemical energy to electricity that are similar to batteries. Like batteries they generate electricity from an electrochemical reaction. However, batteries carry their fuel with them, while fuel cells require fuel to be added. In a typical solid oxide fuel cell, as shown in the figure, the fuel is a source of hydrogen. Oxygen from air is turned into an ion, travels through the oxide (a ceramic crystal) and then reacts with hydrogen to form water. In the reaction an electron is released, which provides the electric current. One key to making a fuel cell efficient is good transport of the oxygen ion through the ceramic electrolyte. So how does the oxygen ion get through a solid material? The answer is defects. The oxygen ion can "hop" through the electrolyte by moving into vacancies. This is why solid oxide fuel cells work best at higher temperatures. As you saw from Equation (4.5.1) for vacancy concentration, there are more vacancies at higher temperatures. This means there are more opportunities for the oxygen ion to hop, so it can move more easily and generate more electricity. A typical solid oxide fuel cell operating temperature is 500° C–1000° C. Of course running at higher temperatures also uses up more energy, so research is ongoing to reduce the operating temperature.

4.8 Determining Crystal Structure: Diffraction

LEARN TO: **Use Bragg's law to relate crystal structure and diffraction peaks.**

Now that we know something about the structure of crystals, how do we actually determine that structure? A powerful and common way to do it is by using x-ray *diffraction*. X-rays were discovered in 1895 by Wilhelm Roentgen, a professor of physics in Bavaria. The theory of x-ray diffraction by crystals was first proposed in 1912 by Paul Ewald as part of his Ph.D. thesis; at the time he was only 24 years old. In 1913 the father–son team of William Henry Bragg and William Lawrence Bragg created what has become known as *Bragg's law* (which we will show below). In 1915 they became the only father–son team to share a Nobel Prize (although other pairs of fathers and sons have won separate prizes). William L. Bragg was only 25 when he won the Nobel Prize, another example of someone in his early twenties making a major contribution to science. This is actually not that unusual; many of the greatest scientific discoveries have been developed by people in their twenties. The experimental setup for x-ray diffraction is shown schematically in Figure 4.8.1.

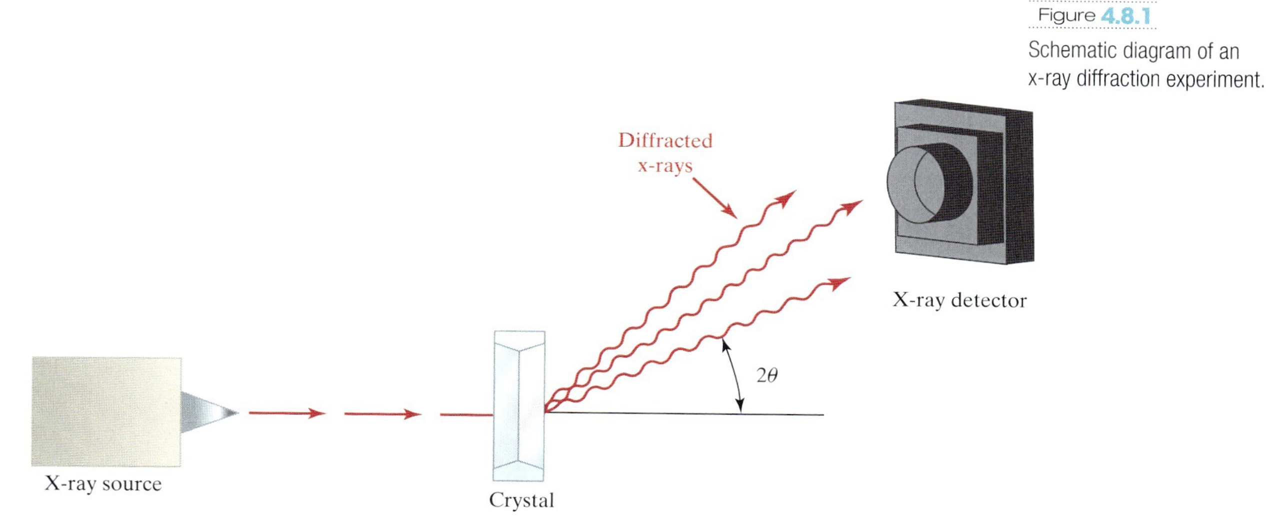

Figure 4.8.1

Schematic diagram of an x-ray diffraction experiment.

An x-ray source creates a beam of x-rays, which hits the sample. The sample causes the x-rays to be deflected, or scattered, in different directions. The angle between the original beam direction and the scattered beam direction is called the *scattering angle* and is given the symbol 2θ. An x-ray detector measures the intensity of the x-rays at the different values of the scattering angle. The power of x-ray diffraction is that it provides a lot of information which can be shown in many different ways. One way is to collect the scattered x-rays on a piece of film. Doing this can provide information about the overall arrangement of the atoms. Figures 4.8.2 shows two examples. Figure 4.8.3 shows the x-ray diffraction pattern of DNA similar to one taken by Rosalind Franklin in 1953. James D. Watson and Francis Crick were shown this diffraction pattern and used it as part of their evidence to show that DNA is a double helix. The history of the papers describing the structure of DNA is somewhat complicated, but essentially Watson and Crick published their paper first with a paper co-authored by Franklin appearing alongside it. Franklin did not share the Nobel Prize at least in part because she had died by the time the prize was awarded in 1962. At the time of their discovery Crick was 36 and Watson was 25, yet another example of a relatively young person making a discovery that would change the world.

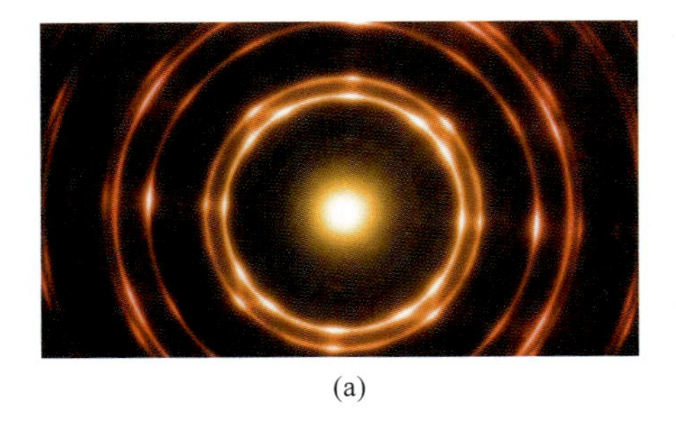

(a)

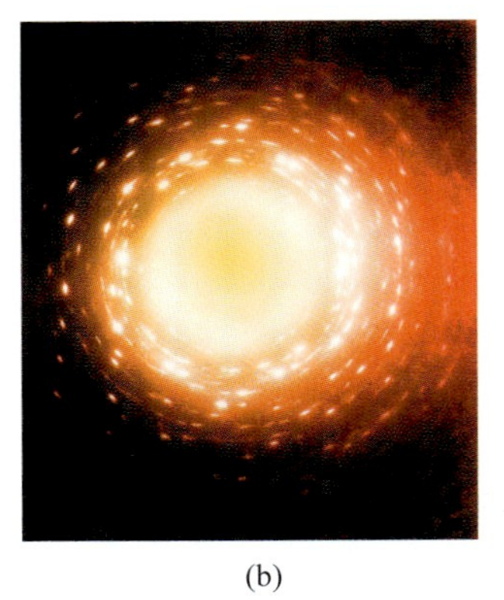

(b)

Figure 4.8.2

X-ray diffraction patterns of (a) chromium and (b) polyethylene crystal. (Science Source/Photo Researchers, Inc.)

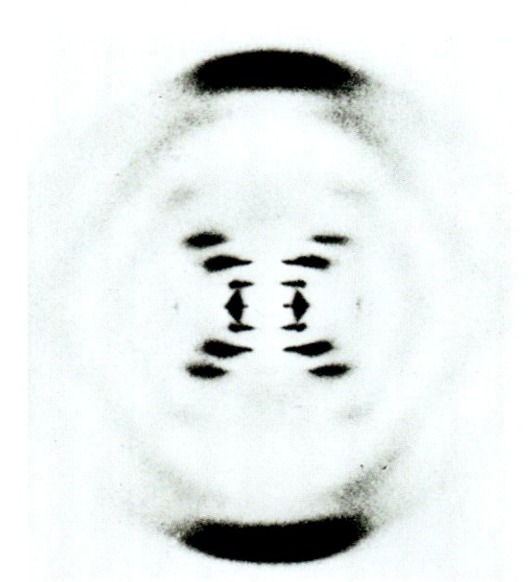

Figure 4.8.3

X-ray diffraction pattern of DNA showing evidence that DNA forms a double helix. (Science Photo Library/ Custom Medical Stock Photography)

Another version of x-ray diffraction is to use a detector that measures x-rays at a single location, and to scan that detector across a range of scattering angles. This type of system is called a powder camera. The result of this experiment is a diffraction pattern like the one shown in Figure 4.8.4. In this figure you can see that x-rays are scattered only at certain scattering angles. Bragg's law allows us to determine at which angles the x-rays are scattered and to relate those angles to the structure of the material.

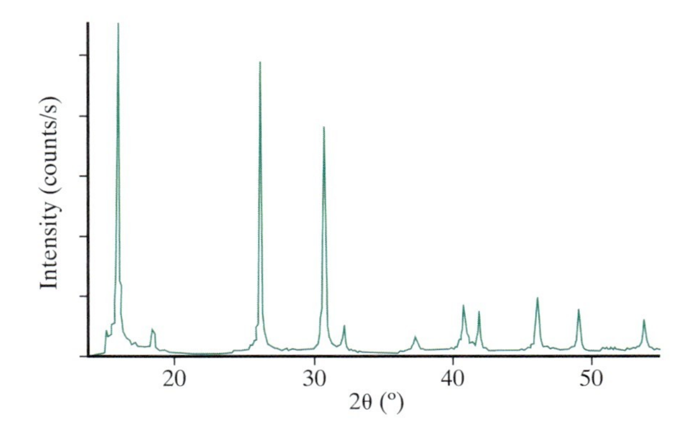

Figure 4.8.4

X-ray diffraction pattern of SiC taken with a powder camera.

To understand how Bragg's law relates scattering angle and structure, we will derive it for a crystal. Figure 4.8.5 shows how the x-rays interact with crystalline planes. In this figure we are looking at the planes edge-on, and so d is the spacing between planes. We imagine there are only two x-ray photons; one that hits the atom on the top plane (photon 1) and one that hits the atom just below it (photon 2). Because photons can be considered as both particles and waves, we will consider the requirement for constructive interference between the two photon waves. Figure 4.8.6 shows that photon 2 travels further than photon 1 by a distance of

$$2d \sin\theta \tag{4.8.1}$$

In order for constructive interference to occur, this extra distance must be equal to an integer number of wavelengths so that the waves of photons 1 and 2 are in phase. Therefore we can say that constructive interference occurs only when

$$n\lambda = 2d \sin\theta \tag{4.8.2}$$

Equation (4.8.2) is Bragg's law. In this equation, n is an integer, which we will generally assume is equal to 1; λ is the wavelength of the x-rays; d is the distance between the planes; and θ is half the scattering angle. That last definition is very important to remember. Because of the way the experiment is set up, if we say "scattering angle" we are talking about 2θ.

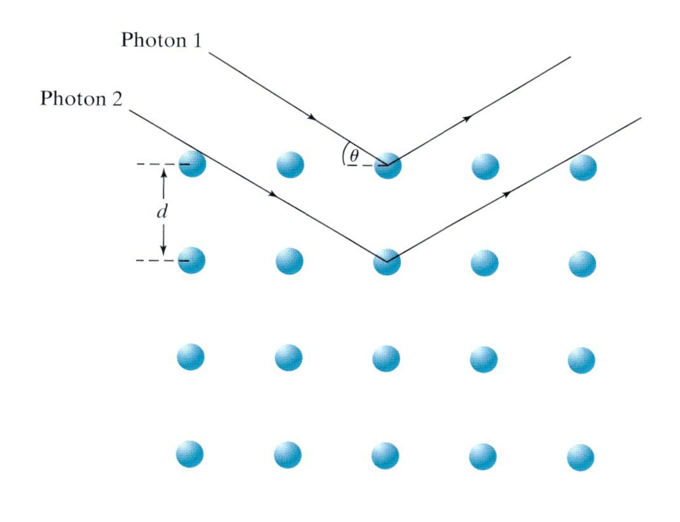

Figure **4.8.5**

Illustration of two x-ray photons interacting with atoms on different planes of the same family in a crystal. In this figure we are looking at the planes edge-on, and d is the distance between the planes.

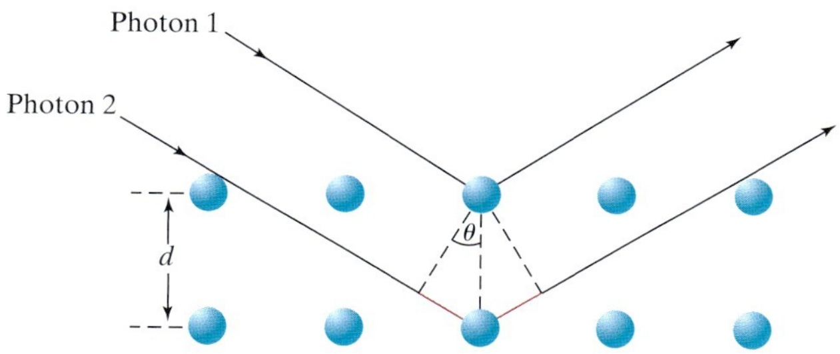

Figure **4.8.6**

Illustration of the additional distance traveled by photon 2 versus photon 1. The red line is the extra distance that photon 2 travels.

What this equation tells us is that we will only detect x-rays scattered at certain angles depending on the spacing between crystallographic planes. At those angles there is constructive interference and at all other angles there is destructive interference. Thus, by knowing the scattering angles at which diffraction occurs (that is, at which there is constructive interference) we can determine the structure of the crystal. This is a simplified derivation of Bragg's law. A more complete derivation uses something called reciprocal space.

In order to determine the crystal structure from x-ray diffraction, we need a little more information. For cubic systems, we can relate d and the lattice parameter for the plane (hkl) with the following equation:

$$d = \frac{a}{\sqrt{h^2 + k^2 + l^2}} \tag{4.8.3}$$

Similar equations exist for noncubic systems, but they are more complicated. There are also selection rules, which account for the presence of parallel planes with the same planar density. For example, if you look at (100) and (200) in FCC you will see that they have the same planar density. Because of this there is destructive interference from these planes. To account for all the possibilities like this that exist for cubic crystals, scattering will occur only if certain selection rules are met. The selection rules are as follows:

- BCC: $h + k + l$ must be even.
- FCC: h, k, l must be all even or all odd (0 counts as even).

Using Bragg's law and the selection rules, we can make simple correlations between crystal structure and x-ray diffraction. To use these, we generally need to know either the crystal structure or the specific plane that the diffraction is coming from. To start with something completely unknown and determine the crystal structure from x-ray diffraction requires more advanced techniques, such as Rietveld refinement. Rietveld refinement uses least squares fitting to match a calculated diffraction pattern for an assumed crystal structure to the measured diffraction pattern. When these two patterns match as closely as possible, you can say that the assumed crystal structure is likely to be the actual crystal structure. Diffraction is also possible with other forms of radiation, such as electrons and neutrons. Because of the wave nature of these subatomic particles, Bragg's law can be used in the same ways for x-rays. You will now practice using Bragg's law and identify what you can learn about a material from x-ray diffraction.

Guided Inquiry: Diffraction

4.8.1 What is the distance between (112) planes in iron?
Don't get confused by the selection rules. You can always calculate a distance between planes even if the selection rule is not met. The selection rules are only to see if scattering is observed from those planes.

4.8.2 At what angle will scattering be observed from the (112) plane of iron using an x-ray wavelength of 0.154 nm?

4.8.3 If scattering is observed at a scattering angle (2θ) of 38.4° from the (111) plane of aluminum, what is the distance between (111) planes in aluminum? The x-ray wavelength is 0.154 nm.

4.8.4 Based on your answer to question 4.8.3, what is the lattice parameter of aluminum?

4.8.5 Based on your answer to question 4.8.4, what is the atomic radius of aluminum?
Does your answer match the value in Table 4.2.1?

4.8.6 What quantities regarding a crystal can be calculated from knowing the scattering angle for a given plane?

4.8.7 What quantities regarding a crystal can be calculated from knowing the atomic radius?

4.8.8 If you are given a scattering angle, plane, and two possible materials, use your answers from questions 4.8.6 and 4.8.7 to create a strategy for determining which of the two materials it actually is.

4.8.9 Diffraction is observed from (110) at a scattering angle of 40.60°. The x-ray wavelength is 0.154 nm. Use your answer from question 4.8.8 to determine whether this is cesium or molybdenum.

Concept Check **4.8.1**

- **At what angle will scattering occur for the (111) plane of iron if the x-ray wavelength is 0.154 nm?**

EXAMPLE PROBLEM 4.8.1

At what angle will scattering occur for the (211) plane of chromium if the x-ray wavelength is 0.154 nm?

The first thing you should always do is check the selection rules. Chromium is BCC, so $h + k + l$ must be even. For this problem $2 + 1 + 1 = 4$, which is even, so the selection rule is satisfied and scattering will occur. Now we just need to figure out the angle from Bragg's law:

$$n\lambda = 2d\sin\theta$$

We assume n equals 1. Rearranging this equation to solve for the angle, we get the following:

$$\sin\theta = \frac{\lambda}{2d}$$

We still need to determine d. For cubic crystals:

$$d = \frac{a}{\sqrt{h^2 + k^2 + l^2}}$$

Now we need the lattice parameter, a. Since chromium is BCC,

$$a = \frac{4R}{\sqrt{3}}$$

The radius of chromium is 0.128 nm. Now we can solve all of these equations:

$$a = \frac{4(0.128 \text{ nm})}{\sqrt{3}} = 0.296 \text{ nm}$$

$$d = \frac{0.296 \text{ nm}}{\sqrt{2^2 + 1^2 + 1^2}} = 0.120 \text{ nm}$$

$$\sin\theta = \frac{0.154 \text{ nm}}{2(0.120 \text{ nm})} = 0.642$$

$$\theta = 39.9°$$

Since the scattering angle is 2θ, the answer is 79.8°.

Summary

This chapter is about the first level of structure in materials: the arrangement of atoms in crystals. By understanding how atoms are packed in unit cells you can calculate basic materials properties, such as atomic packing factor and density. This is true for simple crystals, such as the FCC and BCC metals, or more complicated ionic crystals in which there are atoms on both the lattice sites and the interstitial sites. Both types of crystals also have defects: point, line, and planar defects. For ionic crystals we really looked only at the point defects, but the other types of defects also exist in ionic crystals. The big difference between metallic crystals and ionic crystals is the need to consider the charge on the atoms.

Other parts of the chapter may have seemed more theoretical—for example, things like Miller indices, linear density, and planar density. Although you should now know how to calculate them, why we care is probably not clear. These will come up again in Chapter 9 where we will learn how to predict mechanical properties based on crystal structures.

The last part of the chapter was an introduction to experimental techniques for looking at crystals using x-ray diffraction. X-ray diffraction is a very powerful tool that is used extensively in investigating materials. In addition to the basic calculations you learned how to do, it can be used for determining the crystal structure of materials (such as figuring out if it is BCC, FCC, or something else), looking at changes in crystal structure that occur upon heating, and doing real-time measurements of deformation at the atomic level.

Key Terms

Atomic packing factor
Body-centered cubic
Bragg's law
Burgers vector
Close-packed direction
Close-packed plane
Coordination number
Crystal
Diffraction
Dislocation
Edge dislocation
Face-centered cubic
Family of directions
Family of planes
Frenkel defect
Glass
Grain
Grain boundary
Interstitial impurity
Interstitial site
Lattice parameter
Linear density
Miller indices
Mixed dislocation
Planar density
Plane of highest density
Polycrystal
Scattering angle
Schottky defect
Screw dislocation
Simple cubic
Single crystal
Substitutional impurity
Unit cell
Vacancy

Problems

Skill Problems

4.1 For the crystal structure shown below, answer the following:

a. How many light green atoms are there per unit cell?

b. How many blue atoms are there per unit cell?

c. How many black atoms are there per unit cell?

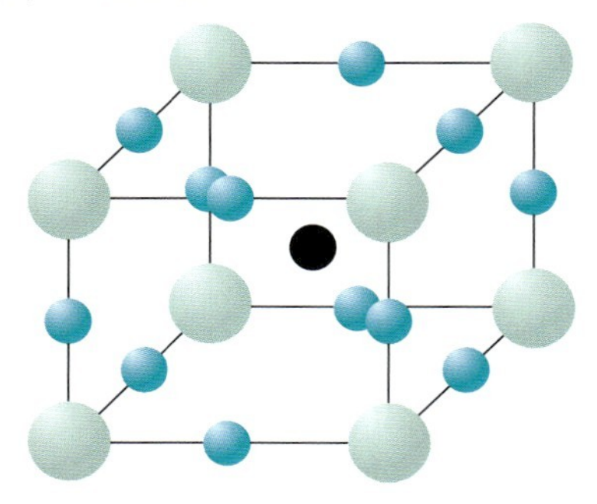

4.2 For the crystal structure shown below, answer the following:

a. How many light green atoms are there per unit cell?

b. How many blue atoms are there per unit cell?

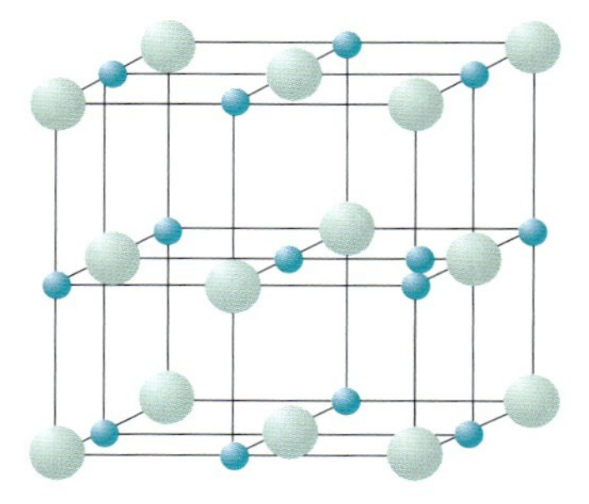

4.3 For the crystal structure shown below, answer the following:

a. How many light green atoms are there per unit cell?

b. How many blue atoms are there per unit cell?

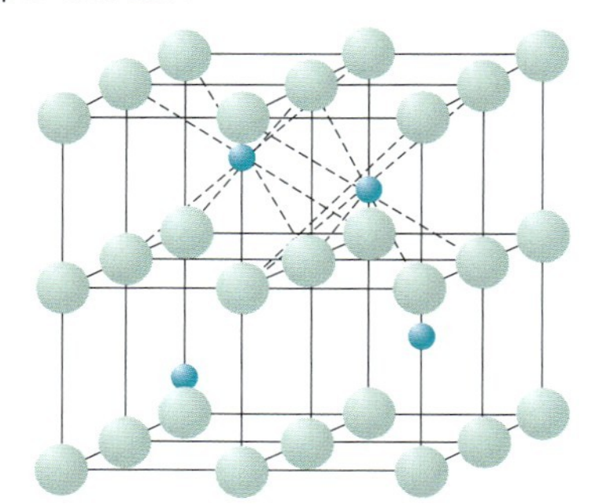

4.4 The crystal structure shown below is the unit cell for a high-temperature superconductor. Answer the following for this material:

a. How many barium atoms are there per unit cell?
b. How many yttrium atoms are there per unit cell?
c. How many copper atoms are there per unit cell?
d. How many oxygen atoms are there per unit cell?

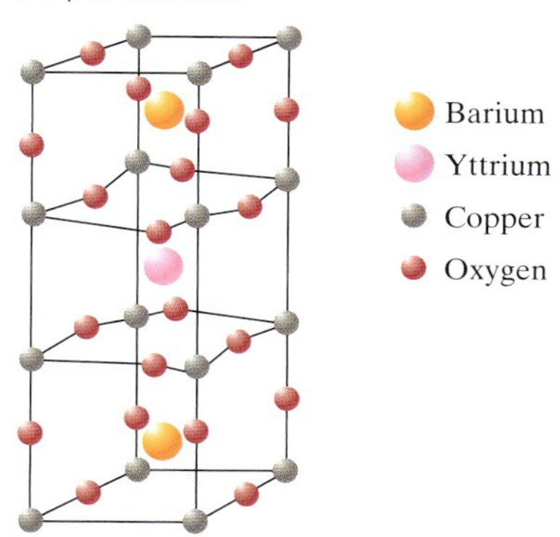

4.5 Calculate the atomic packing factor for a FCC unit cell.

4.6 Calculate the atomic packing factor for a BCC unit cell.

4.7 Calculate the density of lead. Is your answer consistent with the value shown in Table 4.2.1? Should it be?

4.8 Calculate the density of chromium. Is your answer consistent with the value shown in Table 4.2.1? Should it be?

4.9 What are the Miller indices for the directions shown in the figures below?

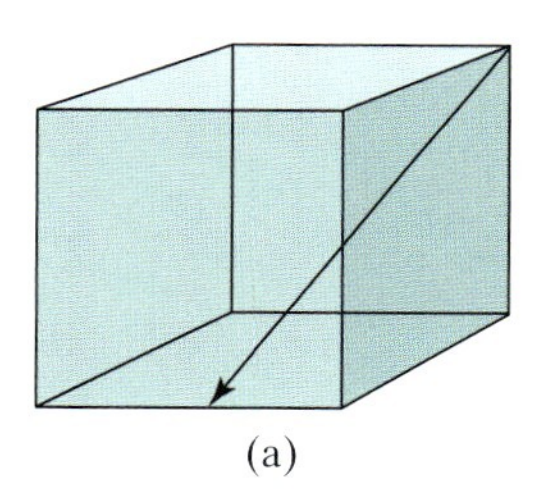
(a)

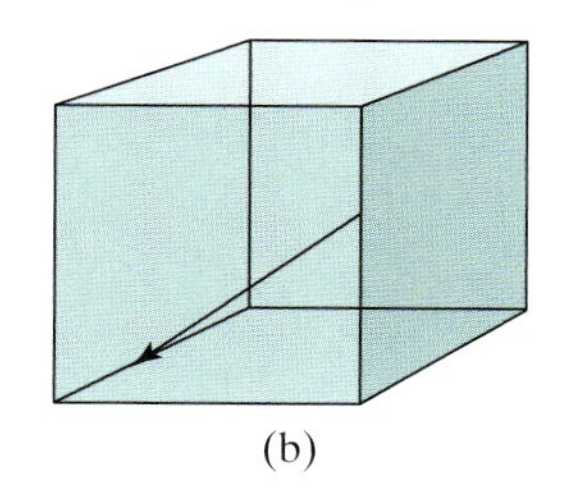
(b)

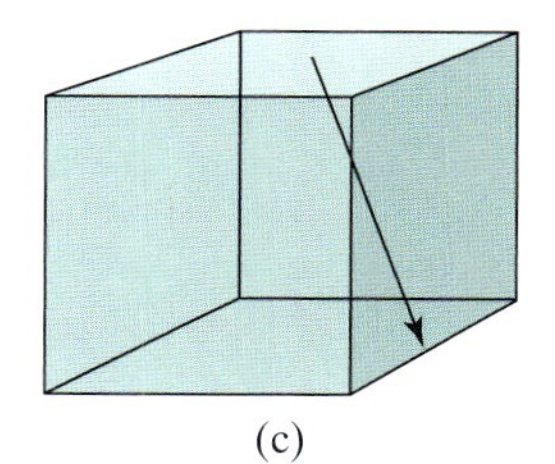
(c)

4.10 What are the Miller indices for the planes shown in the figures below?

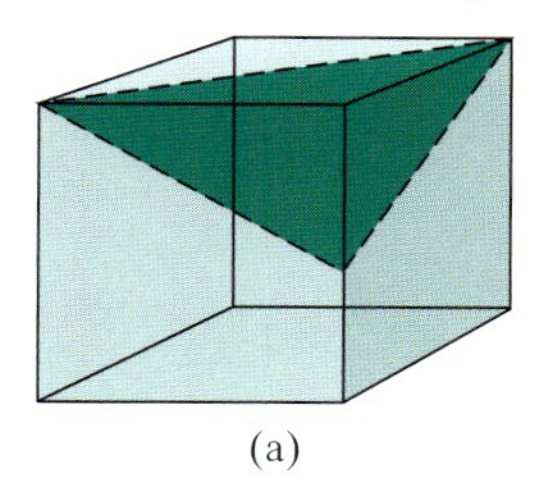
(a)

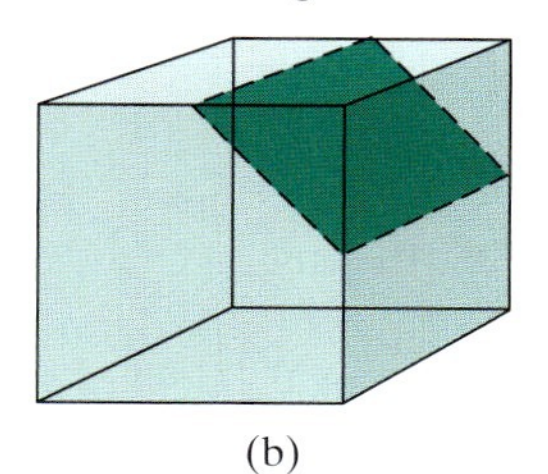
(b)

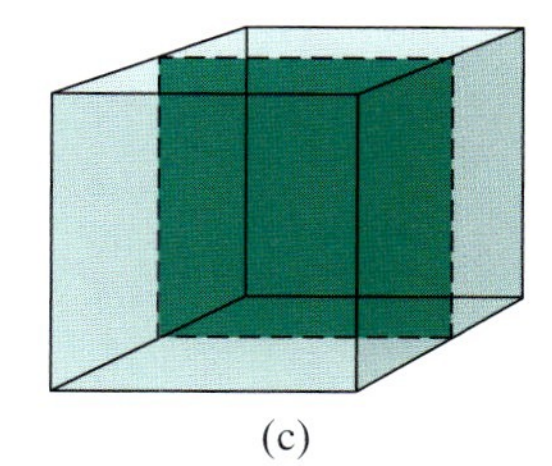
(c)

4.11 Draw the following directions in a cubic unit cell (you may put each in a different unit cell):

a. $[211]$
b. $[11\bar{2}]$
c. $[210]$
d. $[120]$

4.12 Draw the following planes in a cubic unit cell (you may put each in a different unit cell):

a. (001)
b. (121)
c. $(0\bar{1}\bar{1}$
d. $(11\bar{2})$

4.13 Calculate the linear density of the following directions:

a. [100] in FCC
b. [100] in BCC
c. [111] in SC
d. [111] in BCC

4.14 Calculate the planar density of the following planes:

a. (110) in BCC
b. (110) in FCC
c. (111) in SC
d. (111) in BCC

4.15 Is carbon likely to be a substitutional or interstitial impurity in iron? Explain why.

4.16 Is aluminum likely to be a substitutional or interstitial impurity in copper? Explain why.

4.17 Calculate the percentage of vacancies in iron at 1200° C. The activation energy for the formation of vacancies in iron is 1.08 eV/atom.

4.18 What would the temperature need to be to double the percentage of vacancies in iron from the percentage at 1200° C? The activation energy for the formation of vacancies in iron is 1.08 eV/atom.

4.19 What are the magnitude and direction of the Burgers vector for gold?

4.20 What are the magnitude and direction of the Burgers vector for chromium?

4.21 Determine the crystal structure of CsBr.

4.22 Determine the crystal structure of NaCl.

4.23 Determine the crystal structure of CaF_2.

4.24 Some $CaCl_2$ is added as an impurity to a crystal of CaO. If the Cl^- substitutes for the O^{2-}, what additional defect will form, and how many of them will form for each Cl^- impurity atom?

4.25 If a Ca^{2+} interstitial defect is added to a $CaCl_2$ crystal, what additional defect will form, and how many of them will form for each Ca^{2+} impurity atom?

4.26 If a sulfur vacancy is created in a crystal of Na_2S, what additional defect will form, and how many of them will form for each vacancy?

The next three questions refer to tantalum (Ta), which has a BCC crystal structure and an atomic radius of 0.146 nm.

4.27 What is the spacing between {111} planes in tantalum?

4.28 At what scattering angle will diffraction occur for the {111} planes, using an x-ray wavelength of 0.154 nm?

4.29 At what scattering angle will diffraction occur for the {110} planes, using an x-ray wavelength of 0.154 nm?

4.30 X-ray scattering is observed from the (220) plane of palladium at 66.7° when x-rays of wavelength 0.154 nm are used. From this information determine the crystal structure of palladium. The metallic radius of a palladium atom is 0.140 nm.

4.31 You discover a new metal and perform x-ray diffraction on it using a wavelength of 0.154 nm. You find that it is BCC, with scattering occurring from the (202) plane at a scattering angle of 64.2°. What is the atomic radius of this metal?

Conceptual Problems

4.32 As you saw in the Guided Inquiry questions, the relationship between the lattice parameter and the atomic radius for a SC unit cell is $a = 2R$, and for a FCC unit cell it is $a = 2R\sqrt{2}$. Determine this relationship for a BCC unit cell.

4.33 Calculate the density of CsCl. Note that the atoms touch along the body diagonal.

4.34 Consider the noncubic unit cell ($a \neq c$) shown below. For this unit cell, the (001) and (100) planes are not in the same family. Explain why.

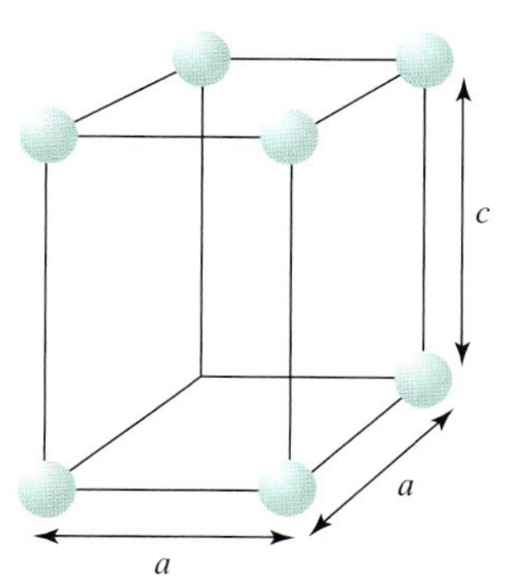

4.35 Your friend claims to have created a perfect, completely defect-free, single crystal of copper. Explain why this is impossible.

4.36 You have discovered a new element. You have determined that its atomic mass is 83.6 g/mol, and from x-ray diffraction you know that it has a BCC crystal structure and scattering is observed from the {110} planes at a scattering angle of 65° when an x-ray wavelength of 0.154 nm is used. What is this element's density in g/cm^3?

Bullet proof vests rely on lightweight polymers that are designed at the molecular level to have good impact absorbing characteristics. (Fotokostic/Shutterstock)

5 The Structure of Polymers

One of the three main types of materials is plastic. The general term "plastic" actually covers a wide range of materials with very different properties. For example, polyethylene can be either opaque and rigid like a milk carton or translucent and flexible like a plastic bag; polystyrene is clear and brittle; epoxy glue starts off as a liquid and then hardens to become a rigid solid. We can also include rubbers such as rubber bands and car tires as a subcategory of plastics. Paints are also plastics—they consist of plastic particles suspended in a liquid. When the liquid dries, the particles fuse together to form a continuous coating.

When we talk about the molecular structure of plastic, we refer to it as a polymer. A *polymer* is a molecule that is a long chain of atoms. Polymers are organic molecules because they are primarily hydrocarbons that derive from petroleum, which in turn came from organic matter (plants and animals) that died millions of years ago. Polymers have very different properties than other organic molecules such as methane, gasoline, or ethanol because of their molecular structure. In fact, polymers provide a great example of how structure and properties are related, so we will spend about half this chapter learning about their structure, and then the other half figuring out how to predict properties from that structure. By the end of this chapter you will:

Know what a polymer is.

Be able to calculate the molecular weight of a polymer.

Understand how polymer structure affects properties.

5.1 Molecular Structure of Polymers

LEARN TO: Describe bonding arrangements in polymers.

To begin, we need to understand the basics of what makes something a polymer. Figure 5.1.1 compares ethane and polyethylene. Polyethylene can be thought of as a group of ethane molecules that have been attached with covalent bonds.

A simpler way to represent a polymer is shown in Figure 5.1.2. The part of the molecule in parentheses is the *repeat unit* (also called the repeat or the mer). The name polymer literally means "many units" in Greek. The original molecule that a polymer is made from is called a monomer ("one unit"). A molecule of several repeats that is not really long enough to be considered a polymer is called an oligomer ("few units"). Similar to a unit cell for a crystal, the repeat unit shows the unit that repeats to make up the entire molecule. You will now examine polyethylene to understand the basic molecular structure of a polymer.

Figure 5.1.1

Molecular structures of (a) ethane and (b) polyethylene. The wavy lines at the ends of the polyethylene structure indicate that the molecule is much longer than what is shown.

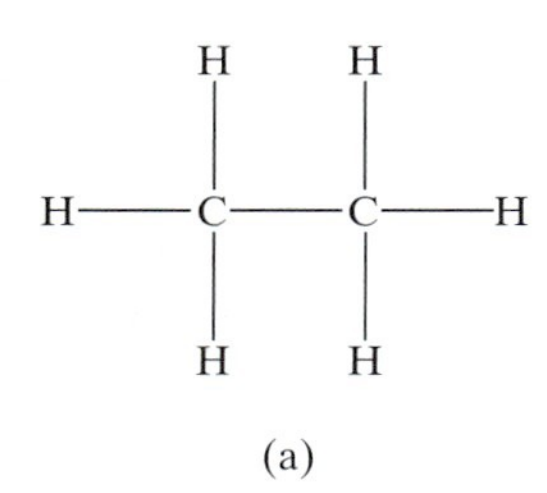

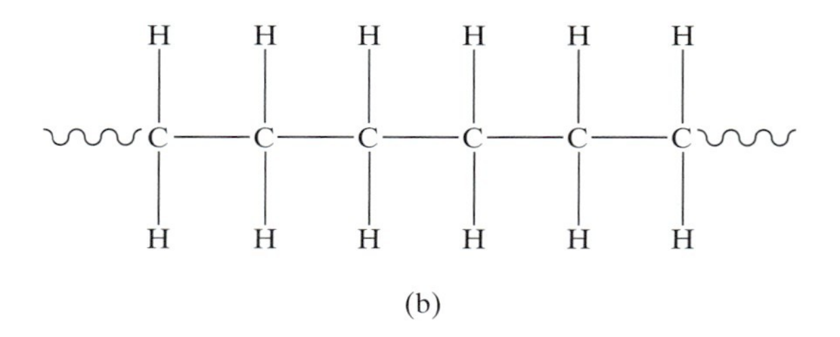

Figure 5.1.2

Simplified molecular structure of polyethylene.

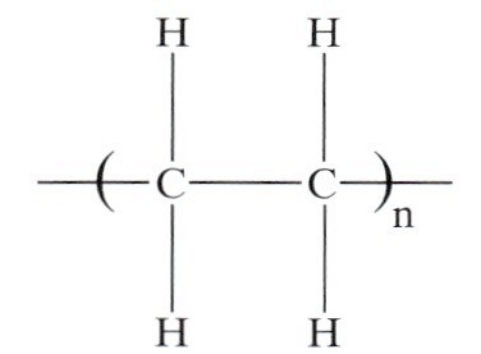

Guided Inquiry: Polymer Molecular Weight

5.1.1 What is the molecular weight of one repeat unit of polyethylene?
When calculating molecular weights use the isotopic average atomic weights, not the integer values.

5.1.2 If the molecular weight of a polyethylene molecule is 28,000 g/mol, the value of n is 1,000. If the molecular weight of a polyethylene molecule is 56,000 g/mol, the value of n is 2,000. Write a mathematical equation that relates the repeat unit molecular weight, the polyethylene molecule molecular weight, and n.

5.1.3 Write a description of what the value of n means.

Concept Check 5.1.1

- **A polyethylene molecule and a polypropylene molecule have the same value of n but different molecular weights. How is this possible?**

The value of n is called the *degree of polymerization*, or DP. Typical values for the degree of polymerization of a polymer range from 100 to hundreds of thousands. Table 5.1.1 provides the repeat units for a variety of polymers. **Use this table to answer the questions that follow.**

TABLE 5.1.1 Structures and names of common polymers

Structure	Full Name	Abbreviation or Common Name
$\left(-\overset{H}{\underset{H}{C}}-\overset{H}{\underset{H}{C}}-\right)_n$	Polyethylene	PE
$\left(-\overset{H}{\underset{H}{C}}-\overset{H}{\underset{CH_3}{C}}-\right)_n$	Polypropylene	PP
$\left(-\overset{H}{\underset{H}{C}}-\overset{H}{\underset{Cl}{C}}-\right)_n$	Poly(vinyl chloride)	PVC
$\left(-\overset{H}{\underset{H}{C}}-\overset{H}{\underset{⬡}{C}}-\right)_n$	Polystyrene	PS
$\left(-\overset{O}{\overset{\Vert}{C}}-\overset{H}{N}-\left[CH_2\right]_5-\right)_n$	Polycaprolactam	Polyamide 6 or Nylon 6
$\left(-\overset{O}{\overset{\Vert}{C}}-\left[CH_2\right]_4-\overset{O}{\overset{\Vert}{C}}-\overset{H}{N}-\left[CH_2\right]_6-\overset{H}{N}-\right)_n$	Poly(hexamethylene adipate)	Polyamide 66 or Nylon 66
$\left(-O-CH_2-CH_2-O-\overset{O}{\overset{\Vert}{C}}-⬡-\overset{O}{\overset{\Vert}{C}}-\right)_n$	Poly(ethylene terephthalate)	PET

Note: The ring structure shown as ⬡ represents a section of the molecule made up of 6 carbon atoms, and hydrogen atoms wherever the ring is not bonded to something else. So in PS, the ring represents a portion of the molecule with a chemical formula of C_6H_5, and in PET it represents a portion of the molecule with a chemical formula of C_6H_4.

Guided Inquiry: Degree of Polymerization

5.1.4 For polypropylene with a DP of 3500, what is the polymer molecular weight?

5.1.5 For Nylon 6 with a polymer molecular weight of 150,000 g/mol, what is the DP?

5.1.6 Does the DP need to be an integer? Explain why or why not.

Concept Check 5.1.2

- **What is the DP for polypropylene with a molecular weight of 35,000 g/mol?**
- **What are the units for DP?**

Application Spotlight PET from Renewable Resources

(Photo courtesy of Lifeprints Photography)

If you look at a water bottle it might say something like "30% from plant-based sources." But there is no difference between that bottle and one made from 100% oil-based sources. How is this possible? PET is made from two chemicals, ethylene glycol and terephthalic acid. There are two ways to make ethylene glycol. Starting with crude oil, you can refine and reform it to make ethylene, which is then chemically converted to ethylene glycol. Alternatively, you can distill corn to make ethanol, which can first be converted to ethylene, and then to ethylene glycol. Since ethylene glycol makes up 30% of the PET repeat unit, if the ethylene glycol is made from this corn route the polymer is 30% plant-based, even though the chemical structure is exactly the same! Companies are now starting to develop processes to also make terephthalic acid from renewable resources. If they are successful, we could have PET bottles that are 100% plant-based.

EXAMPLE PROBLEM 5.1.1

What is the DP of polystyrene with a molecular weight of 100,000 g/mol?

To do this, we use the following relationship:

$$MW = M(\text{repeat}) \times DP$$
$$DP = MW/M(\text{repeat})$$

For polystyrene, the molecular weight of the repeat unit is as follows:

$$M(\text{repeat}) = 8 \times 12.011 \text{ g/mol} + 8 \times 1.008 \text{ g/mol} = 104.15 \text{ g/mol}$$
$$DP = 100{,}000 \text{ g/mol} / 104.15 \text{ g/mol} = 960.15$$

However, DP must be an integer because we can't have a partial repeat unit. Therefore the DP is 960.

One of the great things about polymers is that an infinite number of different ones can be made. Table 5.1.1 shows only a few. Hundreds of other monomers can be used. In addition, we can mix several different types of monomers together to make a *copolymer*. Figure 5.1.3 shows the two primary types of copolymers that can be made. It is important to note that these copolymers are not simple mixtures in the same way metallic alloys are. To make an alloy, you can simply melt a metal, add the impurities you want, and cool it down. But to make a copolymer you can't take a polymer and mix another monomer or polymer with it. Instead, you have to mix the monomers together before you make the polymer. After you mix the monomers, a chemical reaction occurs that creates covalent bonds between the different monomers. The result is a new molecule that is different from a molecule that would be made from either monomer alone. Copolymers are made to get polymers with different properties. Polystyrene homopolymer is a glass at room temperature; it is rigid and transparent. But car tires are made from a random copolymer of styrene and another monomer called butadiene, resulting in a rubber called SBR (which stands for styrene-butadiene rubber). Making a copolymer out of styrene, butadiene, and acrylonitrile results in nitrile rubber, which has good oil resistance.

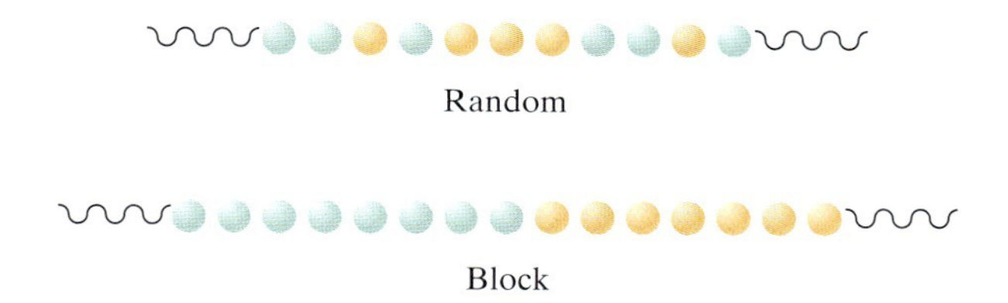

Figure 5.1.3 Schematic structure of random and block copolymers made out of two different monomers, represented by the light green and yellow circles.

Another way to change the properties of a polymer is to change its overall molecular shape. The three main shapes are shown in Figure 5.1.4. In a *linear polymer*, the repeat units are all lined up in a straight chain. In a *branched polymer*, there are *branches* off the main chain. These branches are made of the same repeat units as the main chain. Sometimes students confuse the terms "*side-group*" and "branch." Figure 5.1.5 shows the difference. A "side-group" is a small molecular entity that is part of the repeat unit. A "branch" is a series of repeat units that are hanging off the main chain. In a *network* or *crosslinked polymer*, individual linear polymer molecules are connected by covalent bonds. In a crosslinked polymer every chain is connected to every other chain through a path of covalent bonds. You will now examine how the molecular shape affects some properties of a polymer.

Figure 5.1.4

Possible molecular shapes for polymers. For clarity individual repeat units are not shown. Each line represents an entire polymer molecule.

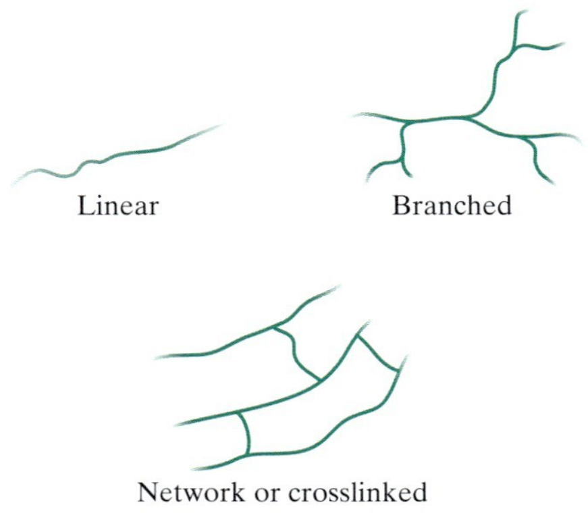

Linear

Branched

Network or crosslinked

Figure 5.1.5

(a) Linear and (b) branched polystyrene, illustrating the difference between a side-group and a branch. The side-group is the phenyl ring that is part of the repeat unit in both structures. A branch is a string of repeat units that are hanging off the main chain.

(a)

Side-group

(b)

Branch

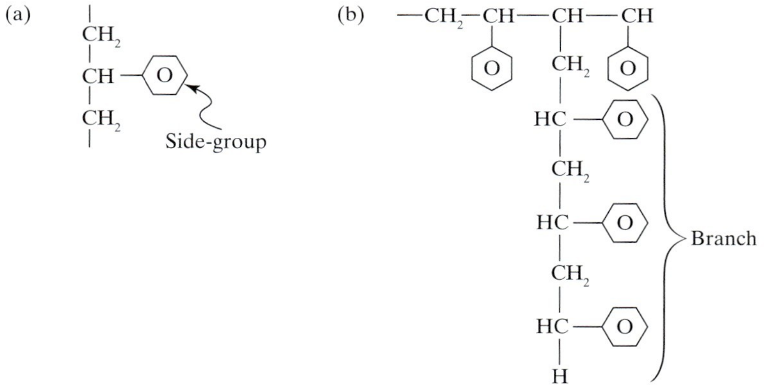

Guided Inquiry: Chain Shape

5.1.7 In which type of structure in Figure 5.1.4 are the chains connected by non-bonding interactions? In which are they connected by covalent bonds?

5.1.8 Of the three shapes shown in Figure 5.1.4, which can be recycled? Explain why.
Once you break a covalent bond, you can't put it back together.

5.1.9 Which polymer will more easily form a crystal; linear or branched? Explain why.

Concept Check 5.1.3

- **Cured epoxy does not become liquid when heated, it only burns. What does this tell you about the molecular shape of epoxy?**

Another important aspect of the shape of some polymers is their *tacticity*. Tacticity occurs in polymers that have all carbon–carbon single bonds in their main chain, such as polypropylene, polystyrene, and poly(vinyl chloride). Tacticity refers to the arrangement of side-groups, as shown in Figure 5.1.5. You will now examine how tacticity affects some properties of a polymer.

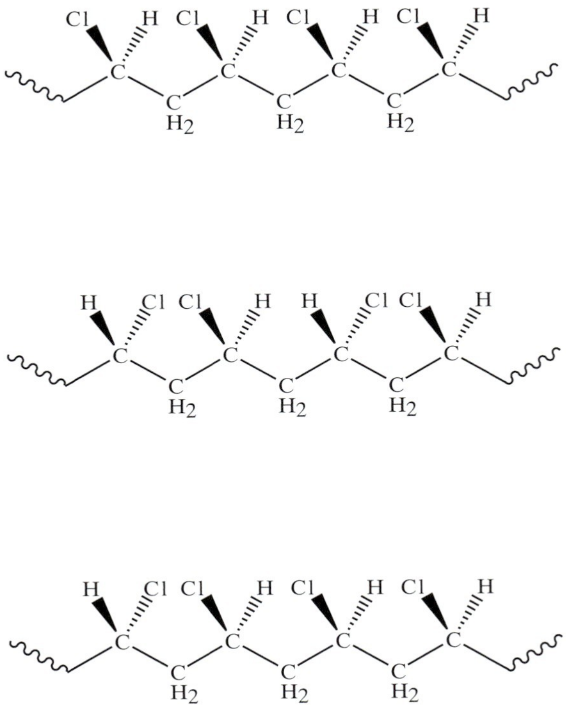

Figure 5.1.6

Possible tacticities for polymers, showing the three-dimensional arrangement of the polymer chains. A line represents a bond in the plane of the paper. A solid wedge represents a bond coming out of the plane of the paper. A dashed wedge represents a bond going back into the plane of the paper.

Isotactic;
All of the side-groups are on the same side of the main chain.

Syndiotactic;
Side-groups are on alternating sides of the main chain.

Atactic;
Placement of side-groups is random.

Guided Inquiry: Tacticity

5.1.10 In which structure is the chlorine atom always on the same side of the chain? In which is the chlorine atom on alternating sides?

5.1.11 Considering the three-dimensional structure of poly(vinyl chloride), describe what is actually meant by "same side" and "alternating sides."

5.1.12 Of the three tacticities, which least easily forms a crystal? Why?

5.1.13 Will PET have different tacticities? Explain your answer.

Concept Check **5.1.4**

- **Which type of polypropylene would form a crystal more easily; isotactic or atactic?**

If you don't get the Concept Check correct, make sure you understand before moving on.

5.2 Molecular Weight

LEARN TO: **Calculate polymer molecular weight (MW) and polydispersity index. Predict the effect of MW and MW distribution on properties.**

In the last activity, you saw how to calculate the molecular weight of a polymer by knowing the molecular weight of the repeat unit and the degree of polymerization. However, in reality a piece of plastic contains a mixture of polymer molecules with different molecular weights. Therefore, we need to be able to calculate an average molecular weight for a mixture of polymer chains. First we need to remember how to do a simple weighted average.

Guided Inquiry: Average Molecular Weight

5.2.1 You have a sample of plastic which you know contains 10 moles of chains that have a molecular weight of 5,000 g/mol, and 5 moles of chains with a molecular weight of 50,000 g/mol. What is the average molecular weight of this mixture?

Before we talk any further about molecular weight, let's imagine you and some friends decide to pool your money and give everyone an equal share. You have one penny and two nickels, one friend has two quarters, and another friend has one penny, one nickel, and two dimes. That's nine coins total, so on average each of you gets three coins. You end up with two pennies and one nickel. When you complain that not only do you have less money than your friends, you have less money than when you started, your friends say it is fair because you each got the same number of coins. Obviously you are right, but the average value you should have used was 29 cents, not 3 coins.

That story is silly, but it makes the point that there are different kinds of averages, and that different properties are correlated with different types of averages. The value of money is not accurately captured with a simple average based on the number of coins. Instead, we need an average that counts the quarter more than the other coins. In the same way, there are different molecular-weight averages we can use with polymers that count the heavier chains more.

The average molecular weight you just calculated in question 5.2.1 is called the *number average molecular weight*, because the average is weighted by the number of molecules that are present for each molecular weight. Mathematically this average can be written as follows:

$$\overline{M}_n = \frac{\sum_i N_i M_i}{\sum_i N_i} \tag{5.2.1}$$

where N_i is the number of moles of chains with molecular weight M_i.

The number average molecular weight is not the only average that you can calculate. The *weight average molecular weight* is calculated by weighting the average based on the mass of the molecules that are present for each molecular weight. Mathematically, this average can be written as follows:

$$\overline{M}_w = \frac{\sum_i w_i M_i}{\sum_i w_i} = \frac{\sum_i (N_i M_i) M_i}{\sum_i (N_i M_i)} = \frac{\sum_i N_i M_i^2}{\sum_i N_i M_i} \tag{5.2.2}$$

where w_i is the total mass of all the chains with molecular weight M_i.

Another important quantity to calculate is the *polydispersity index*, or PDI, which is calculated as follows:

$$PDI = \frac{\overline{M}_w}{\overline{M}_n} \tag{5.2.3}$$

Now you will practice using these equations to calculate the different types of molecular weight for a polymer.

Guided Inquiry: Molecular Weight Calculations

5.2.2 You have a sample of plastic which you know is 15 mol% polymer molecules with molecular weight 500,000 g/mol, and 85 mol% polymer molecules with molecular weight 150,000 g/mol. For this mixture, what numerical values will you use for N to calculate the average molecular weights?

5.2.3 For this mixture, what numerical values will you use for M to calculate the average molecular weights?
This is where most students make a mistake in the calculation; mixing up which values to use for M and which to use for N.

5.2.4 Calculate the number average molecular weight, weight average molecular weight, and PDI for this mixture.

5.2.5 Can $\overline{M}_w$ be less than $\overline{M}_n$? Explain why or why not.
Think of the coin example. Can the average value of the coins for each person be less than the average number of coins each person gets?

5.2.6 What does it mean about the molecular weights of the different polymer molecules in a mixture if PDI equals 1?

Concept Check **5.2.1**

- **What is the number average molecular weight for the mixture of question 5.2.2?**
- **What is the weight average molecular weight for the mixture of question 5.2.2?**

EXAMPLE PROBLEM 5.2.1

Calculate the number average molecular weight, weight average molecular weight, and polydispersity index for the following sample of polystyrene:

Mol% of chain	DP
20.0	1,000
10.0	3,000
70.0	6,000

We solve this using Equations (5.2.1–5.2.3), but we need to identify values for *N* and *M*. *N* is the number of chains, which is equivalent to the mol% of chains if we assume we have a total of 100 moles. *M* we need to get from DP by knowing the repeat unit molecular weight for polystyrene:

$$M(\text{repeat}) = 8 \times 12.011 \text{ g/mol} + 8 \times 1.008 \text{ g/mol} = 104.15 \text{ g/mol}$$

Since MW = DP × *M*(repeat), we can create a new table with all the information we need.

N_i	DP	M_i (g/mol)
20.0	1,000	104,150
10.0	3,000	312,450
70.0	6,000	624,900

Now we can calculate the quantities we want.

$$\overline{M}_n = \frac{(20)(104{,}150 \text{ g/mol}) + (10)(312{,}450 \text{ g/mol}) + (70)(624{,}900 \text{ g/mol})}{20 + 10 + 70} = 489{,}500 \text{ g/mol}$$

$$\overline{M}_n = \frac{(20)(104{,}150 \text{ g/mol})^2 + (10)(312{,}450 \text{ g/mol})^2 + (70)(624{,}900 \text{ g/mol})^2}{(20)(104{,}150 \text{ g/mol}) + (10)(312{,}450 \text{ g/mol}) + (70)(624{,}900 \text{ g/mol})} = 582{,}800 \text{ g/mol}$$

$$\text{PDI} = \frac{582{,}800 \text{ g/mol}}{489{,}500 \text{ g/mol}} = 1.19$$

The molecular weight of a polymer can have an important effect on its properties. We can illustrate this by considering a bowl of cooked spaghetti. Compare a bowl of regular-sized spaghetti to a bowl of spaghetti that has been cut into short pieces. The difference between the spaghetti in these two bowls is equivalent to the difference between polymers with different molecular weights. By understanding this difference you can now predict how molecular weight affects properties.

Figure 5.2.1
Eating regular-sized spaghetti (left) versus eating chopped-up spaghetti (right). (Photo courtesy of Heidi M. Douglas)

Guided Inquiry: MW and Properties

5.2.7 Which is harder to pull apart; regular-sized long spaghetti or chopped-up spaghetti?

5.2.8 Which would you expect to be stronger; a polymer with a high molecular weight or one with a low molecular weight?

5.2.9 Which would flow more easily; a polymer with a high molecular weight or one with a low molecular weight?

5.2.10 You are working at a plastics company, and your boss wants you to modify an existing polymer so it will be stronger and flow more easily for processing. How would you change the molecular weight distribution to meet both of these requirements?

Concept Check 5.2.2

- **Which has a higher strength; polyethylene of 100,000 g/mol or 200,000 g/mol?**
- **Which flows more easily; polyethylene of 100,000 g/mol or 200,000 g/mol?**

The PDI provides a single number that describes the distribution, but in some cases we need more information. Figure 5.2.2 shows some examples of different molecular weight distributions. As you discussed in answering question 5.2.10, changing the distribution can change the properties. In fact, it would be possible to have two polymers with the same PDI but different distributions, and therefore different properties. So while the PDI can be a useful guide, it is not the only factor to consider.

One final very important point to make—when we talk about the molecular weight of a polymer we are talking about the length of the chain, which means DP. If we want to change a property by increasing the molecular weight, we are actually referring to increasing DP, not increasing the molecular weight of the repeat unit. This is because, as with the bowls of spaghetti, properties depend on the length of the molecules, not on how heavy the individual segments are.

Figure 5.2.2

Three possible molecular weight distributions for a polymer; (a) symmetric distribution, (b) asymmetric distribution, (c) bimodal distribution.

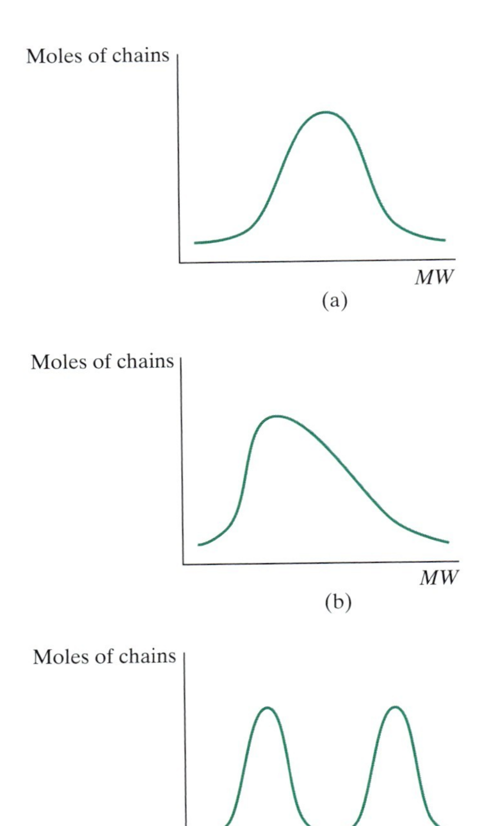

5.3 Polymer Crystals

LEARN TO: Describe the morphology of polymers.
Predict trends in crystallinity.

When we talked about the structure of metals we learned that they are crystalline. Polymers are different. Many polymers are completely amorphous, or glassy, meaning they don't have a regular crystalline structure. Many people will say that a glass is a frozen liquid. This does not mean it will ever flow like a liquid. What it means is that if you took a snapshot of a liquid, it would have the same random structure as a glass. Think about long polymer molecules, and how difficult it would be to arrange them into a regular, repeating crystal structure. The tangled polymer chains can't easily move around and form a crystalline lattice.

Some polymers are *semicrystalline*, which means that a sample of the polymer has two phases, an amorphous phase and a crystalline phase. However, no polymers are 100% crystalline; it's just too hard to get those long molecules arranged in that way. Figure 5.3.1 shows a schematic of the morphology of a semicrystalline polymer. How can a long polymer chain even fit into a unit cell at all? It doesn't. One chain will pass through many unit cells. Figure 5.3.2 shows the unit cell of polyethylene, which has 5 different molecules passing through it. You should also notice that for the crystalline phases in Figure 5.3.1 the chains are folded back and forth within a single crystal. Fortunately the properties of polymers don't depend on the crystal structure like for metals, so we can pretty much ignore the specific crystalline lattice structures of polymers.

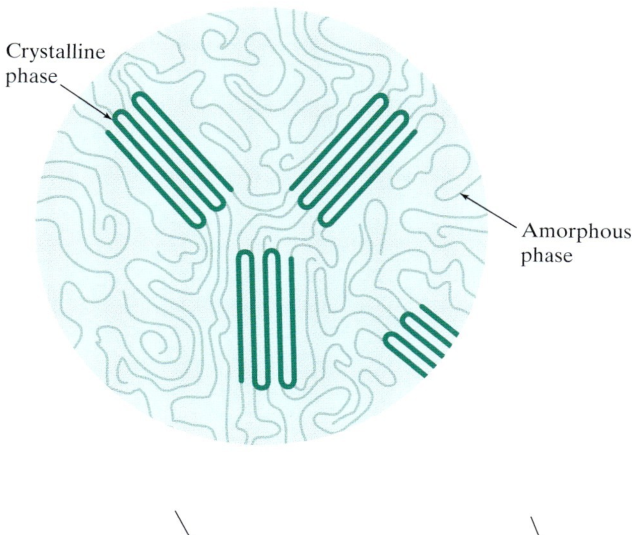

Figure 5.3.1
The two-phase structure of semicrystalline polymers.

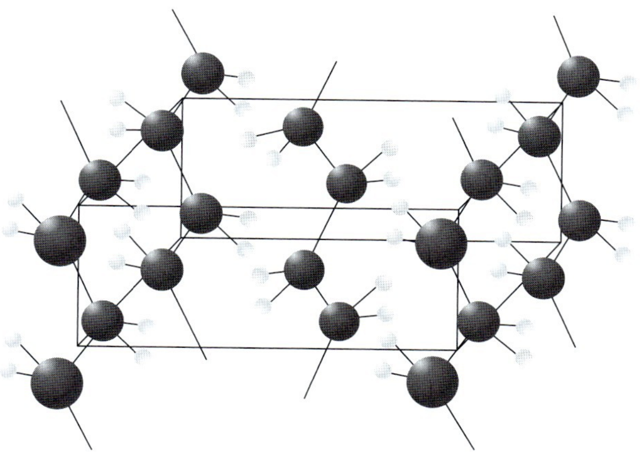

Figure 5.3.2
Unit cell of polyethylene.

Polymer properties are determined partially by the relative amounts of amorphous and crystalline phases. The percentage of the crystalline phase present can be calculated from densities by the following equations:

$$\text{vol\%crystallinity} = \frac{\rho_s - \rho_a}{\rho_c - \rho_a} \times 100 \tag{5.3.1}$$

$$\text{wt\%crystallinity} = \frac{\rho_c(\rho_s - \rho_a)}{\rho_s(\rho_c - \rho_a)} \times 100 \tag{5.3.2}$$

where

ρ_s = density of the semicrystalline polymer
ρ_c = density of the 100% crystalline version of the polymer
ρ_a = density of the 0% crystalline version of the polymer

There are no Guided Inquiry questions on how to use these equations. Make sure you try the problems at the end of the chapter to get practice.

Example Problem 5.3.1 shows you how to use these equations. You may wonder how we can get a value for ρ_c if we can never get a 100% crystalline polymer. We will leave that to the problems at the end of this chapter, which will show you how to get ρ_c.

EXAMPLE PROBLEM 5.3.1

Calculate the weight percent and volume percent crystallinity of polypropylene with a density of 0.92 g/cm³. The density of fully amorphous polypropylene is 0.855 g/cm³ and the theoretical density of fully crystalline polypropylene is 0.946 g/cm³.

From the data provided we know that

$\rho_s = 0.92 \text{ g/cm}^3$

$\rho_a = 0.855 \text{ g/cm}^3$

$\rho_c = 0.946 \text{ g/cm}^3$

Now we can easily use Equations (5.3.1) and (5.3.2) to calculate the crystallinity:

$$\text{vol\%crystallinity} = \frac{\dfrac{0.92\text{g}}{\text{cm}^3} - 0.855\text{g/cm}^3}{\dfrac{0.946\text{g}}{\text{cm}^3} - 0.855\text{g/cm}^3} \times 100 = 71.4\%$$

$$\text{wt\%crystallinity} = \frac{0.946\text{g/cm}^3\left(\dfrac{0.92\text{g}}{\text{cm}^3} - 0.855\text{g/cm}^3\right)}{0.92\text{g/cm}^3\left(\dfrac{0.946\text{g}}{\text{cm}^3} - 0.855\text{g/cm}^3\right)} \times 100 = 73.4\%$$

This means that 71.4% of the volume of the sample is crystalline, but because the crystalline regions have a higher density than the amorphous regions, 73.4% of the mass of the sample is crystalline.

Other than doing a calculation, we can also look at two different polymers and decide which would be more crystalline by considering which would be easier to pack together. **The next questions will help you to learn the factors that affect crystallinity of polymers.**

Guided Inquiry: Polymer Crystallinity

5.3.1 Which would be easier to pack into a box; wooden cubes or wooden pyramids?

5.3.2 You just cut down some bushes and want to stack them all up in a neat pile. What would you do to the bushes so they all stack up neatly?

5.3.3 In general, identify three things about an object that could make it difficult to pack a group of those objects into a neat pile.

5.3.4 For each of the following pairs of polymers, predict which one is more crystalline and explain why:

Remember, this is all about how easily you can pack the molecules together.

a. Polyethylene or polystyrene.
b. Polyethylene with many long branches or polyethylene with a few short branches.
c. Atactic polypropylene or isotactic polypropylene.
d. Polypropylene or poly(ethylene terephthalate) (PET).
e. Polyethylene or a random copolymer of 50% ethylene repeat units and 50% propylene repeat units.
f. PET melted and rapidly cooled to room temperature or PET melted, cooled to 150° C, held there for 1 hour, and then cooled to room temperature.

Concept Check 5.3.1

- **For each of the following pairs, identify which is more crystalline:**

a. Polyethylene or polystyrene.
b. Polyethylene with long branches or polyethylene with short branches.
c. Atactic polypropylene or isotactic polypropyelene.
d. Polypropylene or PET.
e. Polyethylene or a 50/50 ethylene/propylene random copolymer.
f. Rapidly cooled PET or PET held at 150° C for 1 hour.

5.4 Glass Transition and Melting of Polymers

LEARN TO: **Define glass transition and melting temperatures.**
Predict trends in glass transition and melting temperatures.

Think back to an experiment you may have done in high school chemistry and physics. If you measure the temperature of ice water, what happens to the temperature as the ice melts? The temperature remains constant at 0° C as long as any ice is present. This is a characteristic of a *melting temperature*—the temperature remains constant as melting occurs. If we were to plot volume versus temperature for the melting of a solid, we would get the graph shown in Figure 5.4.1.

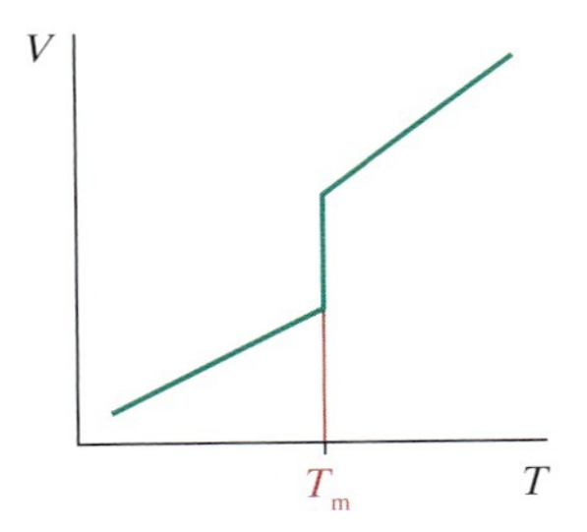

Figure **5.4.1**
Volume versus temperature for a melting transition.

A *glass transition* is also the temperature at which a material changes from a solid to a liquid, but it is different from a melting temperature. The volume-versus-temperature plot for a glass transition is shown in Figure 5.4.2. Instead of a sudden increase in volume, for a glass transition there is just a change in slope.

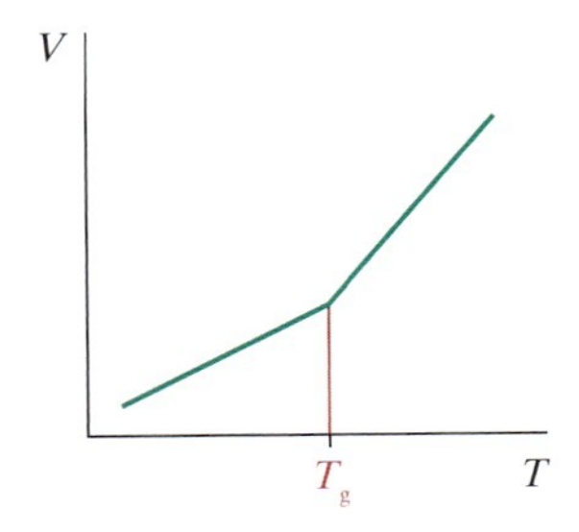

Figure **5.4.2**
Volume versus temperature for a glass transition.

So how do we know whether a material has a melting temperature or a glass transition temperature? That is easy: crystals melt, glasses (amorphous materials) have a glass transition (T_g). This means that amorphous polymers have a T_g but no melting temperature. Semicrystalline polymers have both a T_m and a T_g. The volume-versus-temperature curve for a semicrystalline polymer is shown in Figure 5.4.3. As shown in this figure, the T_g is always lower than the T_m.

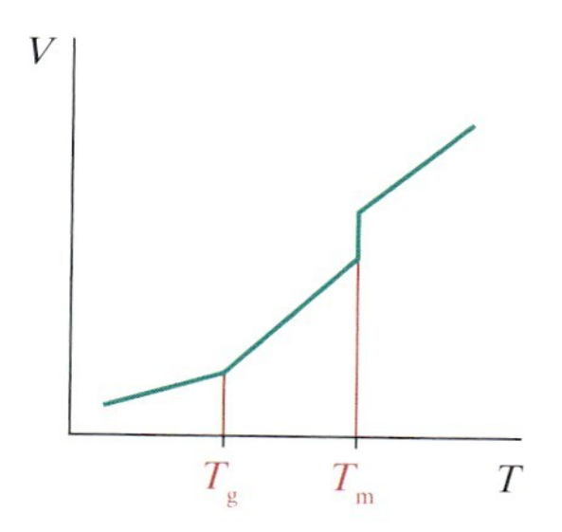

Figure 5.4.3

Volume versus temperature for a semicrystalline polymer.

Just like we did in the previous section for crystallinity, we can compare two polymers and decide which one has the higher T_g or T_m. In this case it is all about chain mobility—the easier it is for the chains to move, the lower will be the T_g or T_m. The chain mobility is related to the available free volume. Free volume is the space between the chains. Figure 5.4.4 shows a schematic of the free volume. Keep in mind that free volume is not holes or pores in the solid—it is the space at the molecular level. A fully dense polymer will have free volume. One of the most interesting experiments I have ever heard of is the use of positrons (antimatter) to measure free volume in polymers. The lifetime of the positrons is related to the size and amount of free volume.

As you answer the next questions, use the idea of chain mobility to determine which of the pair will move more easily.

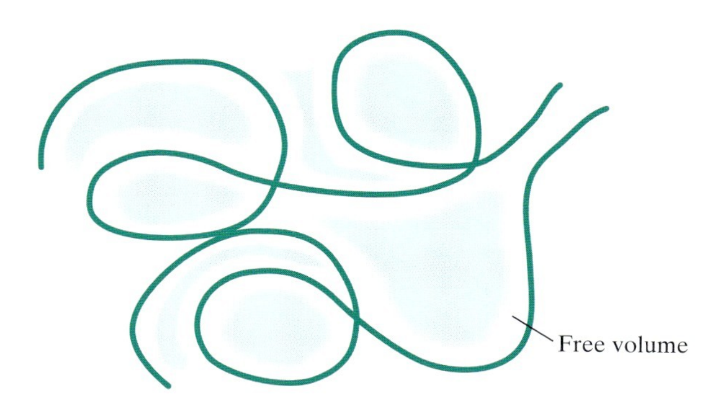

Figure 5.4.4

Illustration of free volume in a polymer.

Guided Inquiry: Glass Transition and Melting

5.4.1 Which is easier to scoop out of a pot; spaghetti or macaroni?

5.4.2 If you have long hair, is it easier to comb your hair when it is dry or when it is wet and conditioned?

5.4.3 Which is easier to separate; copper wires that have been smushed together at room temperature or licorice sticks that have been smushed together at room temperature?

5.4.4 Based on your answers to questions 5.4.1–5.4.3, identify three general aspects that make it more difficult to separate objects.

5.4.5 For each of the following, identify which has the lower T_m or T_g and explain why:

Remember, this is all about how easy it is for the molecules to move past each other.

a. Isotactic polypropylene or isotactic polystyrene.
b. Polypropylene or poly(vinyl chloride) (PVC).
c. Polystyrene with M_n of 5,000 g/mol or polystyrene with M_n of 100,000 g/mol.
d. Pure PVC or PVC mixed with some solvent.

Concept Check 5.4.1

- **For each of the following pairs, identify which has the lower T_g or T_m:**
  a. Isotactic polypropylene or isotactic polystyrene.
  b. Polypropylene or PVC.
  c. Polystyrene with $\overline{M}_n$ of 5,000 g/mol or polystyrene with M_n of 100,000 g/mol.
  d. Pure PVC or PVC mixed with solvent.

Summary

The atomic and molecular level structure of polymers is very different from that of metals and ceramics. Although we typically think of metals and ceramics as individual atoms arranged in a crystal or glass, polymers consist of long chains of atoms connected by covalent bonds, with weaker nonbonding interactions between the molecules. Many different aspects of a polymers' chain structure affect properties: whether they are linear, branched, or crosslinked; the type of tacticity (if present); the strength of nonbonding interactions; and molecular weight. All these can affect things like strength, flow, crystallinity, and glass transition temperature.

It is not possible to predict a property, such as crystallinity or glass transition temperature, just by looking at the structure of a single polymer. What you have seen is that you can make relative comparisons between two polymers based on the structure. This is a good example of part of the MSE triangle; understanding structure–property relationships. If you go on to study polymers in more detail you will see the third element of the triangle; processing. For example, the chemical methods used to make polyethylene determine whether it is linear or branched, which leads to differences in crystallinity and properties; high density polyethylene (HDPE) is mostly linear and is used for milk jugs, while low density polyethylene (LDPE) is more branched and is used for plastic bags. Understanding these differences, as you now do, is the fundamental core of materials science and engineering.

Key Terms

Atactic
Branch
Branched polymer
Copolymer
Crosslinked polymer
Degree of polymerization
Glass transition
Isotactic
Linear polymer
Melting temperature
Network polymer
Number average molecular weight
Polydispersity index
Repeat unit
Semicrystalline
Side-group
Syndiotactic
Tacticity
Weight average molecular weight

Problems

Skill Problems

5.1 If a polymer has a molecular weight of 15,000 g/mol and a degree of polymerization of 125, what is the molecular weight of one repeat unit?

5.2 For each of the following, calculate the molecular weight of a single chain:

a. Polypropylene with DP = 2000.
b. Polyethylene with DP = 2000.
c. Polystyrene with DP = 5500.

5.3 For each of the following, calculate the degree of polymerization of a single chain:

a. Nylon 66 with MW = 30,000 g/mol.
b. Nylon 66 with MW = 150,000 g/mol.

5.4 For the two polymers in problem 5.3, which do you expect to have the higher strength? Which do you expect would flow more easily? Explain why.

5.5 Calculate the number average molecular weight, weight average molecular weight, and polydispersity index for the following sample of poly(vinyl chloride):

Mol% of chain	DP
35.0	1,500
15.0	2,000
50.0	5,000

5.6 Calculate M_n, M_w, and PDI for the following sample of polypropylene:

Mol% of chains	DP
40.0	5,000
40.0	10,000
20.0	30,000

5.7 The theoretical density of 100% amorphous polypropylene is 0.853 g/cm^3, and the theoretical density of 100% crystalline polypropylene is 0.946 g/cm^3. Calculate the volume% crystallinity for samples with densities of 0.860 g/cm^3 and 0.910 g/cm^3.

5.8 The density and associated volume% crystallinity for two nylon 6,6 materials are given below:

Density (g/cm^3)	Crystallinity (vol%)
1.188	67.3
1.152	43.7

a. Compute the densities of totally crystalline and totally amorphous nylon 6,6.
b. Calculate the density of nylon 6,6 with 55.4% crystallinity.

5.9 For each of the following pairs of polymers, indicate which is more likely to crystallize, or if they are the same, or if you can't tell, and give a brief (one-sentence) explanation as to why:

a. Isotactic polypropylene or isotactic polystyrene.
b. Linear polyethylene or branched polyethylene.
c. Polypropylene or PET.
d. An ethylene-propylene alternating copolymer or an ethylene-propylene random copolymer.

5.10 For each of the following pairs of polymers, indicate which has the higher T_g, or if they are the same, or if you can't tell, and give a brief (one-sentence) explanation as to why:

a. Atactic polystyrene of $M_n = 5000$ g/mol or atactic polystyrene of $M_n = 8000$ g/mol.

b. Pure nylon 6,6 or nylon 6,6 with 5 wt% absorbed water.

c. Polypropylene or polystyrene of the same M_n.

d. The following two polymers:

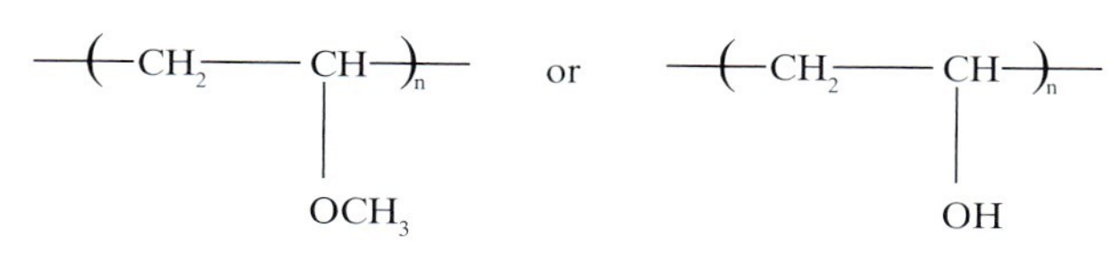

Poly(vinyl methyl ether) Poly(vinyl alcohol)

Conceptual Problems

5.11 Compare the strength of polystyrene with a number average molecular weight of 104,150 g/mol, and PVC with a number average molecular weight of 62,500 g/mol. Which is stronger? Explain why.

5.12 Can polyethylene exhibit tacticity? If it can, draw structures showing the three types of tacticity for it. If not, explain why.

5.13 Can nylon 6 exhibit tacticity? If it can, draw structures showing the three types of tacticity for it. If not, explain why.

5.14 Two common types of glue are epoxy and super glue. Epoxy cannot be dissolved by any solvent, while super glue can be dissolved by acetone. What does this information tell you about the molecular shapes of these two polymers?

The strength of aluminum depends on its morphology. Pure aluminum is very ductile, while high strength aluminum, as would be used for these pipes, is a two-phase system. (ekipaj/Shutterstock)

6 Microstructure—Phase Diagrams

Up to this point we have considered primarily pure substances with an occasional impurity atom thrown in. But in reality, materials used in engineering are complex mixtures that are rarely homogeneous. Their heterogeneity has a strong influence on their properties. Go back and look at the bobby pin experiment described in Section 2.2. The changes in properties during this experiment are caused by changes in heterogeneity. In order to control properties, we need to have some way of talking about these complex mixtures, and even more importantly, we need a way to predict what the mixture actually looks like. By the end of this chapter you will:

> Understand what is meant by the microstructure of a material.
>
> Be able to use phase diagrams to determine the structure of materials.

6.1 Defining Mixtures

> LEARN TO: Given a mixture of substances, identify the components and phases present.
>
> Define solubility limit.
>
> Differentiate between total composition of a mixture, composition of the phases in a mixture, and the relative amounts of phases in a mixture.

From your basic chemistry and physics classes you are familiar with the concept of states of matter. For example, water exists as a gas, liquid, or solid, and the properties of water depend on which state of matter it is in. The same kind of situation occurs for engineering materials, but our concept of "state of matter" needs to be expanded to deal with their complexity. To begin, we will consider three simple mixtures: water with a little bit of sugar in it; water with a lot of sugar in it; and water mixed with oil (see Figure 6.1.1). **In the next questions you will look at the substances and states of matter present for each of these mixtures**.

Guided Inquiry: Mixtures

6.1.1 In the solution of sugar completely dissolved in water, what substances are present?

6.1.2 In the solution of sugar completely dissolved in water, what states of matter are present?
By states of matter we mean solid, liquid, or gas.

6.1.3 In the solution of sugar partially dissolved in water, what substances are present?

6.1.4 In the solution of sugar partially dissolved in water, what states of matter are present?

6.1.5 For a mixture, can you predict the number of states of matter present by knowing the number of substances present? Explain your answer.
Use the answers to the previous questions. Do you see a consistent pattern?

6.1.6 In the mixture of water and oil, what substances and states of matter are present?

6.1.7 What is different about the mixture of water and oil compared to pure water or pure oil?

6.1.8 Is listing the substances and states of matter present sufficient to describe a mixture? Explain why or why not.

Concept Check 6.1.1

- **Nail polish remover is a mixture of two liquids; water and acetone. How is this similar to or different from the oil-and-water mixture?**

(a)

(b)

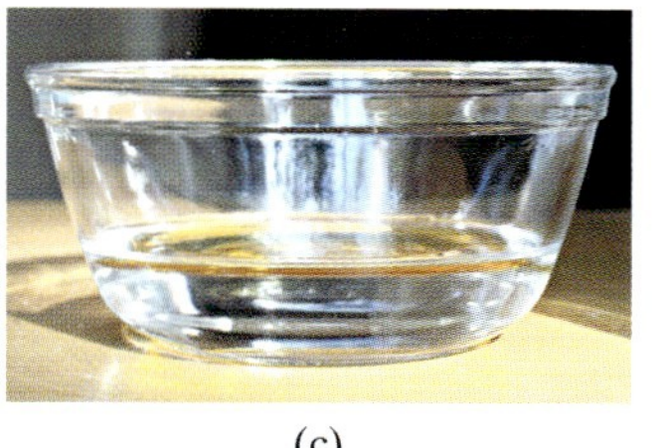
(c)

Figure 6.1.1
Three different kinds of solutions. (a) Sugar completely dissolved in water; (b) Sugar partially dissolved in water; (c) Mixture of oil and water. (Photo courtesy of Lifeprints Photography)

The Guided Inquiry questions you answered to this point should have shown you that the term "state of matter" is insufficient to describe a mixture. After all, the mixture of water and oil has only one state of matter; liquid, yet it is clearly different from pure water or pure oil. In the case of engineering materials we can have several different types of mixtures, which vary in their properties. For example, pure aluminum is a very ductile metal. However, high strength aluminum can be made by creating a material that has particles of Ti_3Al embedded in an aluminum/titanium mixture. Clearly we need some terms that more accurately describe a mixture. The terms we will use are *component* and *phase*. Instead of being given the definitions, you will work out the definitions yourself using the following information:

- Sugar completely dissolved in water has two components and one phase.
- Sugar partially dissolved in water has two components and two phases.
- A mixture of oil and water has two components and two phases.
- Solid sugar has one component and one phase.
- Pure water has one component and one phase.

Guided Inquiry: Components and Phases

6.1.9 Based on Figure 6.1.1 and the information provided previously, provide a definition for the term "component."

6.1.10 Based on Figure 6.1.1 and the information provided previously, provide a definition for the term "phase."
Make sure your definition is consistent with all of the information provided.

6.1.11 Compare sugar completely dissolved in water and sugar partially dissolved in water. What is different about the amounts of the components present?

Concept Check 6.1.2

- How many components are present in water with ice cubes?
- How many phases are present in water with ice cubes?
- How many components are present in sparkling water (water with CO_2 bubbles)?
- How many phases are present in sparkling water?

The properties of materials depend on the components and phases that are present. The components are determined by what we mix together to make the material. However, we need some way to predict what phases will be present. We do this by using a type of graph called a *phase diagram*. A phase diagram is a map showing what phases are present as a function of temperature and the amounts of each component present. Figure 6.1.2 shows the phase diagram for sugar and water. You will now begin to learn how to use phase diagrams.

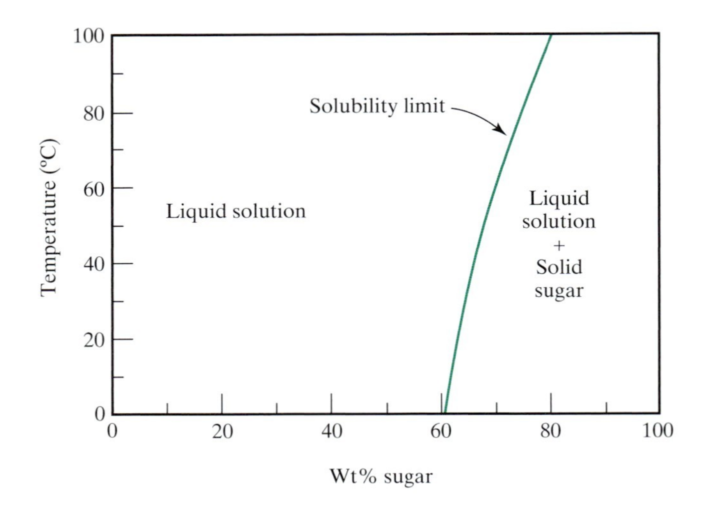

Figure **6.1.2**

Phase diagram for sugar and water.
(Data from Young and Jones, "Sucrose Hydrates—The Sucrose-Water Phase Diagram." *J of Phys Chem*, (1949), 53(9): 1334–1350.)

Guided Inquiry: Phase Diagrams

6.1.12 What are the components for the sugar–water phase diagram?

6.1.13 What phases are present at low concentrations of sugar?

6.1.14 What phases are present at high concentrations of sugar?

6.1.15 What is the significance of the line labeled *solubility limit*?

6.1.16 Propose a method for determining the solubility limit of sugar in water.

Concept Check 6.1.3

- **You have a mixture of 70 wt% sugar/30 wt% water at 20° C. What are two ways you can transform this mixture from two phases to one phase?**

In addition to knowing what components and phases are present in a mixture, we also need to know something about the *microstructure*—the size and arrangement of the phases. Even if the components and phases are the same, materials with different microstructures can have different properties. For example, plain carbon steel that is 0.76 wt% carbon in iron can be very ductile or very strong and brittle, depending on the microstructure.

Two important terms that can become confusing are *phase composition* and *phase amount*. Phase composition describes the amounts of each component present in a particular phase. For example, solution (b) in Figure 6.1.1 has two phases. The phase composition of the liquid phase (assuming room temperature) is approximately 65 wt% sugar and 35 wt% water. The phase composition of the solid phase is 100 wt% sugar. Phase amount is the amount of each phase in the total mixture. So, for example, we might determine that solution (b) in Figure 6.1.1 has phase amounts of 50 wt% liquid phase and 50 wt% solid phase. Another way of looking at it is that phase composition tells you something about one particular phase, while phase amount tells you about all the phases present. Because these concepts can be confusing, this section leads you through some additional examples.

Guided Inquiry: Phase Compositions and Amounts

6.1.17 In Figure 6.1.3, what components are present in alloy A? Alloy D? Alloy E?

6.1.18 What phases are present in alloy A? Alloy D? Alloy E?

6.1.19 Estimate the phase compositions and phase amounts for alloys A, D, and E. *Remember, dark green represents pure aluminum. The lighter the green, the more copper is in the mixture.*

6.1.20 Propose two different methods to determine the phase compositions and phase amounts in metallic alloys.

6.1.21 Draw examples similar to those shown in Figure 6.1.3 that illustrate the structure of the following iron-copper alloys:

a. A single-phase alloy that consists of 25% pure iron and 75% pure copper.

b. A two-phase alloy of 25% α and 75% β; α contains 20% iron, β contains 60% iron.
The Greek letters α and β are simply used as names for different solid phases.

c. A two-phase alloy of 25% α and 75% β; α is pure iron and β contains 60% iron.

Concept Check 6.1.4

- **What are the phase compositions for alloy A?**
- **What are the phase amounts for alloy A?**

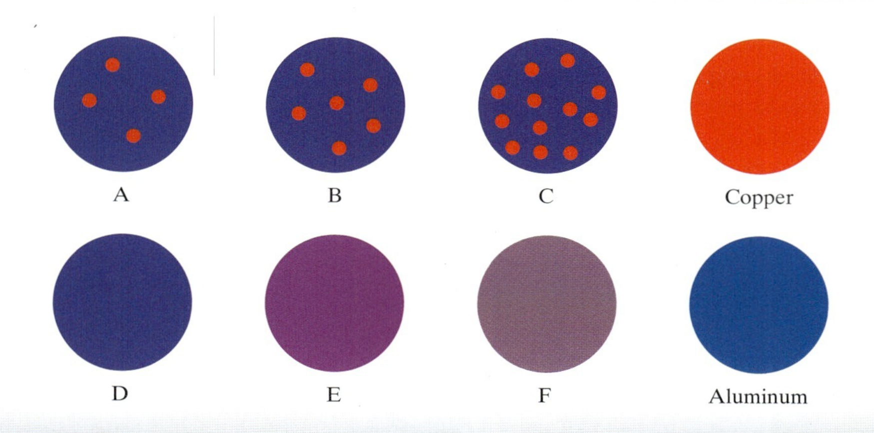

Figure 6.1.3
Some possible microstructures for mixtures of aluminum and copper. In this figure, aluminum is represented as blue and copper as red. Mixtures of the aluminum and copper are purple; the more bluish the purple, the more aluminum is present.

6.2 Isomorphous Binary Phase Diagrams—The Lever Rule

LEARN TO: **From a phase diagram, determine phases present, phase compositions, phase amounts.**

As we saw in the previous section, a phase diagram is a map of the phases present as a function of composition and temperature. Phase diagrams are named by how many components they have, so a *binary phase diagram* has two components. For metallic phase diagrams, solid phases are usually named with Greek letters. So, for example, a solid solution of two metals might be called the α phase. Figure 6.2.1 shows the binary phase diagram for copper and nickel. This is the simplest type of phase diagram, in which there is complete solubility of the two components in each other for all compositions; in other words, at low temperatures there is always only one phase, no matter what the composition. The boundary between the solid phase region and the two-phase region is called the solidus, and the boundary between the liquid phase region and the two-phase region is called the liquidus. We will use this phase diagram to start learning about how to get information from phase diagrams.

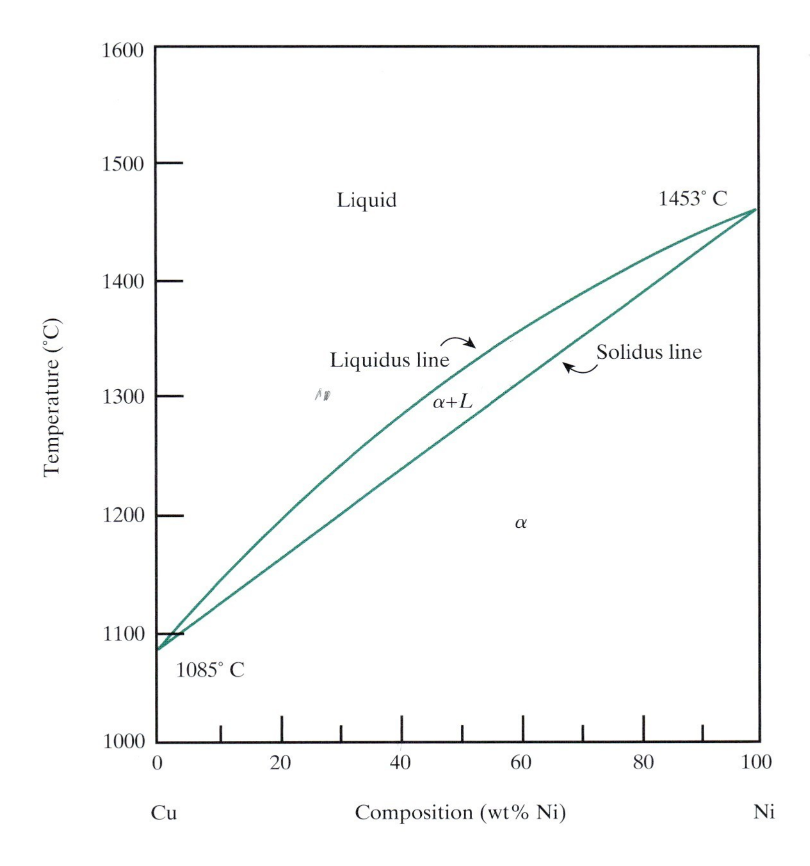

Figure 6.2.1 Binary phase diagram for copper and nickel.

Guided Inquiry: Interpreting Phase Diagrams

6.2.1 What components are present for the phase diagram in Figure 6.2.1?

6.2.2 For a mixture of 20 wt% Ni and 80 wt% Cu at 1300° C, what phases are present?

6.2.3 For a mixture of 50 wt% Ni and 50 wt% Cu at 1300° C, what phases are present?

6.2.4 For a mixture of 80 wt% Ni and 20 wt% Cu at 1300° C, what phases are present?

6.2.5 For each of the situations in questions 6.2.2–6.2.4, identify the amount of the phases present if you can. If you cannot, explain why.

6.2.6 For each of the situations in questions 6.2.2–6.2.4, identify the compositions of the phases present if you can. If you cannot, explain why.

Concept Check 6.2.1

- **What are the phase compositions for 20 wt% Ni and 80 wt% Cu at 1300° C?**
- **What are the phase compositions for 50 wt% Ni and 50 wt% Cu at 1300° C?**
- **What are the phase amounts for 20 wt% Ni and 80 wt% Cu at 1300° C?**
- **What are the phase amounts for 50 wt% Ni and 50 wt% Cu at 1300° C?**

What you probably realized when completing questions 6.2.5 and 6.2.6 was that there is a difference between points in the one-phase region and two-phase region of the phase diagram. For the one-phase region, the phase amount is just 100% of whatever phase is shown on the phase diagram, and therefore the phase composition is just the same as the overall given composition. However, for two-phase regions it is not so easy. How do you know how much of each phase is present and how the components are divided up between the two phases? The answer is found with something called a *tieline*. To learn what a tieline is and how to use it, look at Figure 6.2.2. Let's say we want to know the phase compositions

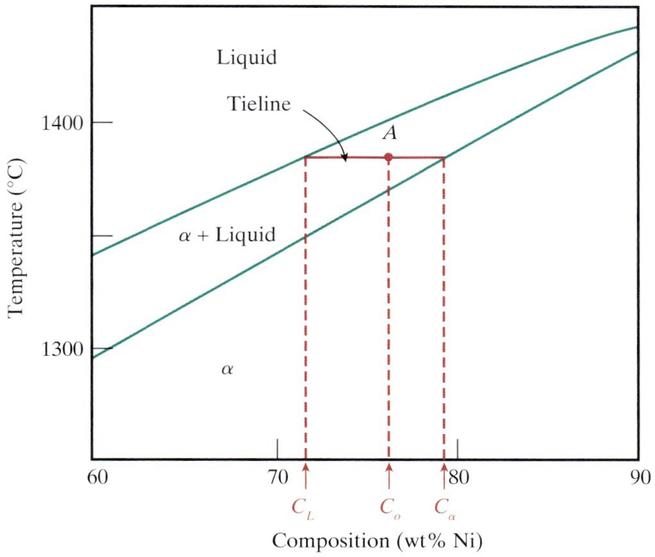

Figure 6.2.2
Illustration of how to construct a tieline.

and amounts for point A. We know by looking at the phase diagram that two phases are present; liquid and solid. To get the phase compositions, we do the following:

1. Draw a line through point A, parallel to the composition axis, that stops at the liquidus and solidus. This is the tieline.
2. The composition of the liquid phase is read off the composition axis from where the tieline hits the liquidus, which is C_L in Figure 6.2.2. So for point A, the composition of the liquid phase is 28 wt% Cu and 72 wt% Ni.
3. The composition of the solid phase is read off the composition axis from where the tieline hits the solidus, which is C_α in Figure 6.2.2. So for point A, the composition of the solid phase is 21 wt% Cu and 79 wt% Ni.

For phase amounts we also use the tieline, but it is a little more complicated. We need to use something called the *lever rule*. Here is what we do:

1. Figure out the length of the entire tieline. In Figure 6.2.2 this would be $C_\alpha - C_L$.
2. Figure out the length of the tieline that goes from the solidus to point A. This would be $C_\alpha - C_0$.
3. The amount of the **liquid** phase is the fraction of the tieline that goes from the **solidus** to point A. In terms of steps 1 and 2, this becomes

$$wt\%\ liquid = \frac{C_\alpha - C_0}{C_\alpha - C_L} \times 100 \tag{6.2.1}$$

The amount of solid present would then just be 100 – wt% liquid. It is important to note that the amount of **liquid** comes from the length of the tieline that is on the same side as the **solidus**. This may seem backward, but it actually makes sense if you look at the phase diagram in Figure 6.2.1. At high temperatures in the two-phase region, there should be a lot more liquid than solid, and if you look at the tieline you will see that at high temperatures the length of the tieline is longer on the side toward the solidus.

Now that you know how to use a tieline, you should be able to determine the phase compositions and phase amounts in the two-phase region. **The next question will give you some practice doing that**.

Guided Inquiry: The Lever Rule

6.2.7 Complete your answers to questions 6.2.5 and 6.2.6 for those situations that you could not answer previously.

Concept Check 6.2.2

- **What are the phase compositions for 35 wt% Ni and 65 wt% Cu at 1250° C?**
- **What are the phase amounts for 35 wt% Ni and 65 wt% Cu at 1250° C?**

EXAMPLE PROBLEM 6.2.1

At a temperature of 1225° C, determine the phase compositions and phase amounts at overall compositions of 30 wt% Ni/70 wt% Cu and 40 wt% Ni/60 wt% Cu.

The phase diagram for this system is shown in Figure 6.2.1. The point corresponding to 1225° C and 30 wt% nickel is in the two phase region, so the phases present are solid and liquid. To determine the compositions and amounts we need to use the tieline. Drawing the tieline and estimating the corresponding compositions shows us that the tieline intersects the liquidus at 27 wt% nickel and intersects the solidus at 38 wt% nickel. These are estimates, so if you get something slightly different, that is OK. From the intersection of the tieline with the liquidus, we get the composition of the liquid; from the intersection of the tieline with the solidus, we get the composition of the solid, so:

Liquid is composed of 27 wt% nickel and 73 wt% copper
Solid is composed of 38 wt% nickel and 62 wt% copper

To get the amounts of each phase, we use the *lever rule*, remembering that the amount of solid is determined by the length of the tieline on the liquidus side of the overall composition:

$$wt\%\ solid = \frac{30 - 27}{38 - 27} \times 100 = 27\%$$

$$wt\%\ liquid = 100 - 27 = 73\%$$

So overall, there is 27 wt% solid and 73 wt% liquid.

The point corresponding to 40 wt% nickel and 1225° C is in the single-solid-phase region. This means the only phase present is solid. Since everything is in one phase, the composition of the solid is the same as the composition that was given, so the solid consists of 40 wt% nickel and 60 wt% copper. And since only solid is present, the amount of solid is 100 wt%.

6.3 Isomorphous Binary Phase Diagrams—Microstructure

LEARN TO: **Predict the microstructure for both equilibrium and nonequilibrium cooling from a binary isomorphous phase diagram.**

Phase diagrams are *equilibrium phase diagrams*, meaning they tell us about the state of the system with the lowest free energy. However, we can also learn something about the microstructure from phase diagrams. The microstructure depends both on the phases that are present, as shown on the phase diagram, and on how the phases form as the material is cooled from the liquid to the solid. At each step of the cooling process, the microstructure evolves from the microstructure that was present at the higher temperature, and so it must be consistent with the previous microstructure. Generally, the process of solidification is, first, that small particles form, and then grow larger until no liquid is left. Figure 6.3.1 shows the microstructure of a Cu/Ni mixture at one point in the two-phase region. You will now use information from the phase diagram to see how this microstructure changes as the material is cooled through the liquid/solid two-phase region.

Guided Inquiry: Microstructure

6.3.1 If the microstructure shown in Figure 6.3.1 is cooled to a lower temperature so that the particles grow, draw a picture of the resulting microstructure.

6.3.2 The microstructure you drew is cooled further to the one-phase solid region, and so the particles continue to grow until no liquid is left. Draw a picture of that microstructure.

6.3.3 What would you call the type of microstructure that you drew for question 6.3.2?
Remember the lesson on crystal defects.

6.3.4 Using your knowledge of how to use the tieline and your answers to questions 6.3.1 and 6.3.2, draw the microstructure at each of the points *A*, *B*, *C*, *D*, and *E* in the phase diagram shown in Figure 6.3.2.
Point B is just below the liquidus, and point D is just above the solidus.

Concept Check 6.3.1

- **Figure 6.3.1 shows three solid particles forming in the liquid. If five particles formed instead, how would the final microstructure be different?**

Figure 6.3.1

Microstructure at one point in the two-phase region of the Cu–Ni phase diagram. There are particles of solid (light grey) in the liquid (dark grey).

Figure 6.3.2

Portion of the copper–nickel phase diagram.

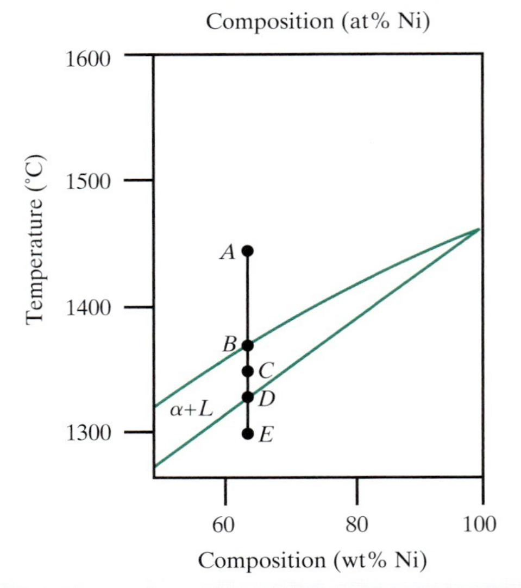

6.4 Eutectic Phase Diagrams—Microstructure

LEARN TO: **Draw the microstructures developed on cooling for a eutectic phase diagram.**

Phase diagrams as simple as the one shown in Figure 6.2.1 are actually pretty rare. Most phase diagrams are more complicated, with limited solubility of the two components in each other. Figure 6.4.1 shows the Pb–Sn phase diagram.

This type of phase diagram is called a *eutectic phase diagram*. There are a few special things to notice about it. First, at low temperatures you will see that the two components have limited solubility in each other. For example, at 100° C the solubility limit of tin dissolved in lead is only about 5 wt%; if any more tin is added, two phases will form. This is just like our original example of sugar and water, except that instead of a liquid phase and a solid phase, for lead and tin there are two solid phases. As a shorthand, for metallic phase diagrams we name the different solid phases from left to right across the phase diagram with Greek letters. There is no special meaning to the α phase on the Pb–Sn phase diagram. It's just easier to say "α solid" than "the solid that has a composition ranging from pure lead to 18.3 wt% lead and 81.7 wt% tin."

Also notice what happens as you cool down from high temperatures at 30 wt% Sn. At this composition you start in the liquid phase, go to a two-phase region of solid plus liquid, then go to a two-phase region of two solids. However, there is one special composition at 61.9 wt% Sn; at this composition when you cool down you go directly from one liquid

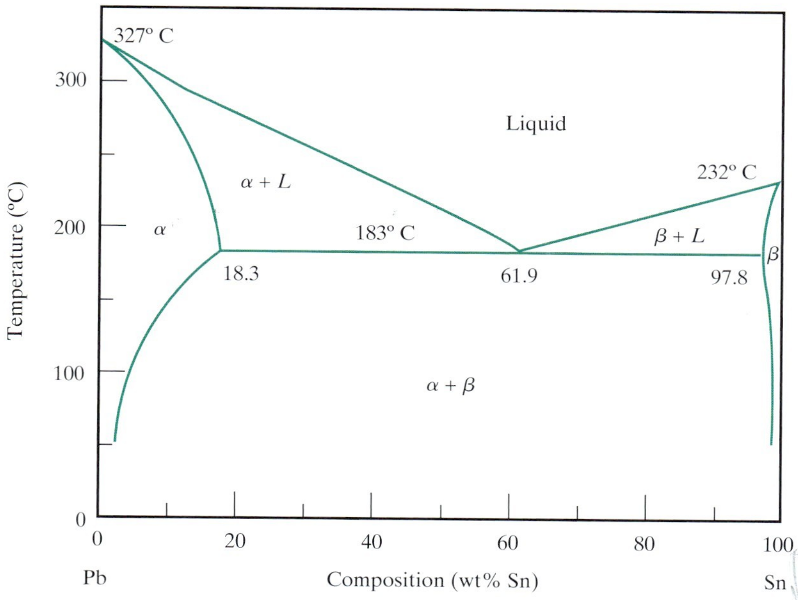

Figure 6.4.1

The lead–tin phase diagram.

phase to two solid phases. The composition at which that occurs is called the *eutectic composition*, while the temperature at which it occurs is called the *eutectic temperature*. At this eutectic point, the following *eutectic reaction* occurs:

$$L \rightleftharpoons \alpha + \beta$$

This reaction defines a eutectic phase diagram. A diagram in which this reaction is present is a eutectic phase diagram. The eutectic point is one specific kind of invariant point on a phase diagram. As we will see later, there are other types of invariant points.

Guided Inquiry: Eutectic Phase Diagrams

6.4.1 What phases are present at 10 wt% Sn and 200° C?

6.4.2 What phases are present at 40 wt% Sn and 100° C?

6.4.3 Draw a picture showing the microstructure for question 6.4.1.

6.4.4 Explain the difference between pure Pb and the α phase.

6.4.5 Identify all compositions on the lead–tin phase diagram for which melting occurs at a single temperature. What are those temperatures?

Don't just look for intersections of lines on the phase diagram. Draw lines showing the path the material takes during cooling to see where melting occurs at a single temperature.

6.4.6 Why is lead–tin solder made up of 60 wt% tin and 40 wt% lead?

Concept Check **6.4.1**

- **On the lead–tin phase diagram, is there a single melting temperature at a composition of 18.3 wt% tin?**

Because of the presence of two solid phases, microstructure development in eutectic phase diagrams can be more complicated than for the simple phase diagrams we saw in the last section. The first important thing to remember is this: **If we are only dealing with a single solid phase and a single liquid phase, the microstructure develops in the same way as we have already seen**. It's when the second solid phase appears that something different happens.

The most important point for us to consider is the eutectic point. As a liquid of eutectic composition (61.9 wt% Sn for the Pb–Sn phase diagram) is cooled past the eutectic temperature (183° C for the Pb–Sn phase diagram), the eutectic microstructure is formed.

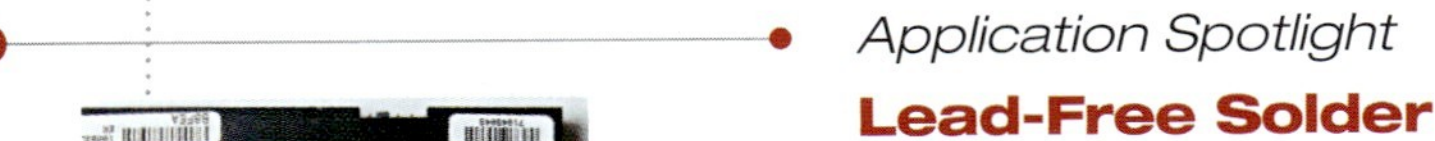

Application Spotlight

Lead-Free Solder

(Photo courtesy of Lifeprints Photography)

As you probably know, soldering is the process of using a low-melting metal to join together two other pieces of metal. Soldering is used most commonly in plumbing and in electronics. But what is solder made of? For electronics, the most common solder has been lead–tin alloy. Various compositions are used, but the best joints are obtained with alloys that are 63 wt% tin/37 wt% lead. Take a look at Figure 6.4.1 to see why—this is close to the eutectic composition, and so this mixture has a single melting temperature. When the solder cools, it solidifies instantly, creating a strong joint. For other compositions (such as 50/50) solidification occurs over a range of temperatures. If the joint moves during this solidification, cracks can develop and the joint is weaker. While lead–tin solder has good properties, lead is toxic, and regulations have been established, especially in the European Union, restricting its use. Many lead-free solders are now available. For example, alloys of tin, silver, and copper have a eutectic temperature of 217° C. The search for lead-free solders continues, and researchers rely largely on understanding the phase diagrams for these alloys to obtain low eutectic temperatures.

This microstructure is shown in Figure 6.4.2; it consists of alternating layers of the α and β phases, which are called lamellae. These lamellae form because of the need for solidification to occur quickly at a single temperature. In order to limit the distance over which the atoms need to move, the lamellae form as shown in Figure 6.4.3. The overall composition of the mixture is 61.9 wt% tin, but the α solid is only 18.3 wt% tin. Therefore, tin atoms have to move away from the regions that form the α solid. For the same reason, the lead atoms have to move away from the regions that form the β solid. In order to keep the distance over which they move short, lamellae form as shown. Keep this process in mind, because it will become important in the next chapter when we talk about the microstructure of steel.

The final important thing to think about is the rate of atomic motion in solids. Compared to a liquid, motion in solids is very slow because the host atoms are fixed in place and not moving. As a result, the atoms cannot move very far in a solid when phase transformations occur.

You will now put these concepts into practice by predicting the microstructures that form at different points on a eutectic phase diagram.

Figure 6.4.2

Eutectic microstructure of lead and tin. (ASM Handbook, Vol. 3: *Alloy Phase Diagrams*. Hugh Baker, Ed., pp. 1.1–1.29. Copyright © 1992 ASM International®. All rights reserved.)

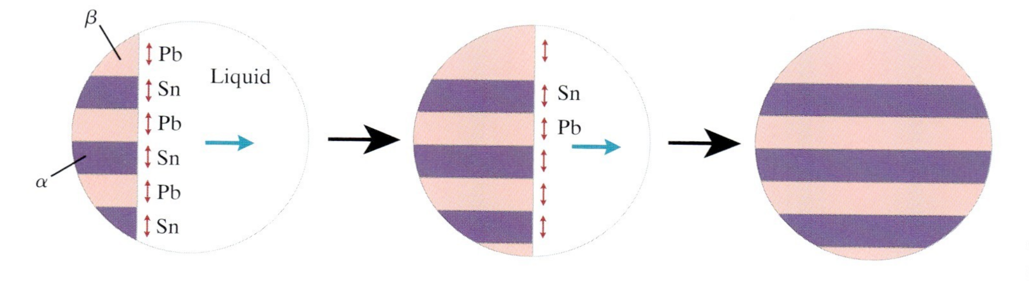

Figure 6.4.3

Development of the eutectic microstructure upon cooling through the eutectic point. The red arrows show the direction of diffusion for the lead and tin atoms, the blue arrow shows the direction of growth for the solid.

Guided Inquiry: Eutectic Microstructure

6.4.7 Draw the microstructures at points *A*, *B*, and *C* in Figure 6.4.4.

6.4.8 What phases are present at point *C* in Figure 6.4.4? At point *D*?

6.4.9 Would you expect atomic motion to be fast or slow in a solid compared to a liquid?

6.4.10 In a solid, in a given amount of time would you expect the atoms to move a long or a short distance?

6.4.11 In Figure 6.4.5, in which image do the atoms move a shorter distance to get from one particle to another?

6.4.12 Based on your answers to questions 6.4.9–6.4.11, would you expect phases that grow from a solid phase to consist of a few large particles or many small particles? *Think of how far the atoms have to move to get to a particle.*

6.4.13 Based on your answer to question 6.4.12, draw the microstructure at point *D*. *Do you expect the atoms to move a long or a short distance? Which image in Figure 6.4.5 do you expect it to look like?*

6.4.14 Circle the eutectic point in Figure 6.4.4.

6.4.15 If a liquid is cooled through the eutectic point, what microstructure forms?

6.4.16 Draw the microstructures at points *E* and *F*.

6.4.17 Draw the microstructure at points *G* and *H*.

6.4.18 What is the composition of the liquid at point *H*?

Draw the tieline for point H. Is there something special about the composition of the liquid?

6.4.19 What will the microstructure of this liquid be, once it solidifies?

6.4.20 Based on your answer to question 6.4.15, draw the microstructure at point *I*.

Don't look ahead! Try to figure it out in your group.

Concept Check 6.4.2

- **What is different about the microstructures at points *I* and *J* in Figure 6.4.4?**

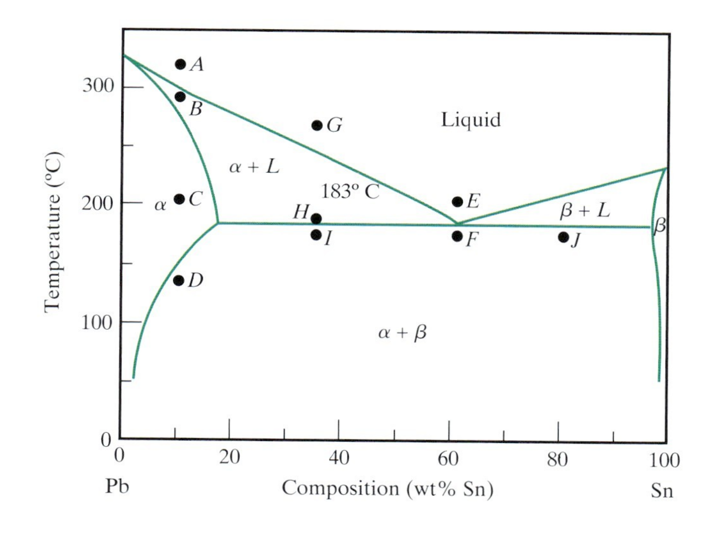

Figure 6.4.4

Phase diagram to use with questions 6.4.8–6.4.18.

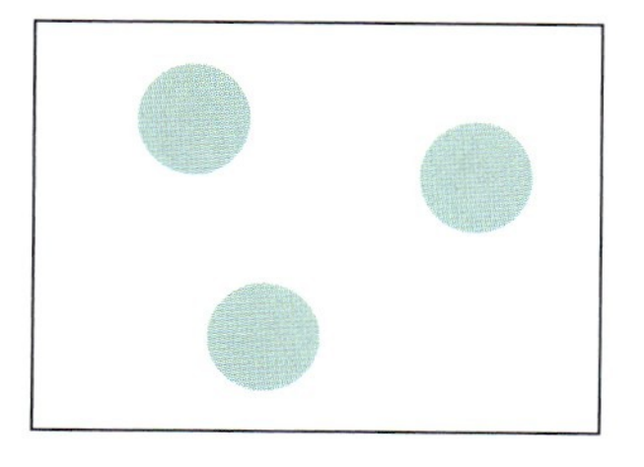

Figure 6.4.5

Two possible microstructures in a material.

6.5 Eutectic Phase Diagrams—Microconstituents

LEARN TO: From a eutectic phase diagram, calculate the amounts of microconstituents present.

Because eutectic phase diagrams result in such complicated microstructures, we need a specific way of talking about them. The term we use is *microconstituent*. Microconstituents are specific parts of microstructures. Figure 6.5.1 shows the microstructure of a eutectic alloy. The different microconstituents are:

- Primary α: This is the large particles of α (the *primary solid*) that form above the eutectic temperature.
- *Eutectic solid*: This is the solid that consists of lamellae of α and β.
- Eutectic α: The lamellae of α.
- Eutectic β: The lamellae of β.

Note that in this case there is no primary β. By looking at the phase diagram and thinking about how the microstructure forms during cooling, you should see why you can have either primary α or primary β, but not both.

From the lever rule we can actually determine the amounts of each of the individual microconstituents. However, before we try to use the phase diagram to do that, we need an overall strategy for doing it. **Questions 6.5.1–6.5.7 should be done only by estimating the amounts shown in the pictures of the microstructure**. Don't try to use the lever rule until you get to question 6.5.8.

Guided Inquiry: Microconstituents

6.5.1 What is the amount of α present at point *l*?
Estimate the amount based on the images of the microstructure in Figure 6.5.1.

6.5.2 What is the amount of primary α at point *m*?

6.5.3 What is the amount of liquid at point *l*?

6.5.4 What is the amount of eutectic solid at point *m*?

6.5.5 What is the amount of β at point *m*?

6.5.6 What is the amount of eutectic β at point *m*?

6.5.7 Based on your answers to questions 6.5.1–6.5.6, describe a general approach to determining the following quantities in a eutectic alloy just below the eutectic temperature:
a. Amount of primary α.
b. Amount of primary β.
c. Amount of eutectic solid.
d. Amount of eutectic β.
e. Amount of eutectic α.
Look at the relationships you used to answer questions 6.5.1–6.5.6.

6.5.8 Use your procedure from question 6.5.7 to determine the phase compositions, phase amounts, and microconstituent amounts for an alloy of 50 wt% Pb and 50 wt% Sn at 182° C.

Concept Check **6.5.1**

- **How much primary α is present for an alloy of 70 wt% Pb and 30 wt% Sn at 182° C?**
- **How much eutectic α is present for an alloy of 70 wt% Pb and 30 wt% Sn at 182° C?**
- **How much eutectic β is present for an alloy of 70 wt% Pb and 30 wt% Sn at 182° C?**

Figure 6.5.1

Lead–tin phase diagram showing the microstructure that develops during cooling from the liquid for a composition of 35 wt% tin.

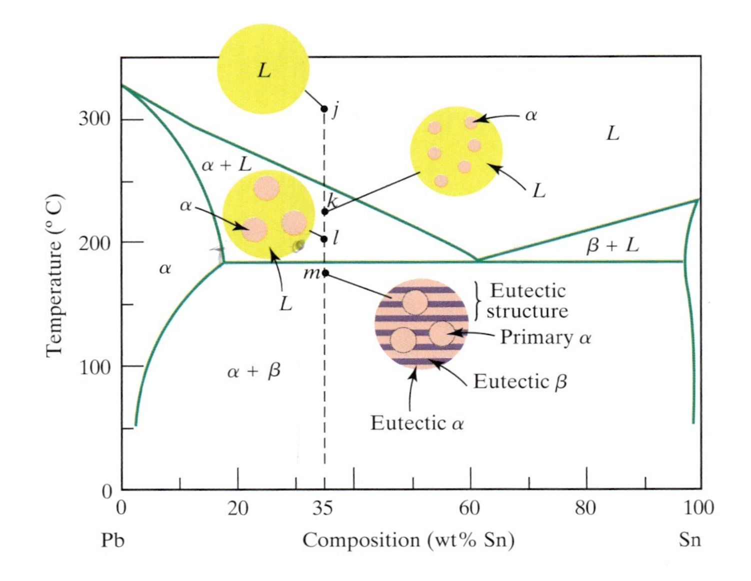

EXAMPLE PROBLEM 6.5.1

Determine the phase compositions, phase amounts, and microconstituent amounts for an alloy of 30 wt% Sn and 70 wt% Pb at 182° C.

The point corresponding to 30 wt% Sn and 182° C is in the $\alpha + \beta$ region of the phase diagram, just below the eutectic temperature. To get the phase compositions and amounts we just do a normal tieline calculation as we would do in the two-phase region of any phase diagram. The tieline to use is shown in red in the phase diagram below. Since 182° C is very close to the eutectic temperature, we will use the compositions as given on the tieline instead of trying to estimate. From this we get:

Composition of $\alpha = 18.3$ wt% Sn and 81.7 wt% Pb

Composition of $\beta = 97.8$ wt% Sn and 2.2 wt% Pb

$$\text{Amount of } \alpha = \frac{97.8 - 30}{97.8 - 18.3} \times 100 = 85.3 \text{ wt\%}$$

Amount of $\beta = 100 - 85.3 = 14.7$ wt%

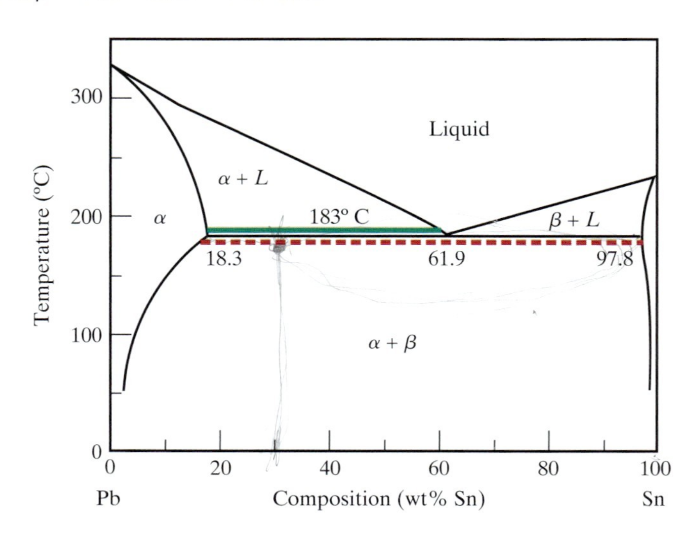

Now for the microconstituents. In order to determine the amounts of these, it is helpful to visualize the microstructure. Following the description in Section 6.4, primary α forms in the $\alpha + L$ region, and eutectic α and β form when the liquid left just above the eutectic temperature is cooled down to below the eutectic temperature. This means the microstructure looks like this:

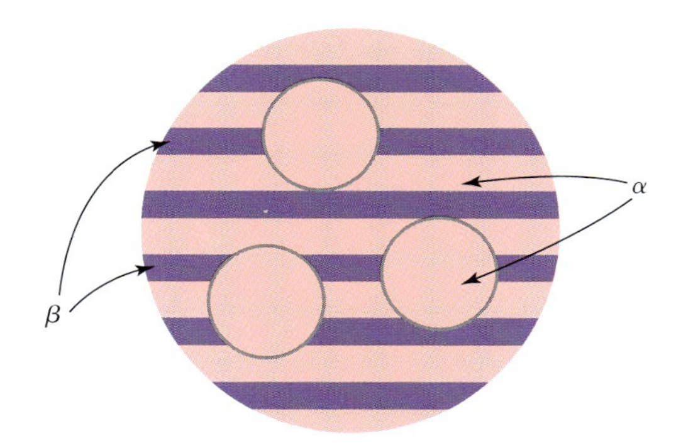

Looking at this microstructure, we see that all of the β is eutectic β, so the amount of eutectic β is the same as the total amount of β, which from above is 14.7 wt%. The α phase is trickier because some of it is primary α and some is eutectic α. The easiest way to deal with this is to note that the amount of solid just above the eutectic is the same as the amount of primary α just below the eutectic. So to get the amount of primary α, figure out how much solid is present just above the eutectic using the green tieline in the phase diagram below.

Amount of primary α at 182° C = amount of solid at 184° C

$$= \frac{61.9 - 30}{61.9 - 18.3} \times 100 = 73.2 \text{ wt\%}$$

We now know how much primary α and eutectic β there is. Looking at the picture of the microstructure, we see the only thing left is eutectic β, so

Amount of eutectic α = 100 − amount of eutectic β − amount of primary α

$= 100 - 14.7 - 73.2 = 12.1$ wt%

Now we can list the amounts of microconstituents:

Amount of primary α = 73.2 wt%

Amount of eutectic α = 12.1 wt%

Amount of eutectic β = 14.7 wt%

6.6 Peritectic Phase Diagrams

LEARN TO: Identify the peritectic point.
Conduct calculations for peritectic phase diagrams.

Another type of phase diagram is the peritectic phase diagram, shown schematically in Figure 6.6.1. It has an invariant point, called a peritectic point. **You will use the questions to figure out what the *peritectic reaction* is.**

Guided Inquiry: Peritectic Phase Diagrams

6.6.1 What phases are present at temperature T_1 and composition x_2?

6.6.2 What phases are present at temperature T_2 and composition x_1?

6.6.3 What phases are present at temperature T_2 and composition x_2?

6.6.4 If an alloy of composition x_1 is cooled from T_1 to T_2, write the reaction that occurs. Can this reaction occur at any other composition? Explain why or why not.

6.6.5 If an alloy of composition x_2 is cooled from T_1 to T_2, write the reaction that occurs. Can this reaction occur at any other composition? Explain why or why not.

6.6.6 Based on your answers to questions 6.6.4 and 6.6.5, circle invariant points on the peritectic phase diagram in Figure 6.6.1.

An invariant point is a point where a phase change occurs at a single temperature.

Concept Check 6.6.1

- How do eutectic reactions differ from peritectic reactions?

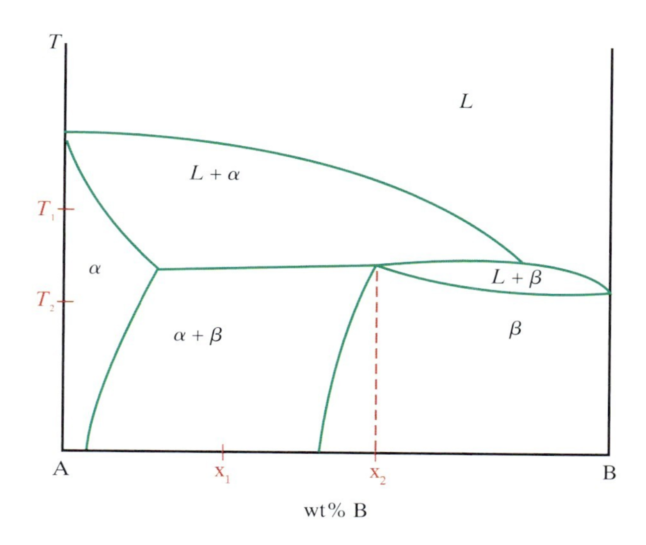

Figure 6.6.1

Peritectic phase diagram. The compositions x_1 and x_2 and the temperatures T_1 and T_2 are for use in the questions.

You probably identified the peritectic point as the point corresponding to point P in Figure 6.6.2; that is how most people define it. However, you should be aware that a few textbooks define it as point P'. You should also note that the phase diagram in Figure 6.6.1 is a little bit distorted, so it is easier for you to see what is going on. Figure 6.6.2 is more accurate, but you should now be able to see how the two are the same. One final thing to be aware of is that the microstructure development is very complicated for peritectic phase diagrams and is beyond the scope of this book. But for basic calculations of phase compositions and phase amounts, just use the lever rule in the same way as you have for isomorphous and eutectic phase diagrams.

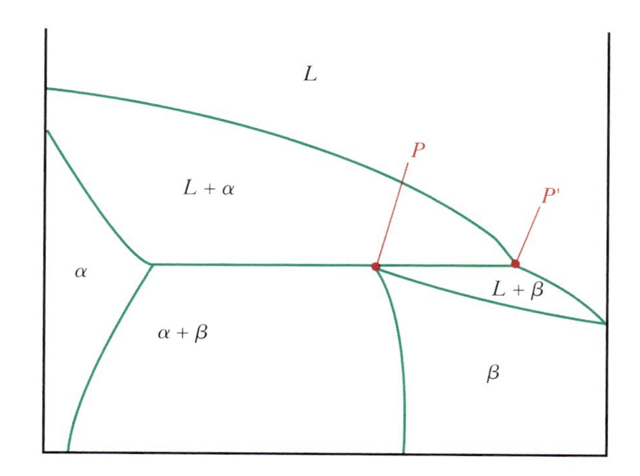

Figure 6.6.2

Peritectic phase diagram showing the possible peritectic points. Most people consider the peritectic point to be at point P, but a few texts define it as P'.

6.7 Intermetallic and Ceramic Phase Diagrams

LEARN TO: **Identify the phases present on phase diagrams.**

So far we have dealt with relatively simple phase diagrams that are of one clear type—for example, eutectic. However, many phase diagrams look more complicated. Even if they do, make sure you follow one simple rule: **don't panic!** All of the rules we have learned so far still apply, no matter how complicated the phase diagram looks. That means if you are looking at a two-phase region, you still use the tieline and the lever rule just as you have already learned to do to determine the phase compositions and phase amounts.

Figure 6.7.1 shows the phase diagram for magnesium and lead. This phase diagram has a *line compound.* A line compound is a single phase of an exact composition, so it appears on the phase diagram as a vertical line rather than a region; the regions on either side of the line compound are two-phase regions. In the Mg–Pb phase diagram, Mg_2Pb is a line compound. Up until now, whenever we saw a line on a phase diagram it was a boundary between a one-phase region and a two-phase region, but with a line compound this is not true.

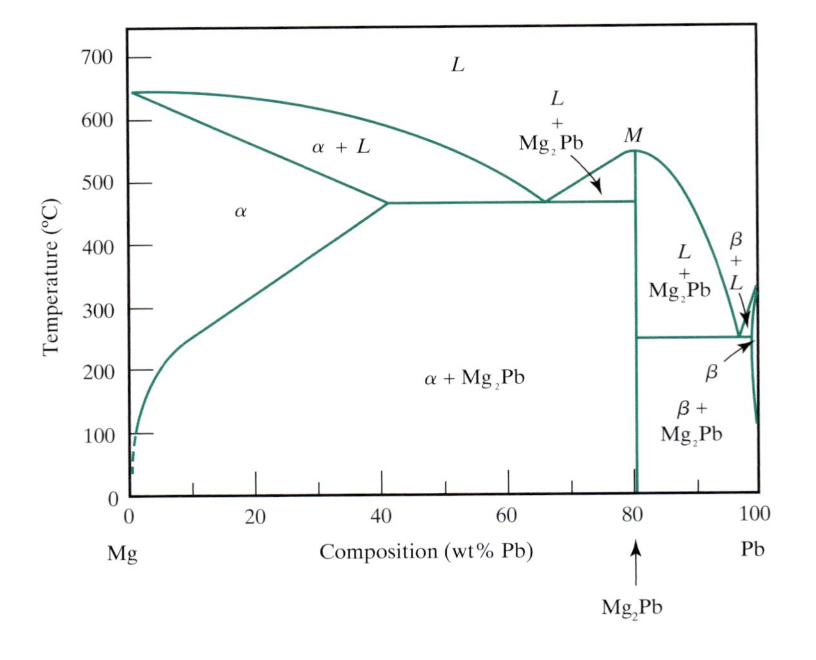

Figure 6.7.1
The magnesium–lead phase diagram.

Another important consideration is how we define components for ceramic phase diagrams. Metal oxides are one type of ceramic. Oxides are typically named by changing the "um" or "on" at the end of the name of the metal to "a." So silicon oxide is silica, germanium oxide is germania, and so on. Let's consider mixtures of zirconia (ZrO_2) and calcia (CaO). We could create a ternary phase diagram of three components: zirconium, calcium, and oxygen. But it is much easier to just make a binary phase diagram, with zirconia and calcia as the components. We can do this because zirconia and calcia are specific chemical compounds.

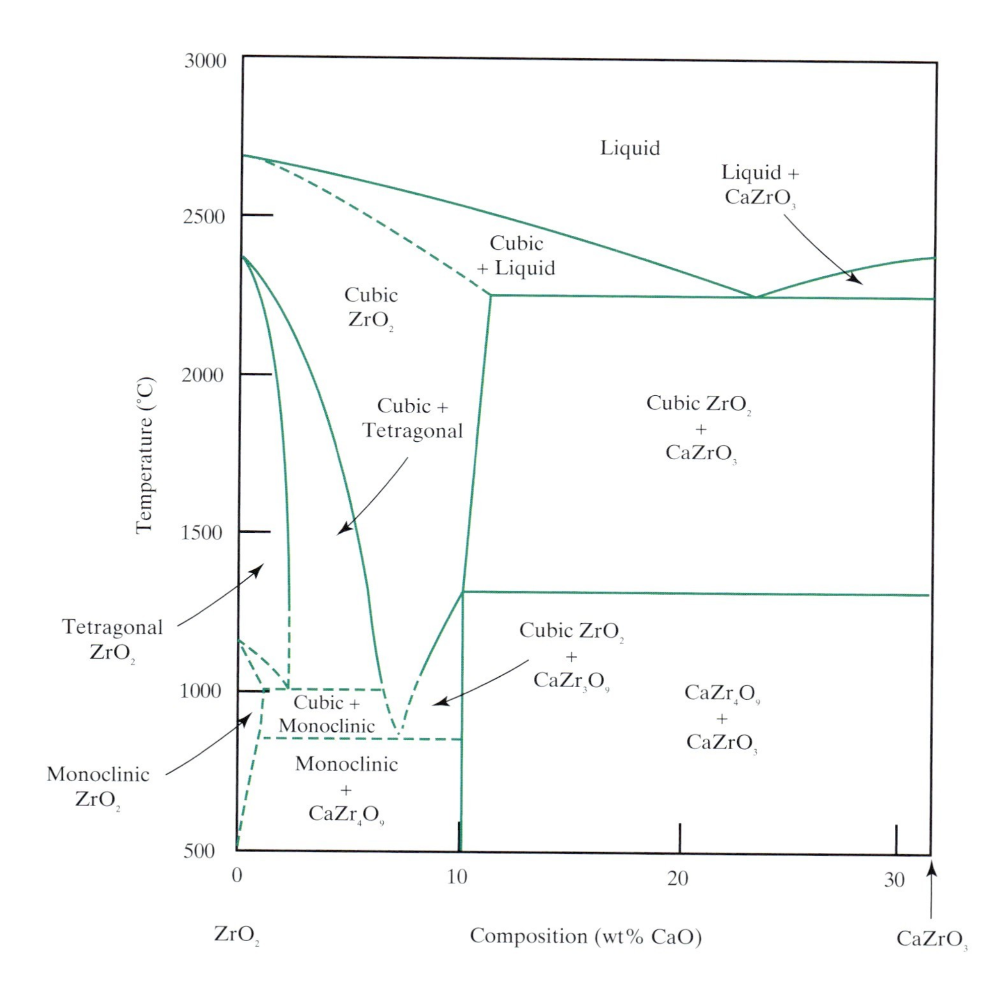

Figure 6.7.2

Calcia–zirconia phase diagram.

And now you know everything you need to know about phase diagrams (at least for this introductory course)! **The questions in this section will give you some more practice in working with phase diagrams**.

Guided Inquiry: Complex Phase Diagrams

6.7.1 In the Mg–Pb phase diagram, what phases are present at 40 wt% Pb and 200° C? What are the compositions of those phases?
You can also work in your group to calculate the phase amounts if you want practice doing that.

6.7.2 What phases are present at 90 wt% Pb and 200° C? What are the compositions of those phases?

6.7.3 Look at the zirconia–calcia phase diagram. What are the components for this phase diagram?

6.7.4 Based on your answer to question 6.7.3, what is one difference between the components for metallic phase diagrams and for ceramic phase diagrams?

6.7.5 On the zirconia–calcia phase diagram, what phases are present at 5 wt% calcia and 1500° C? Other than composition, what distinguishes these two phases?

6.7.6 Cubic zirconia is a cheap substitute for diamond. On the basis of everything you have learned from this book to this point, explain what the structure of cubic zirconia is, based upon its name.

Concept Check 6.7.1

- **Why is it incorrect to say that cubic zirconia (ZrO_2) has a composition of 1 mole of zirconium for every 2 moles of oxygen?**

Summary

The primary goal of this chapter was for you to learn how to use phase diagrams. While it may seem to have gotten complicated, the rules are always the same: when in a two-phase region, use the tieline; the phase compositions are obtained from the ends of the tieline and the phase amounts are obtained from the lever rule. Also remember that the lever rule is really an inverse lever rule: for example, you get the amount of the solid from the proportion of the tieline that is closest to the liquid region of the phase diagram.

You also saw how we can start to get information about microstructure from a phase diagram. But the information we get is really limited. For example, the microstructure development in Section 6.3 depended on the number of particles that initially form in the liquid, but how do you know how many particles will form? That question is the topic of kinetics, which we begin in the next chapter.

Key Terms

Binary phase diagram
Component
Composition
Equilibrium phase diagram
Eutectic composition
Eutectic phase diagram
Eutectic reaction
Eutectic solid
Eutectic temperature
Lamellae
Lever rule
Line compound
Microconstituent
Microstructure
Peritectic reaction
Phase
Phase amount
Phase composition
Phase diagram
Primary solid
Solubility limit
Tieline

Problems

Skill Problems

6.1 For each of the following, list the components and the phases:

a. Sugar dissolved in water.
b. Water with ice cubes in it.
c. Sparkling water (pure water with carbon dioxide bubbles).
d. Oversaturated sugar/water solution, so there is some undissolved sugar sitting on the bottom.

6.2 Give the phases that are present and the phase compositions for the following alloys:

a. 30 wt% Pb and 70 wt% Mg at 425° C.
b. 70 wt% Pb and 30 wt% Mg at 425° C.
c. 90 wt% Pb and 10 wt% Mg at 425° C.
d. 30 wt% Pb and 70 wt% Mg at 550° C.

6.3 Give the amounts of the phases present for each of the alloys in problem 6.2.

6.4 For a copper–silver alloy of composition 25 wt% Ag–75 wt% Cu at 775° C, calculate the following:

a. The amounts of the α and β phases.
b. The amounts of primary α and total eutectic microconstituents.
c. The amounts of eutectic α and eutectic β.

6.5 For a lead–magnesium alloy of composition 50 wt% Pb–50 wt% Mg at 465° C, calculate the following:

a. The amounts of the α, β, and Mg_2Pb phases.
b. The amounts of primary α, primary β, primary Mg_2Pb, and total eutectic microconstituents.
c. The amounts of eutectic α, eutectic β, and eutectic Mg_2Pb.

6.6 For a lead–magnesium alloy of composition 70 wt% Pb–30 wt% Mg at 465° C, calculate the following:

a. The amounts of the α, β, and Mg_2Pb phases.
b. The amounts of primary α, primary β, primary Mg_2Pb, and total eutectic microconstituents.
c. The amounts of eutectic α, eutectic β, and eutectic Mg_2Pb.

6.7 For a lead–magnesium alloy of composition 90 wt% Pb–10 wt% Mg at 255° C, calculate the following:

a. The amounts of the α, β, and Mg_2Pb phases.
b. The amounts of primary α, primary β, primary Mg_2Pb, and total eutectic microconstituents.
c. The amounts of eutectic α, eutectic β, and eutectic Mg_2Pb.

6.8 Draw the microstructure for a lead–magnesium alloy of composition 20 wt% Pb–80 wt% Mg at 600° C.

6.9 Draw the microstructure for a lead–magnesium alloy of composition 50 wt% Pb–50 wt% Mg at 400° C.

6.10 Draw the microstructure for a lead–magnesium alloy of composition 20 wt% Pb–80 wt% Mg at 400° C.

6.11 Draw the microstructure for a lead–magnesium alloy of composition 20 wt% Pb–80 wt% Mg at 200° C.

6.12 Draw the microstructure for a lead–magnesium alloy of composition 90 wt% Pb–10 wt% Mg at 600° C.

6.13 Draw the microstructure at each of the points *A*, *B*, *C*, *D* in the phase diagram given below. Clearly describe any differences in the microstructures if they are not evident in your sketches.

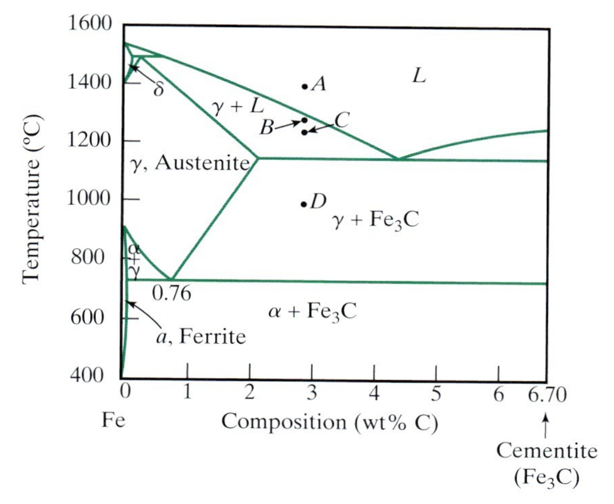

Conceptual Problems

6.14 What is the maximum mass of sugar that could be dissolved in 1 L of water?

6.15 On the magnesium–lead phase diagram, identify all compositions at which melting occurs at a single temperature.

6.16 For the phase diagram given below, within each circle sketch the microstructure at the indicated point. In other words, within circle A sketch the microstructure at point *A* on the phase diagram, etc.

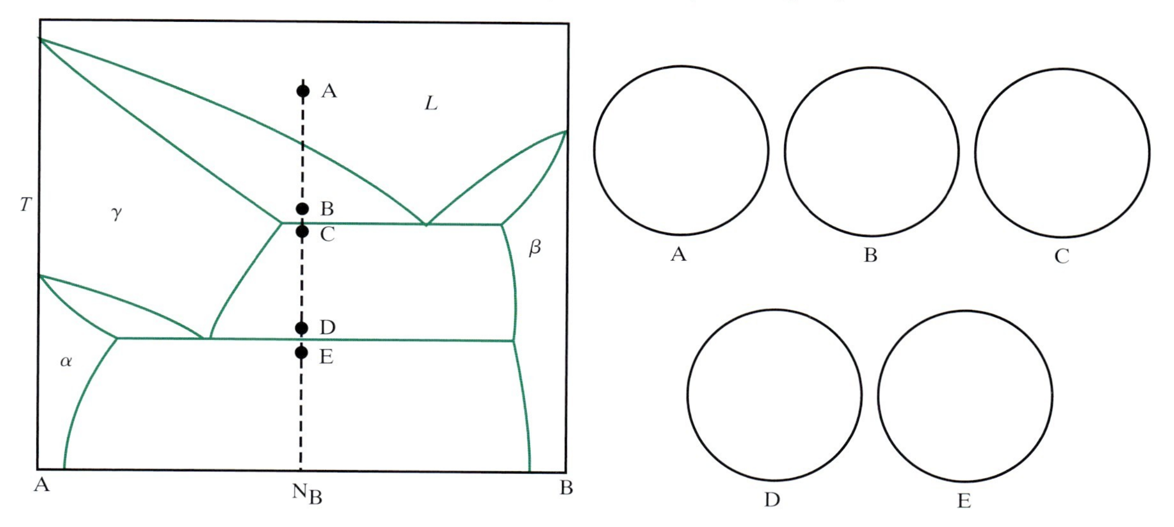

6.17 In the titanium–copper phase diagram, Ti_2Cu is a line compound. What is the composition of this phase?

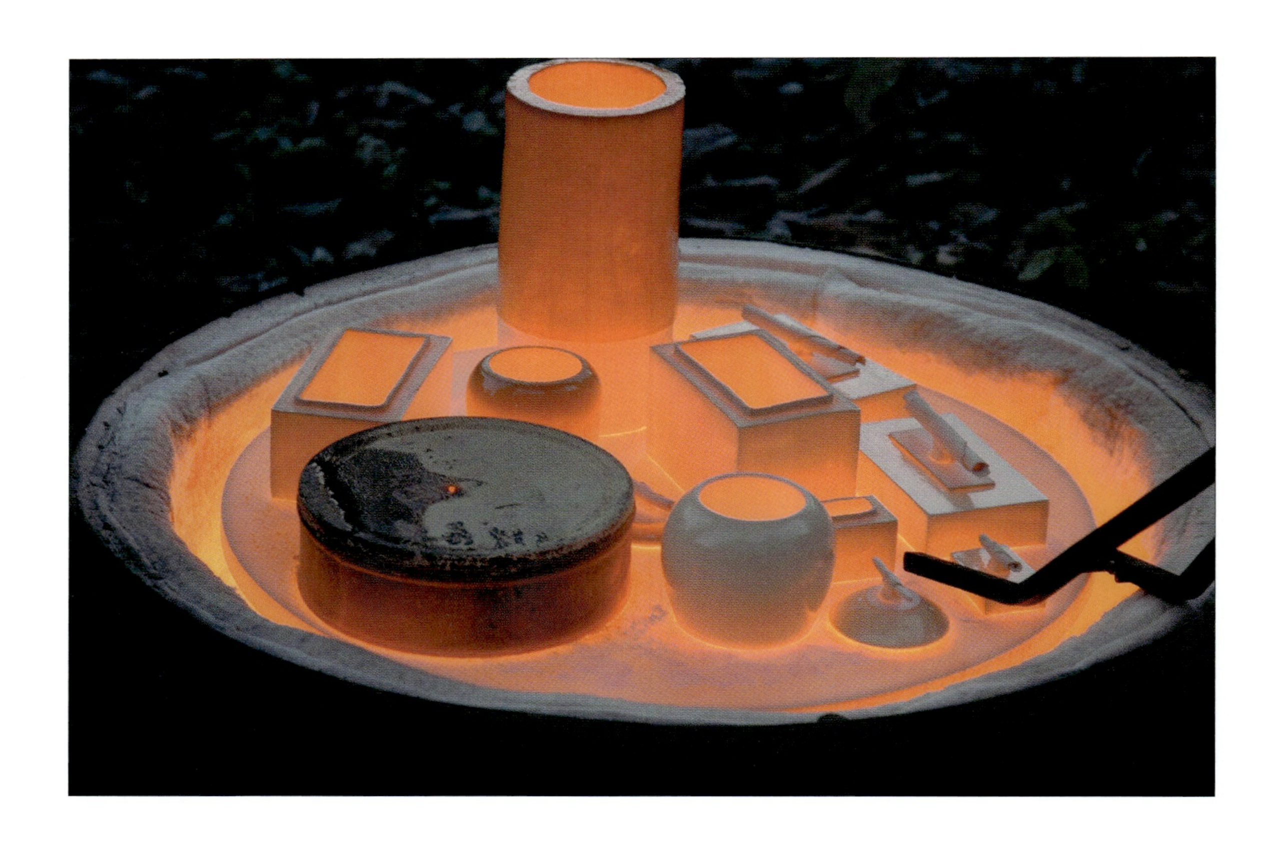

The process of making pottery involves binding ceramic particles together and then fusing them together at high temperatures through a process called sintering. Sintering relies on diffusion to fuse the particles into a solid piece. (Sander Claes/Shutterstock)

7 Diffusion

An important process used in materials engineering is *diffusion*, in which atoms or molecules move through a substance. Note that this is not the same as flow. Flow implies that a bulk fluid is moving through space. In diffusion individual atoms or molecules move through the bulk of another material. Diffusion can occur when gas molecules move through a solid, when atoms of one solid move through another solid, or when the molecules of one liquid move through another liquid. You can observe diffusion in action by putting a drop of food coloring into a glass of water. Even if you don't stir it, the food coloring will diffuse through the entire glass and change the color of the water. Diffusion can be used in processing, for example, to separate gases or to create computer chips. Diffusion can also be a desired part of an engineering system, such as when oxygen diffuses through a ceramic in the oxygen sensor in your car. By the end of this chapter you will:

Understand the mechanisms of diffusion.
Be able to perform diffusion calculations.

7.1 Diffusion Mechanisms

LEARN TO: Define diffusion.
Describe diffusion mechanisms.

In order to understand how diffusion works, we will begin by considering the simple case of nitrogen gas diffusing through a solid aluminum sheet, as shown in Figure 7.1.1. On one side of the metal the gas has pressure P_A and on the other side pressure P_B. There are no holes or pores in the metal sheet; the gas is moving through the spaces between the atoms. By examining how pressure affects diffusion, you will work out why diffusion occurs.

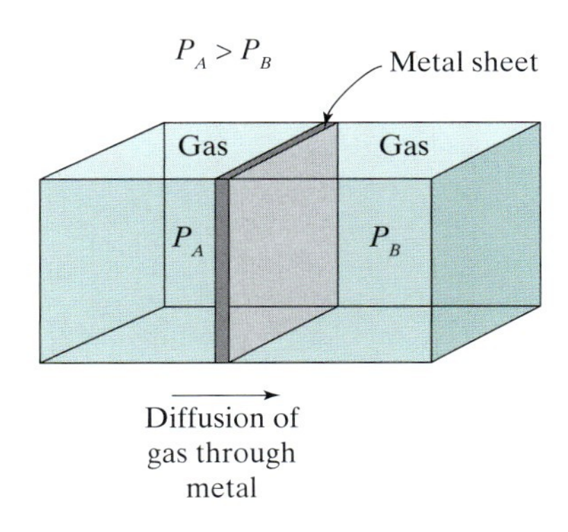

Figure 7.1.1 Example of diffusion of a gas through a solid metal sheet.

Guided Inquiry: Diffusion

7.1.1 Which side has the greater nitrogen pressure, *A* or *B*?
Your instructor may have given you a time limit to complete these questions. Make sure one person in your group keeps track of time and makes sure your group doesn't get bogged down on a question.

7.1.2 Which side has the greater nitrogen concentration; *A* or *B*?

7.1.3 Based on Figure 7.1.1, which way does the nitrogen move through the piece of aluminum?

7.1.4 If Figure 7.1.1 were changed so that the concentration of nitrogen were greater on side *B* than on side *A*, which way would the nitrogen diffuse?

7.1.5 If the figure were changed so that the concentration of nitrogen were the same on sides *A* and *B*, which way would the nitrogen diffuse?

7.1.6 What is the driving force for nitrogen diffusion through the metal?
By "driving force" I mean the thing that causes diffusion. If the driving force were not present, diffusion would not occur.

Concept Check 7.1.1

- **If $P_B > P_A$ in Figure 7.1.1, which way will the nitrogen diffuse?**
- **If $P_B = P_A$ in Figure 7.1.1, which way will the nitrogen diffuse?**
- **Consider two situations, both with $P_A > P_B$. If P_A increases from double P_B to triple P_B; how will the rate of diffusion change?**

Now that you know what diffusion is, the obvious question to ask is: How can atoms or molecules move through a solid? The rest of this section guides you to think about possible diffusion mechanisms.

Application Spotlight **Water Purification**

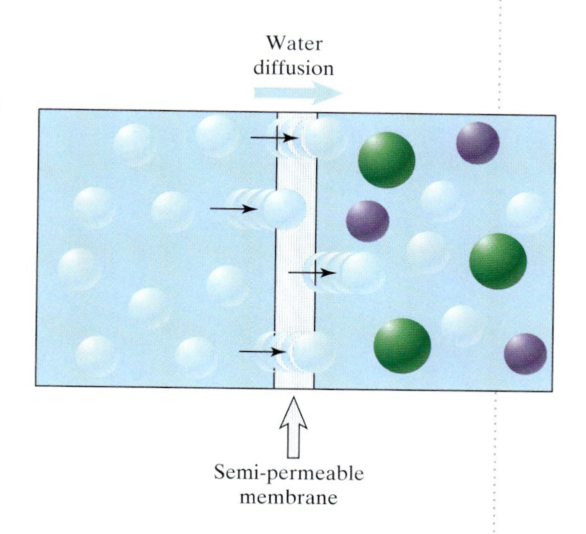

Water is essential for daily life, and for much of human history it has been considered an inexhaustible resource. However, it has become clear recently that sources of fresh water are limited. Some policy analysts have predicted that water will soon replace oil as a commodity over which political and military battles are fought. Even in the developed world, water rights have become a controversial issue. For example, in the United States there have been continuing "water wars" between the states of Florida and Georgia over use of water from rivers that flow through both states.

Desalination is an important technology that can be used to convert seawater to fresh water. Desalination uses diffusion in a process called reverse osmosis. Before we can understand reverse osmosis, we have to understand osmosis. The process of osmosis is shown schematically in the accompanying figure. Here, two compartments are separated by what is called a semipermeable membrane; this membrane will allow water to diffuse through, but not other substances, such as salt. To make osmosis occur, one compartment has pure water and one has salt water. We know that the difference in concentration causes diffusion to occur. Normally we would expect the salt to diffuse from the salt water to the fresh water until the concentrations on each side are equal, but the semipermeable membrane won't let that happen. Instead, the water diffuses from the fresh water to the salt water in an attempt to dilute the salt water and make the concentrations equal. Obviously this can never happen, and there is a limit to how much water will diffuse to the salt-water side. You should see that what we need in order to purify water is the opposite of this osmosis process, so that water molecules move from the salt-water side to the fresh-water side, generating fresh water from salt water. To do this, desalination plants work by applying a pressure that opposes the diffusion process. Pressures of 600–1000 psi are needed for purification of seawater. This pressure forces water diffusion to occur opposite the concentration gradient, resulting in generation of fresh water from seawater.

Guided Inquiry: Diffusion Mechanisms 1

7.1.7 Circle the substitutional impurity in Figure 7.1.2.

7.1.8 Circle the vacancy in Figure 7.1.2.

7.1.9 If the impurity atom were going to diffuse, where could it move to?

7.1.10 Can the impurity atom continue to move further away from its original position based only on what you see in Figure 7.1.2? Is there anywhere else it can go, other than back to where it was?

7.1.11 If the impurity atom were going to move further away from its original position, what would have to be present?

7.1.12 Circle the interstitial atom in Figure 7.1.3.

7.1.13 Circle vacant interstitial sites in Figure 7.1.3.

7.1.14 If the impurity atom were going to diffuse, where could it move to?

7.1.15 Can the impurity atom continue to move further away from its original position based only on what you see in Figure 7.1.3? Is there anywhere else it can go other than back to where it was?

Concept Check 7.1.2

- **How will the rate of diffusion change as the size of the impurity atom increases?**

Figure 7.1.2

Part of a crystal with a substitutional impurity and a vacancy.

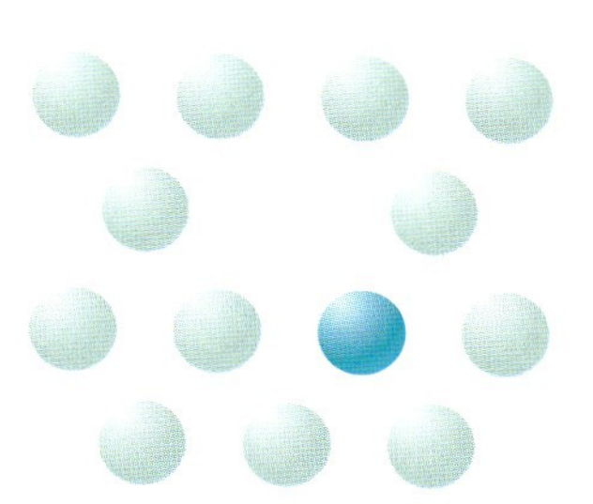

Figure 7.1.3

Part of a crystal with an interstitial impurity.

Figure 7.1.2 illustrates *vacancy diffusion*; it is called this because you can think of it occurring by motion of vacancies—even though it is really atoms that are moving, it looks like the vacancy is moving. Figure 7.1.3 illustrates *interstitial diffusion*; it is called this because diffusion occurs through interstitial sites. A third type of diffusion is called *grain boundary* or *short-circuit diffusion*. In this type of diffusion the diffusing atoms bypass, or "short-circuit," the bulk of the material and travel along grain boundaries. While short-circuit diffusion is fast, the total volume of the grain boundaries in a crystal is very small, so short-circuit diffusion does not contribute much to the total diffusion. For the rest of this section we will consider only vacancy and interstitial diffusion.

Guided Inquiry: Diffusion Mechanisms 2

7.1.16 Go back and look at Figures 7.1.2 and 7.1.3. For which case is the diffusing atom smaller; vacancy diffusion or interstitial diffusion? Which will diffuse more easily; a smaller atom or a larger atom?

7.1.17 Based on your answers to 7.1.10 and 7.1.14, which case has more spaces for the diffusing atom to go to; vacancy diffusion or interstitial diffusion?

7.1.18 Which do you think is harder; vacancy diffusion or interstitial diffusion? Explain why.

Concept Check 7.1.3

- **Which is faster; vacancy diffusion or interstitial diffusion?**

7.2 Diffusion Calculations

LEARN TO: Perform calculations for diffusion.

Figure 7.2.1 shows an example of diffusion that can occur when two metals are placed into contact with each other. Normally we would graph the concentration versus distance, as shown in the bottom row of the figure, and this graph represents what is called the *concentration gradient*, which is the change in concentration as a function of position. We can calculate this concentration gradient, how long it will take diffusion to occur, and what the concentration will be for a particular diffusion situation. First, however, we have to understand the possible diffusion situations that can occur. We will begin by considering two different scenarios, shown in Figure 7.2.2.

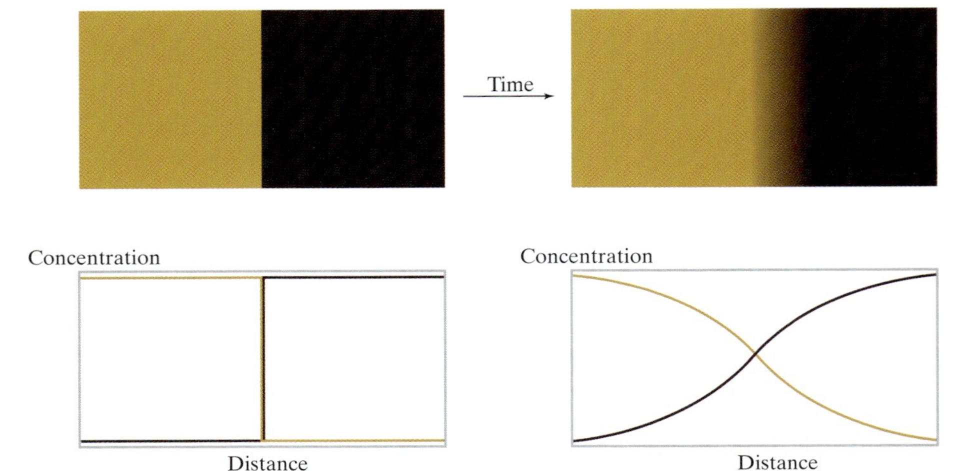

Figure 7.2.1

Example of diffusion resulting when gold and copper are placed into contact. The top row shows the change in concentration in the materials, while the bottom row shows the resulting concentration gradient. Note that the materials would have to be heated to very high temperatures for this to occur in a reasonable amount of time.

Column A

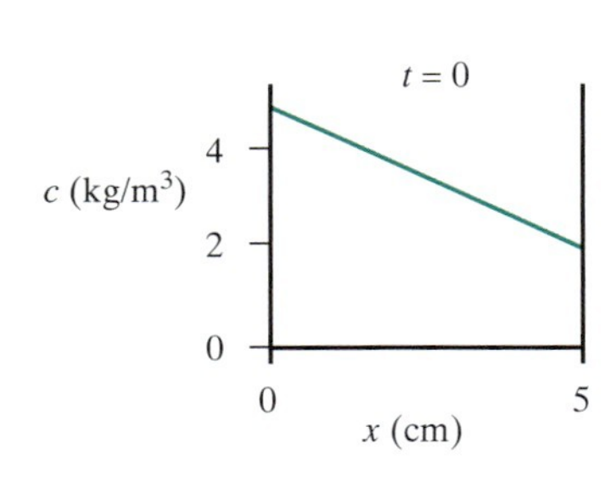

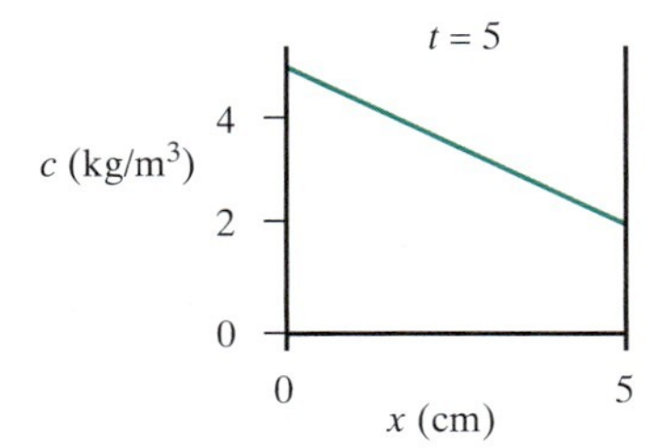

Column B

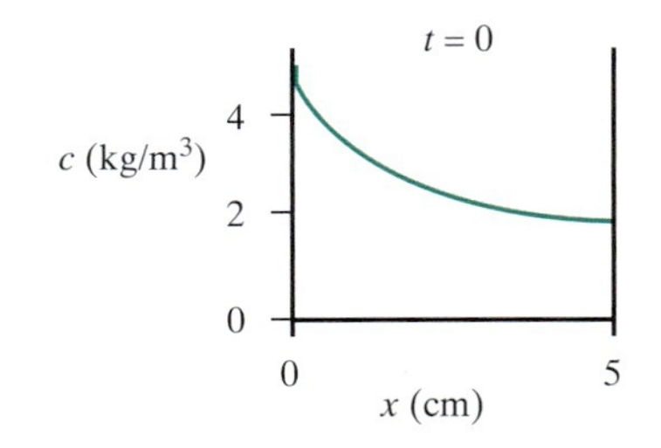

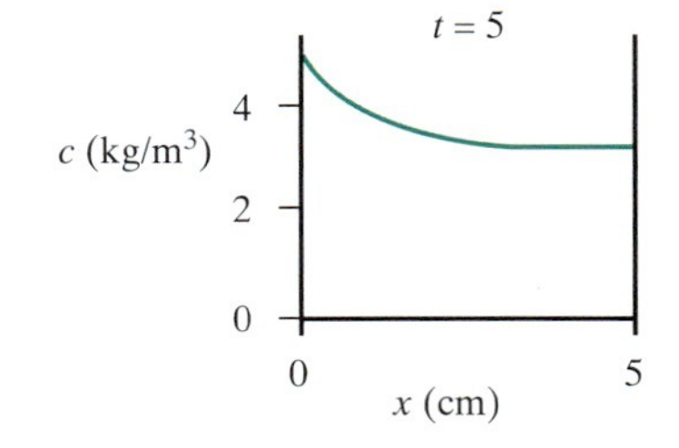

Figure 7.2.2

Two different scenarios for how concentration can change over time due to diffusion.

Guided Inquiry: Types of Diffusion

7.2.1 In column A, what is the value for the concentration at $x = 5$ cm and time = 0?

7.2.2 In column A, what is the value for the concentration at $x = 5$ cm and time = 5 minutes?

7.2.3 In column B, what is the value for the concentration at $x = 5$ cm and time = 0?

7.2.4 In column B, what is the value for the concentration at $x = 5$ cm and time = 5 minutes?

7.2.5 In which column is the concentration independent of time? In which is it dependent on time?

7.2.6 Describe the difference between column A and column B in at least two different ways.
There is no single correct answer. I just want you to have a picture in your head of how these two situations are different.

Concept Check 7.2.1

- **If you want to set up a system for continuous purification of nitrogen, would you want it to operate as shown in column A or in column B?**

The situation shown in column A of Figure 7.2.2 is an example of *steady-state diffusion*. Steady-state diffusion is occurring if any of the following is true:

1. The flux is constant. Flux is the mass of material that diffuses per unit area per unit time. It is sort of equivalent to a flow rate—but remember, diffusion is not the same as flow!
2. The concentration gradient does not change over time.
3. The concentration at any given point does not change over time.

Looking at column A, you should easily see that the numbers 2 and 3 are true. It is probably not obvious that the flux is constant, but since the other two items are true, the flux must be constant. Now take a careful look at column B. Again, it should be easy to see that numbers 2 and 3 are not true, and so column B shows *nonsteady-state diffusion*.

For steady-state diffusion we can calculate the rate of diffusion with *Fick's first law*:

$$J = -D\frac{dc}{dx} \tag{7.2.1}$$

In this equation, J is the flux, which has the units kg/m^2-s. D is the *diffusion coefficient*, which is a material property, c is the concentration, which typically has units of kg/m^3, x is the position, and thus dc/dx is the change in concentration with position, or the concentration gradient. To understand how to use Fick's first law we will go through a problem.

The next questions are based on the following scenario: Carbon is passed through a plate of steel at 700° C. At steady state, the carbon concentration 5 mm below the surface is 1.2 kg/m^3, and the carbon concentration 10 mm below the surface is 0.8 kg/m^3. The total flux through the plate is 2.4 × 10^{-9} kg/ m^2-s. Assume a linear concentration profile.

Guided Inquiry: Fick's First Law

7.2.7 In the problem above, what are the units and value for J?

7.2.8 What are the units for c? *Hint:* Make sure you are using standard SI units.

7.2.9 What are the units for x?

7.2.10 What are the units and value for dc/dx?

7.2.11 What are the units and value for D?
Check your answers to make sure everyone in your group agrees on how to do the calculation.

Concept Check 7.2.2

- **What is the diffusion coefficient for question 7.2.11?**

EXAMPLE PROBLEM 7.2.1

A sheet of polyethylene 2.0-mm thick is being used as an oxygen barrier. If the flux is 2.48×10^{-8} kg/m²-s, what concentration of oxygen must be maintained on the high-pressure side to ensure that the concentration on the low-pressure side is maintained at 0.5 kg/m³? The diffusion coefficient for oxygen in polyethylene at the operating temperature is 4.50×10^{-11} m²/s.

Since the concentrations are being maintained at a constant value, and the flux is given as a constant, this is a steady-state diffusion problem and we can use Fick's first law. The first thing to do is to draw an image that illustrates the situation:

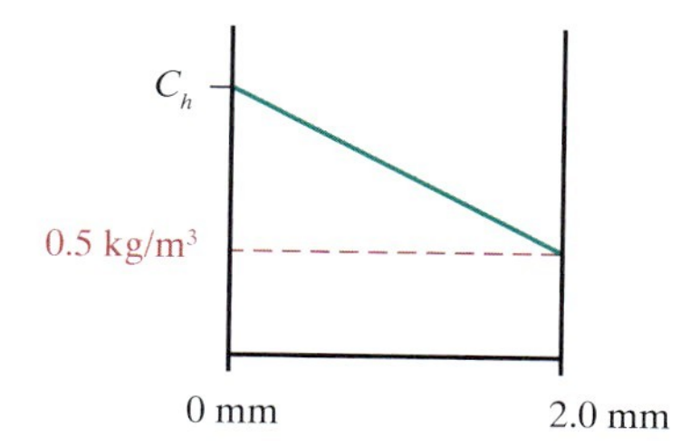

From this image we can see how to set up the calculation:

$$J = -D\frac{dc}{dx} = -D\frac{C_h - C_l}{X_h - X_l}$$

In this equation the unknown variable is C_h, so, rearranging it, we have the following:

$$C_h = C_l - \frac{J(X_h - X_l)}{D} = 0.5\ \text{kg/m}^3 - \frac{(2.48 \times 10^{-8}\ \text{kg/m}^2\text{-s})(0\ \text{m} - 2.0 \times 10^{-3}\ \text{m})}{4.50 \times 10^{-11}\ \text{m}^2/\text{s}} = 1.60\ \text{kg/m}^3$$

This concentration is higher than the concentration on the low-pressure side, which is what we would expect. This example problem shows one of the tricks to using Fick's first law: the negative sign occurs because the high-pressure side is defined as having a position of 0. If you don't do this, you will get strange results, like negative concentrations.

For nonsteady-state diffusion the rate is calculated using *Fick's second law*:

$$\frac{\partial c}{\partial t} = D\frac{\partial^2 c}{\partial x^2} \qquad (7.2.2)$$

One solution to this equation is based on the initial conditions shown in Figure 7.2.3. Under these conditions, the solution to Fick's second law is as follows:

$$\frac{c_x - c_0}{c_s - c_0} = 1 - \text{erf}\left(\frac{x}{2\sqrt{Dt}}\right) \qquad (7.2.3)$$

The meanings of the concentrations on the left side of this equation are shown in Figure 7.2.3, and t is time. The function "erf" is the error function. The error function can be treated like other functions you might have on your calculator, like sin or cos. However, you probably don't have an error-function button (or an inverse error-function button) on your calculator. Table 7.2.1 tabulates values of the error function for you, and you can interpolate from this table to get values not listed. There are also online calculators for both the error function and the inverse error function that you can find through an internet search.

Interpolation assumes that there is a linear relationship between any two points on the table. See Example Problem 7.2.3 if you are not sure how to do this.

Figure 7.2.3

Initial conditions for one solution to Fick's second law.

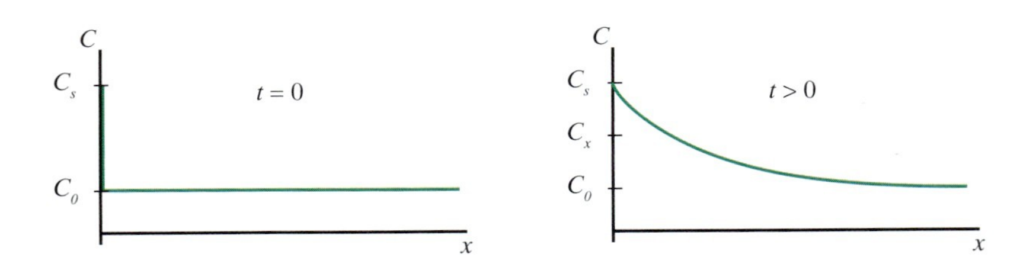

TABLE 7.2.1 Values for the error function

y	erf(y)	y	erf(y)
0.00	0.0000	0.75	0.71116
0.05	0.05637	0.80	0.74210
0.10	0.11246	0.85	0.77067
0.15	0.16800	0.90	0.79691
0.20	0.22270	0.95	0.82089
0.25	0.27632	1.00	0.84270
0.30	0.32863	1.10	0.88021
0.35	0.37938	1.20	0.91031
0.40	0.42839	1.30	0.93401
0.45	0.47548	1.40	0.95229
0.50	0.52050	1.50	0.96611
0.55	0.56332	1.60	0.97635
0.60	0.60386	1.70	0.98379
0.65	0.64203	1.80	0.98909
0.70	0.67780	1.90	0.99279

Fick's second law can be further simplified under the condition that we are maintaining all concentrations constant and trying to change only process conditions. In other words, sometimes we want to get the same value of concentration but to have it at a different position, or to take less time, or to do it at a different temperature. In that situation, the equation simplifies to:

$$\frac{x_1^2}{D_1 t_1} = \frac{x_2^2}{D_2 t_2} \tag{7.2.3}$$

where the subscripts 1 and 2 refer to the two different process conditions. Again, we will do some problems to see how to use these equations.

The next set of questions are based on the following scenario: You are performing a treatment to diffuse copper into the surface of a piece of aluminum. At 600° C, 10 hours of treatment gives a copper concentration of 0.2 wt% at a depth of 0.5 cm. How long will it take to get the same result if you perform the treatment at 500° C? At 500° C the diffusion coefficient is 4.8 × 10^{-14} m^2/s, and at 600° C the diffusion coefficient is 5.3 × 10^{-13} m^2/s.

Guided Inquiry: Fick's Second Law

7.2.12 For the problem on the previous page, what are the values of x_1 and x_2?

7.2.13 What are the values of D_1 and D_2?

7.2.14 What are the values of t_1 and t_2?

7.2.15 Solve for the unknown value. Make sure to include appropriate units.
Do you know why this problem used the simplified version of Fick's second law instead of the full version? Discuss this in your group to make sure everyone understands.

7.2.16 The text says that we can use this equation to choose a different temperature. Which variable depends on temperature?

The next question is based on the following scenario: You are performing a treatment to diffuse copper into the surface of a piece of aluminum. At 600° C, 10 hours of treatment gives a copper concentration of 0.2 wt% at a depth of 0.5 cm. At what temperature would you get the same result if you performed the treatment for 20 hours?

7.2.17 Do you have all the information you need to solve the problem above? If yes, solve it. If not, describe what kind of additional information you need.

Concept Check 7.2.3

- What is the answer for question 7.2.15?
- If a diffusion problem states that the flux is constant, which equation would you use to solve it?
- If a diffusion problem asks how long it will take to get a certain concentration at a certain position, which equation would you use to solve it?

EXAMPLE PROBLEM 7.2.2

Diffusion of carbon into the surface of BCC steel at 600° C for 10 hours results in a carbon concentration of 0.2 wt% at a depth of 1.5 mm below the surface. At what depth will the same concentration be obtained if the diffusion time is increased to 20 hours, leaving the temperature at 600° C?

Since this problem asks for the same carbon concentration but different process conditions, we use the simplified version of Fick's second law:

$$\frac{x_1^2}{D_1 t_1} = \frac{x_2^2}{D_2 t_2}$$

Let's identify the original conditions as condition 1 and the changed conditions as condition 2. We note that the temperatures for the two conditions are the same, so the diffusion coefficient will be the same. This means we can further simplify the equation:

$$\frac{x_1^2}{t_1} = \frac{x_2^2}{t_2}$$

The variables are as follows:

$x_1 = 1.5$ mm

$t_1 = 10$ hours

$x_2 =$ unknown

$t_2 = 20$ hours

We also note that there are no constants that specifically require SI units, so we can just leave all the units as is. Solving for the unknown gives

$$x_2 = \left(\frac{t_2 x_1^2}{t_1}\right)^{1/2} = \left(\frac{(20\text{ hr})(1.5\text{ mm})^2}{(10\text{ hr})}\right)^{1/2} = 2.12\text{ mm}$$

This depth is larger than the depth for the original conditions, which is what we would expect, since the time is longer.

EXAMPLE PROBLEM 7.2.3

A piece of pure iron is exposed to a source of carbon at a concentration of 2.0 kg/m^3. How long will it take for the carbon concentration in the iron to reach a value of 0.2 kg/m^3 at a depth 0.5 mm below the surface? The diffusion coefficient for carbon in iron at the operating temperature is 6.55×10^{-12} m^2/s.

This is not steady state, because the concentration is changing, and there is no initial set of process conditions. Therefore we need to use the full version of Fick's second law:

$$\frac{c_x - c_0}{c_s - c_0} = 1 - \operatorname{erf}\left(\frac{x}{2\sqrt{Dt}}\right)$$

Values for the variables are as follows:

$c_x = 0.2$ kg/m^3

$c_s = 2.0$ kg/m^3

$c_0 = 0$ kg/m^3 (because it is pure iron, so the initial concentration is zero)

$x = 0.5$ mm

$D = 6.55 \times 10^{-12}$ m^2/s

$t =$ unknown

Rearranging the equation gives us the following:

$$\frac{C_x}{C_s} = 1 - \operatorname{erf}\left(\frac{x}{2\sqrt{Dt}}\right)$$

$$\operatorname{erf}\left(\frac{x}{2\sqrt{Dt}}\right) = 1 - \frac{C_x}{C_s} = 1 - \frac{0.2\ \text{kg/m}^3}{2.0\ \text{kg/m}^3} = 0.9$$

Now we need to use the error-function table. What we want is to know the value of y that gives erf(y) = 0.9. Looking at table 7.2.1, we see that there are error-function values of 0.88021 for y of 1.10 and 0.91031 for y of 1.20. Since the values we need are in between these, we will use a linear interpolation:

$$\frac{1.20 - 1.10}{0.91031 - 0.88021} = \frac{y - 1.10}{0.9 - 0.88021}$$

$$y = 1.166$$

This is between 1.10 and 1.20, which we would expect.

Now we know that erf(1.166) = 0.9, so

$$\frac{x}{2\sqrt{Dt}} = 1.166$$

$$t = \left(\frac{x}{(1.166)2\sqrt{D}}\right)^2 = \left(\frac{0.5 \times 10^{-3}\ \text{m}}{(1.166)2\sqrt{6.55 \times 10^{-12}\ \text{m}^2/\text{s}}}\right)^2 = 7018\ \text{s} = 1.95\,\text{hr}$$

The temperature dependence of the diffusion coefficient is given by the Arrhenius equation:

$$D = D_0 e^{-E_d/RT}$$

where R is the ideal gas constant, T is the temperature, and D_0 and E_d are constants that can be looked up for a given pair of materials. Table 7.2.2 has some values for these constants. Note that the values depend on which material is the host and which material is diffusing. So, for example, the constants are different for nickel diffusing in copper compared to copper diffusing in nickel.

TABLE 7.2.2 Preexponential factors (D_0) and activation energies (E_d) for diffusion

Material	D_0 (m^2/s)	E_d (kJ/mol)
Cu in Ag	1.20×10^{-4}	193
Cu in Al	1.50×10^{-5}	126
Cu in Ni	2.7×10^{-5}	256
Zn in Ag	5.40×10^{-5}	174
Zn in Cu	3.40×10^{-5}	191
Ni in Cu	2.70×10^{-4}	236
Ni in Fe	7.70×10^{-5}	280
C in BCC Fe	2.00×10^{-6}	84
C in FCC Fe	2.00×10^{-5}	142
N in Fe	3.00×10^{-7}	76
Al in Al_2O_3	2.8×10^{-3}	477
O in Al_2O_3	1.9×10^{-1}	636
Mg in MgO	2.49×10^{-5}	330
O in MgO	4.3×10^{-9}	344
Ni in MgO	1.8×10^{-9}	202
O in SiO_2	2.7×10^{-8}	111
CO_2 in PET	6.0×10^{-5}	51
CO_2 in Polyethylene	2.0×10^{-4}	38
CO_2 in PVC	4.2×10^{-2}	64
O_2 in PET	5.2×10^{-5}	47
O_2 in Polyethylene	6.2×10^{-4}	41
O_2 in PVC	4.1×10^{-3}	54

The next questions are based on the following scenario: You are performing a treatment to diffuse copper into the surface of a piece of aluminum. At 600° C, 10 hours of treatment gives a copper concentration of 0.2 wt% at a depth of 0.5 cm. At what temperature would you get the same result if you performed the treatment for 20 hours?

Guided Inquiry: Temperature Dependence

7.2.18 In the problem at the bottom of the previous page, which material is diffusing and which is the host material?

Getting this right is important. D_0 and E_d can be different for the same pair of materials depending on which is the host. Look at Table 7.2.2 and compare the values for Cu diffusing in Ni versus Ni diffusing in Cu.

7.2.19 What values and units of D_0 and E_d would you use in the Arrhenius equation for this problem?

7.2.20 In the Arrhenius equation, what are the value and the units for R?

Make sure these units are consistent with the units for E_d.

7.2.21 What is the value for the known temperature that you would use in the Arrhenius equation?

Make sure these units are consistent with the units for E_d and R.

7.2.22 Calculate the value for D at the known temperature.

7.2.23 Calculate the value for D at the unknown temperature.

7.2.24 Calculate the unknown temperature.

Concept Check 7.2.4

- **What are the correct value and units to use for the known temperature in the Arrhenius equation for question 7.2.21?**
- **What is the unknown temperature in question 7.2.24?**

EXAMPLE PROBLEM 7.2.4

You are hardening a tool made of BCC iron by increasing the carbon concentration at its surface by heating the tool in the presence of a carbon-rich gas (for example, methane). A heat treatment at 600° C for 100 minutes results in a carbon concentration of 0.75 wt% at a position 0.5 mm below the surface. How long would it take to obtain a carbon concentration of 0.75 wt% at a position 0.5 mm below the surface if the heat treatment were conducted at 900° C?

Since this problem asks for the same carbon concentration but different process conditions, we use the simplified version of Fick's second law:

$$\frac{x_1^2}{D_1 t_1} = \frac{x_2^2}{D_2 t_2}$$

Let's identify the original conditions as condition 1 and the changed conditions as condition 2. We note that the depths for the two conditions are the same. This means we can further simplify the equation to the following:

$$D_1 t_2 = D_2 t_2$$

For the times we have the following:

$t_1 = 100$ min

$t_2 =$ unknown

We also note that we are given temperatures but we need diffusion coefficients in this equation. We can use the Arrhenius equation to get D at each temperature:

$$D = D_0 e^{-E_d/RT}$$

From Table 7.2.2 we can see that for C diffusing in BCC Fe, D_0 is 2.00×10^{-6} m²/s and E_d is 84 kJ/mol.

$$D_1 = (2.00 \times 10^{-6}\ \mathrm{m^2/s})\, e^{\left(-\frac{84\times 10^3\ \mathrm{J/mol}}{(8.314\ \mathrm{J/mol-K})(873\ \mathrm{K})}\right)} = 1.88 \times 10^{-11}\ \mathrm{m^2/s}$$

$$D_2 = (2.00 \times 10^{-6}\ \mathrm{m^2/s})\, e^{\left(-\frac{84\times 10^3\ \mathrm{J/mol}}{(8.314\ \mathrm{J/mol-K})(1173\ \mathrm{K})}\right)} = 3.63 \times 10^{-10}\ \mathrm{m^2/s}$$

Now we have all the values to solve for t_2. Note that the units will cancel out, so we can leave the units of t in minutes:

$$t_2 = \frac{D_1 t_1}{D_2} = \frac{(1.88 \times 10^{-11}\ \mathrm{m^2/s})(100\ \mathrm{min})}{(3.63 \times 10^{-10}\ \mathrm{m^2/s})} = 5.18\ \mathrm{min}$$

This is shorter than the original condition, which is what we would expect, since the temperature is higher.

Summary

Diffusion is a very important concept that has applications all across materials science and engineering. At a macroscopic level you learned how to do diffusion calculations using Fick's first and second laws. It is very important to understand when to use each of these, and also when you can use the simplified version of Fick's second law. Remember, Fick's first law is used for steady-state diffusion, Fick's second law is used for nonsteady-state diffusion, and the simplified Fick's second law is for nonsteady-state diffusion when the concentration remains constant.

At the microscopic level we looked at the two mechanisms of diffusion; vacancy and interstitial. Although we didn't go into this in detail, you should understand that knowledge of these mechanisms provides another avenue to understanding structure–property–processing relationships and controlling the behavior of materials. For example, let's say you want to improve the production rate of a membrane being used to purify nitrogen by diffusion. By understanding the diffusion mechanism involved, you could modify the structure of the membrane to change the diffusion rate.

One last word is needed about the driving force for diffusion, which we described as the concentration gradient. Strictly speaking, this is not correct. The true driving force for diffusion is the chemical potential gradient. For this text you don't need to know what the chemical potential is, and in most cases the chemical potential gradient is the same as the concentration gradient. But in some cases it isn't, such as the reverse osmosis process described in the Application Spotlight. Another example is the process by which a one-phase system becomes two phases, which is the topic for the next chapter.

Key Terms

Concentration gradient
Diffusion
Diffusion coefficient
Fick's first law
Fick's second law
Grain boundary diffusion
Interstitial diffusion
Nonsteady-state diffusion
Short-circuit diffusion
Steady-state diffusion
Vacancy diffusion

Problems

Skill Problems

7.1 A thin sheet of iron 0.5 mm thick has a nitrogen concentration on one side of 0.5 kg/m^3 and a nitrogen concentration on the other side of 3.0 kg/m^3. The diffusion coefficient is 3.44×10^{-15} m^2/s. What is the flux of nitrogen through the iron?

7.2 For the situation in problem 7.1, if the concentration on the low-pressure side is increased to 1.0 kg/m^3, by how much will the flux change? Solve this problem without actually calculating the flux.

7.3 For the situation in problem 7.1, if the thickness of the aluminum sheet is decreased to 0.25 mm, by how much will the flux change? Solve this problem without actually calculating the flux.

7.4 The flux of oxygen through a 4-mm thick sheet of PVC at steady state is 5.47×10^{-8} kg/m^2-s at a temperature of 100° C. The concentration on the high-pressure side is 3.2 kg/m^3. What is the concentration on the low-pressure side?

7.5 The flux of oxygen through a 4-mm thick sheet of PVC at steady state is 5.47×10^{-8} kg/m^2-s at a temperature of 100° C. The concentration on the low-pressure side is 0.7 kg/m^3. What is the concentration of oxygen in the PVC at a depth 1 mm from the high-pressure side?

7.6 Diffusion of carbon into BCC iron for 12 hours at a temperature of 500° C results in a carbon concentration of 0.2 wt% at a depth 2 mm from the surface. How long would it take to get a concentration of 0.2 wt% at a depth 3 mm from the surface at 500° C?

7.7 Diffusion of carbon into BCC iron for 12 hours at a temperature of 500° C results in a carbon concentration of 0.2 wt% at a depth 2 mm from the surface. At what depth would the concentration be 0.2 wt%, if the diffusion occurred for 20 hours at 500° C?

7.8 Diffusion of carbon into BCC iron for 12 hours at a temperature of 500° C results in a carbon concentration of 0.2 wt% at a depth 2 mm from the surface. At what depth would the concentration be 0.2 wt%, if the diffusion occurred for 6 hours at 800° C?

7.9 A piece of pure aluminum is placed in contact with a piece of pure copper at 600° C. How long will it take for diffusion to result in a concentration of 200 kg/m^3 of copper in the aluminum at a depth of 0.3 mm? Pure copper has an effective concentration of 8,940 kg/m^3.

7.10 A piece of pure aluminum is placed in contact with a piece of pure copper at 600° C. If diffusion occurs for 20 hours, at what depth will the concentration of copper in the aluminum be 400 kg/m^3? Pure copper has an effective concentration of 8,940 kg/m^3.

7.11 A piece of pure aluminum is placed in contact with a piece of pure copper. At what temperature should diffusion occur for the concentration of copper in aluminum to be 600 kg/m^3 at a depth of 0.2 mm after 10 hours? Pure copper has an effective concentration of 8,940 kg/m^3.

7.12 A piece of pure aluminum is placed in contact with a piece of pure copper at 600° C. What will the concentration of copper in aluminum be at a depth of 0.5 mm after 15 hours? Pure copper has an effective concentration of 8,940 kg/m^3.

Conceptual Problems

7.13 What is the mechanism for diffusion of nickel in copper? Explain the reason for your answer.

7.14 What is the mechanism for diffusion of carbon in copper? Explain the reason for your answer.

7.15 Which will diffuse faster in iron; zinc or carbon? Explain the reason for your answer.

7.16 A thin sheet of aluminum that has dimensions 0.5 mm $\times$ 150 cm $\times$ 150 cm is being used as a membrane to purify nitrogen. If the flux of nitrogen through the membrane is 8.2×10^{-7} kg/m^2-s, what is the production rate of nitrogen in g/hr?

7.17 Identify two ways to increase the production rate of nitrogen for the membrane of problem 7.16. Explain why these will increase the production rate.

7.18 Based on the Arrhenius equation, does a higher activation energy result in a faster or slower increase in the rate of diffusion with temperature?

7.19 Compare the activation energies for copper diffusing in aluminum and copper diffusing in nickel. Based on these activation energies, for which one will the diffusion rate increase faster with temperature?

7.20 In this chapter we have expressed concentrations in kg/m^3, but in practice we usually express concentrations in terms of wt%. If the concentration of carbon in iron is 1 wt%, what is the concentration of carbon in kg/m^3?

Samurai swords are known for their strength and durability. They are made through a complex process of heating and shaping to modify the steel microstructure. (Anneka/Shutterstock)

8 Microstructure—Kinetics

Chapter 7 was based on thermodynamic equilibrium; in other words, phase diagrams represent the lowest free energy state of a mixture. This chapter is about *kinetics*, how fast different microstructures form and the ways in which they evolve. It's important to keep in mind that there may be kinetic barriers to achieving an equilibrium state. Even if the phase diagram says that certain phases should exist, we can use a processing method that doesn't give those phases enough time to form, and thus freeze in a nonequilibrium structure. By the end of this chapter you will:

Understand the mechanisms of phase transformations.
Be able to predict the microstructures that form for different processing conditions.

8.1 Nucleation and Growth

LEARN TO: Describe the steps of phase transformations.
Describe the factors affecting the rate of phase transformations.

Later in this chapter we will talk a lot about steel as an example of how we can use processing methods to change the microstructure and the properties. But before we get to that we need to understand some general aspects of *phase transformations*. Phase transformations occur by a two-step process. In order to understand this, think of a liquid that is cooled to below its melting temperature and then solidifies. The first step in this process is called *nucleation*. During nucleation small particles of the solid, called *nuclei*, form in the liquid. The second step is *growth*, during which the particles get bigger. Even though the examples in this section are all for solidification from a liquid, the same concepts also apply to solid–solid transformations—for example, the formation of a β phase from an α phase.

The process of phase transformations is governed by changes in the free energy of the system. We need to consider two aspects of the free energy: the bulk free energy and the surface free energy. Phase transformations occur when the total free energy of the system (bulk plus surface) is reduced. The first step is to understand how the bulk and surface components affect the total free energy. **To illustrate this, the first questions consider the simple case of a sphere.**

Guided Inquiry: Nucleation

8.1.1 What is the first step in the solidification process? What happens during that process?

8.1.2 At a temperature below the melting temperature, which form of the material is more stable; the liquid or the solid?

8.1.3 At a temperature below the melting temperature, which form of the material has the lower free energy; the liquid or the solid?

8.1.4 If a nucleus is formed, there is a surface formed between the nucleus and the surrounding liquid. Does this surface have a higher energy or a lower energy than the bulk solid?

8.1.5 What is the equation for the volume of a sphere?

8.1.6 What is the equation for the surface area of a sphere?

8.1.7 Compare your answers to questions 8.1.5 and 8.1.6. If you increase the radius of the sphere, which increases by a larger amount; the volume or the surface area?

8.1.8 Based on your answer to question 8.1.7, if you increase the radius of a nucleus, which will increase by a larger amount; the free energy of the bulk nucleus (see question 8.1.3) or the free energy of the surface (see question 8.1.4)?
Consider only the absolute values; do not worry about whether the energy changes are negative or positive.

8.1.9 Based on your answer to question 8.1.8, if you increase the radius of the nucleus, will the total free energy change (bulk plus surface) increase or decrease?
Now you need to consider whether each type of energy is positive or negative. Remember that things that are favorable have negative free-energy changes.

Concept Check 8.1.1

- **Which has greater free energy; two spheres, each with a volume of 2 cm^3, or one sphere with a volume of 4 cm^3?**

Your answer to question 8.1.9 was probably that the free energy decreases because the volume of a sphere increases more than the surface area. However, the situation is actually more complicated than this. Figure 8.1.1 shows how the free energy changes as the radius of the nucleus changes. The line in red shows how the free energy from the volume of the sphere decreases, the line in green shows how the free energy from the surface energy increases, and the line in blue is the total change in free energy, which is the sum of the two. **For the next set of questions you should consider the total change in free energy (the blue line).**

Figure 8.1.1

Free energy versus radius of a nucleus for the bulk free energy (red), surface free energy (green), and total free energy of the nucleus (blue).

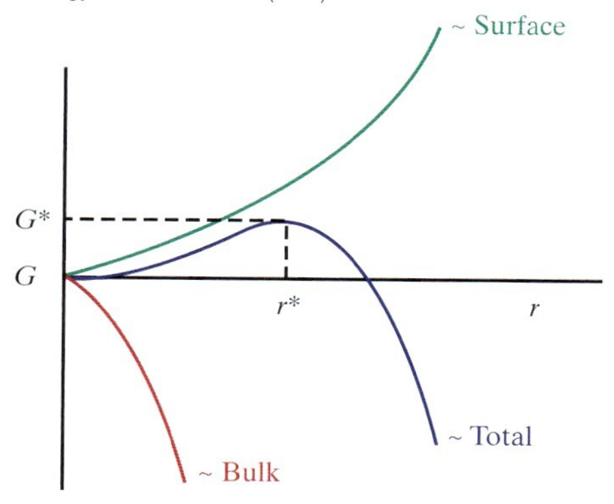

Guided Inquiry: Nucleus Stability

8.1.10 If a nucleus with a radius less than r^* grows larger, what will happen to its free energy?

Consider the total free energy as shown in Figure 8.1.1.

8.1.11 How can a nucleus with a radius less than r^* decrease its free energy?

8.1.12 If a nucleus with a radius greater than r^* grows larger, what will happen to its free energy?

8.1.13 How can a nucleus with a radius greater than r^* decrease its free energy?

8.1.14 Based on your answers to questions 8.1.10–8.1.13, what is the critical radius size for a nucleus to allow it to grow larger?

Concept Check 8.1.2

- If a nucleus with a radius less than r^* spontaneously forms, what will then happen to it?

Nucleation will occur at different rates depending on the temperature. We can understand the temperature dependence by thinking about the stability of the liquid below the melting point. In Figure 8.1.1 there is an energy barrier, given by G^*, which is the *activation energy* for nucleation. You can think of this activation energy as an energy "wall" that needs to be climbed over for nucleation to occur. This activation energy depends on temperature according to

$$G^* \propto \frac{1}{(T_m - T)^2} \tag{8.1.1}$$

The important part of this equation is that the activation energy is inversely proportional to the undercooling, $T_m - T$. This means that as the temperature gets lower, the activation energy gets lower, which makes it easier for a nucleus to form and thus makes nucleation faster. Figure 8.1.2 shows how this energy barrier changes with temperature. Qualitatively, we can think of it as being caused by how close we are to the melting temperature. Very close to the melting temperature, the material is not very far from where it would rather be a liquid, so there is only a small driving force to form the solid and the nucleation rate is low. At very low temperatures, however, the material is very far from where it would rather be a liquid, so there is a very high driving force to form a solid and the nucleation rate is high.

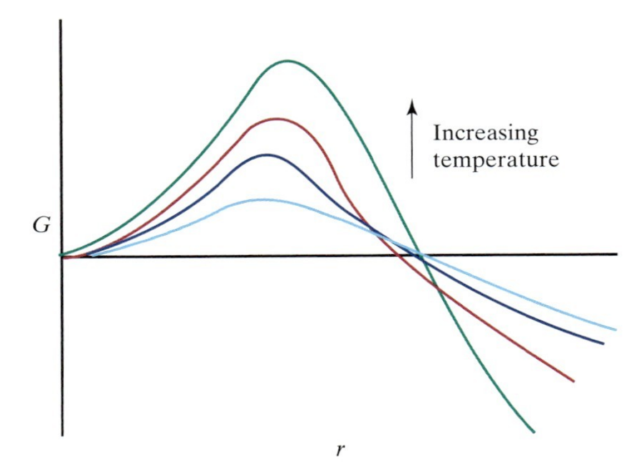

Figure **8.1.2**

Change in activation–energy barrier for nucleation with temperature.

So far we have considered only nucleation. But to get a complete picture of phase transformations we also need to consider growth. A nucleus will form by random fluctuations, during which a bunch of atoms come together to form a nucleus. For the nucleus to get bigger, other atoms need to diffuse to the nucleus and add on to make the solid particle bigger. Because this process depends on diffusion, we can use the temperature dependence of diffusion to describe the temperature dependence of the growth step:

$$\text{Rate} = Ae^{-E_a/kT} \tag{8.1.2}$$

where E_a is the activation energy for growth. Because diffusion is faster at higher temperatures, growth will be fast at high temperatures and slow at low temperatures, just the opposite of nucleation. The overall transformation rate, then, depends on some kind of balance between nucleation and growth.

We can represent the overall nucleation and growth process with a graph like that shown in Figure 8.1.3. At short times nucleation is occurring. Because the nuclei are so small, the overall amount of solidified material during nucleation is pretty close to zero. Only when growth occurs is there any substantial amount of solidification, so we usually consider the beginning of the growth phase to mark the start of the transformation process. If we get graphs like Figure 8.1.3 at different temperatures we can put them together into an *isothermal transformation diagram* like Figure 8.1.4. The solidification curve at temperature T_1 in Figure 8.1.3 provides the points t_0, t_{50}, and t_{100}. Solidification curves like Figure 8.1.4 at different temperatures are used to create the overall isothermal transformation diagram. This diagram shows how the time for a transformation to occur changes depending on the temperature. To use this diagram, pick a temperature at which solidification is occurring. By drawing a line at that temperature across the diagram you can determine how long it will take the material to solidify. **The rest of the questions in this section will help you understand how to interpret this diagram. You will also use this information to start to understand how processing affects structure, which in turn affects properties. This relationship is at the heart of materials science and engineering.**

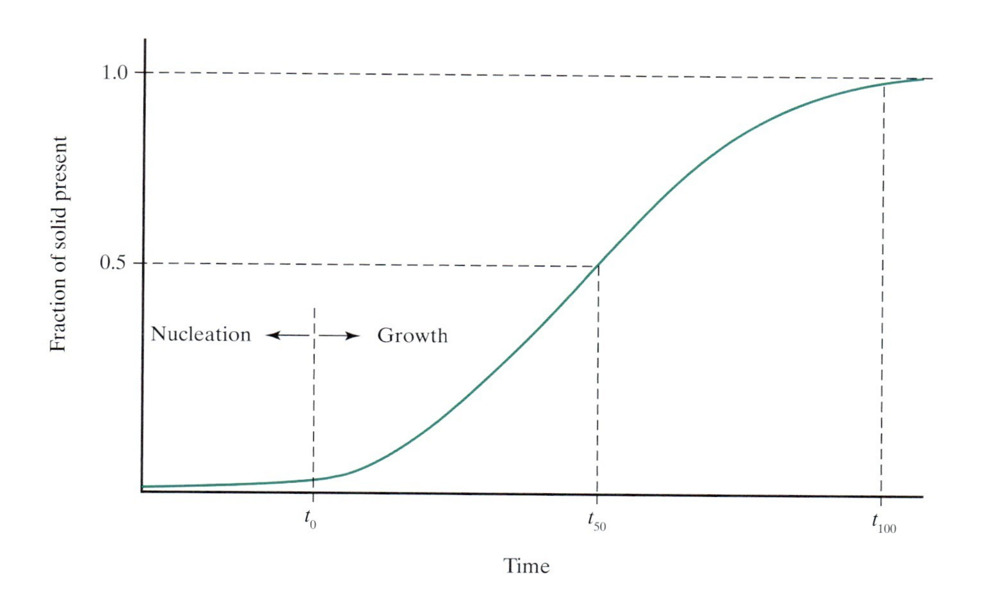

Figure 8.1.3

Fraction of solidified material versus time for formation of a solid from a liquid below the melting temperature at a temperature of T_1. The times t_0, t_{50}, and t_{100} can be used to obtain points for the isothermal transformation diagram in Figure 8.1.4.

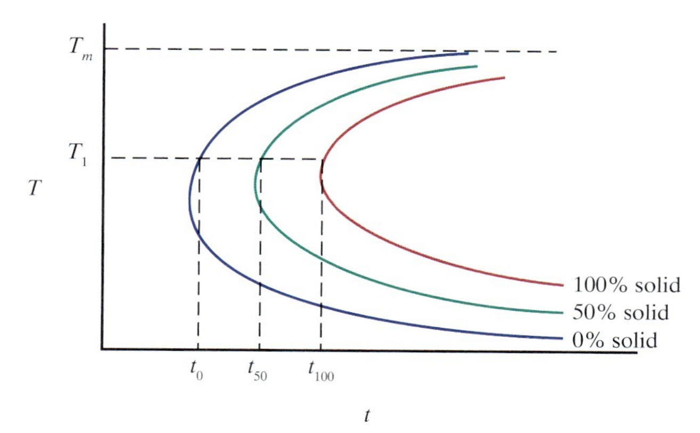

Figure 8.1.4

Isothermal transformation diagram for solidification. The blue line represents the start of the solidification, the green line represents when half the liquid has transformed to a solid, and the red line represents when the liquid has just finished transforming to all solid. The solidification curve at temperature T_1 in Figure 8.1.3 provides the points t_0, t_{50}, and t_{100}.

Guided Inquiry: Nucleation and Growth

8.1.15 At a high temperature close to (but still below) the melting temperature, is nucleation fast or slow? Is growth fast or slow?

Remember, if it takes a long time, that means the rate is slow.
This is not shown on Figure 8.1.4. That figure is the combined effect of nucleation and growth.

8.1.16 At a high temperature close to (but still below) the melting temperature, is the overall rate of the transformation (nucleation plus growth) fast or slow? Explain why.

8.1.17 At a low temperature, is nucleation fast or slow? Is growth fast or slow?

8.1.18 At a low temperature, is the overall rate of the transformation fast or slow? Explain why.

8.1.19 At what temperatures do you think the overall transformation rate is fast?

8.1.20 At high temperatures, are many nuclei formed, or few? What about at low temperatures?

8.1.21 Based on your answers to question 8.1.20, how can you get a material with large grains?

You may want to take a look at Section 6.3.

Concept Check **8.1.3**

- **At what temperatures is the overall transformation rate the fastest?**
- **What solidification temperatures will result in copper with the largest grain sizes?**

8.2 Heterogeneous Nucleation

LEARN TO: Calculate contact angle.
Compare homogeneous and heterogeneous nucleation.

The nucleation process described in Section 8.1 is called homogenous nucleation, meaning that the solid nuclei spontaneously form in the middle of the liquid. We can also have heterogeneous nucleation, in which the nuclei form on a solid surface. If you've ever made rock candy you've seen heterogeneous nucleation in action. To make rock candy, you create a supersaturated solution of sugar at high temperature and suspend strings into it. As the solution cools, nuclei form on the string (and the walls of the container) and then grow to form sugar crystals. Add some flavoring to the solution and you can have a sweet snack!

We begin an exploration of heterogeneous nucleation by first considering how liquids wet a surface. If you put a droplet of water on a solid, it will bead up more or less, depending on what the solid is. We can actually calculate the shape of the droplet by knowing the surface tensions (or surface energies) involved. Figure 8.2.1 shows two examples of a liquid droplet on a solid surface. The surface tensions shown are for the solid–gas interface (γ_{SG}), the liquid–gas interface (γ_{LG}), and the solid–liquid interface (γ_{SL}). The droplet takes a certain shape, defined by the contact angle θ, based on the relative magnitudes of these surface tensions. **Through the next set of questions you will examine how the surface tensions relate to the contact angle.**

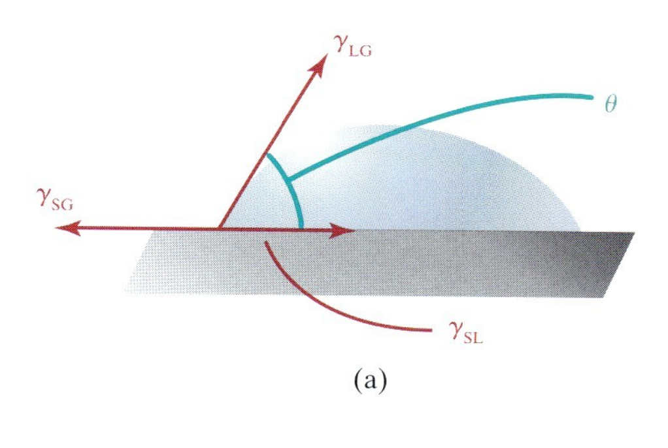

γ_{LG}
θ
γ_{SG}
γ_{SL}
(b)

Figure 8.2.1
Droplets of a liquid sitting on a surface. In both parts the solid is the same but the liquid is different. The contact angle θ is the angle between γ_{LG} and γ_{SL}.

Guided Inquiry: Contact Angle

8.2.1 In which part of Figure 8.2.1 is there a greater solid–liquid surface tension?

8.2.2 In which part of Figure 8.2.1 is there a greater contact angle?

8.2.3 If the solid–liquid surface tension becomes larger, what happens to the contact angle?

8.2.4 If the solid–liquid surface tension becomes smaller, what happens to the contact angle?

8.2.5 Create an equation relating the surface energies to the contact angle.
Treat the surface tensions as forces in equilibrium along the horizontal direction.

8.2.6 Draw a picture of the liquid droplet on the surface if the contact angle is 0°.

8.2.7 Draw a picture of the liquid droplet on the surface if the contact angle is 180°.

Concept Check 8.2.1

- **If γ_{SG} is equal to γ_{SL}, what is the contact angle?**

The example above was for a liquid droplet on a solid surface, but the same principle applies for heterogeneous nucleation, where there is a solid "droplet" on a solid surface of a different type. But why would heterogeneous nucleation occur? **The next questions address that issue. To answer these questions, use the data from Table 8.2.1, which shows the surface tension of various solids at the interface with air. You also need to know that the surface tension of water in contact with air is 73 mN/m.**

TABLE 8.2.1 Surface tensions of various solids in contact with air

Material	Surface tension (γ_{SG}) (mN/m)
Poly(tetrafluoroethylene) (PTFE, Teflon®)	19.1
Polystyrene	40.6
Nickel	120
Silicon oxide	2672

Guided Inquiry: Heterogeneous Nucleation

8.2.8 Which has a higher surface tension when in contact with air; water or PTFE?

8.2.9 When a droplet of water is placed on PTFE, which surface would rather be exposed to the air?

Remember, high energy is unfavorable.

8.2.10 Based on your answer to question 8.2.9, when a droplet of water is placed on PTFE, do you expect it to bead up or spread out?

8.2.11 Which has a higher surface tension when in contact with air; water or silicon oxide?

8.2.12 When a droplet of water is placed on silicon dioxide, which surface would rather be exposed to the air?

8.2.13 Based on your answer to question 8.2.12, when a droplet of water is placed on silicon oxide, do you expect it to bead up or spread out?

8.2.14 Draw pictures of a droplet of water on each of the solids listed in Table 8.2.1, showing how the droplet shape is different for each solid.

8.2.15 Copper is melted in a silicon oxide container and allowed to solidify. The surface tension of solid copper is 1360 mJ/N and the surface tension of liquid copper is 1320 mN/m. Which solid, copper or silicon oxide, would rather be exposed to the liquid copper?

8.2.16 When copper is melted in a silicon oxide container and allowed to solidify, where do the nuclei form?

8.2.17 Use the concept of surface tension to explain why heterogeneous nucleation is favored over homogeneous nucleation.

8.2.18 Propose a method to ensure that during the solidification of copper, nucleation will occur uniformly throughout the liquid.

Concept Check 8.2.2

- **Which is faster; heterogeneous nucleation or homogeneous nucleation?**

8.3 Equilibrium versus Nonequilibrium Cooling

LEARN TO: **Draw the microstructures that form on equilibrium and nonequlibrium cooling.**

The microstructural development you looked at in Chapter 6 assumes that the system remains at equilibrium throughout the entire process. In a practical sense, this means that the atoms can move around as much as they need to in order to make the compositions of the solid and liquid phases match what is given by the tieline. But think about how difficult it is for atoms to move in a solid compared to a liquid. It is more realistic to assume that the atoms in a solid cannot move. This means that once a solid of a certain composition forms, it will not change. This is actually not completely true, as we saw in Chapter 7, and in later parts of this chapter we look at microstructural changes that occur in solids. But for now, we will assume that atoms in a solid cannot move at all, and that atoms in a liquid can move as much as they need to.

To understand the microstructure that forms during nonequilibrium cooling we will go back to the Cu–Ni phase diagram, which is shown in Figure 8.3.1. In Section 6.3 you used this phase diagram to understand the microstructure that forms during equilibrium cooling. Figure 8.3.2 shows the first steps of the solidification process for nonequilibrium cooling. This structure results because as the material cools, the solid forms based on the tieline at each temperature. In Figure 8.3.2 the first solid that forms, at the center of the particles,

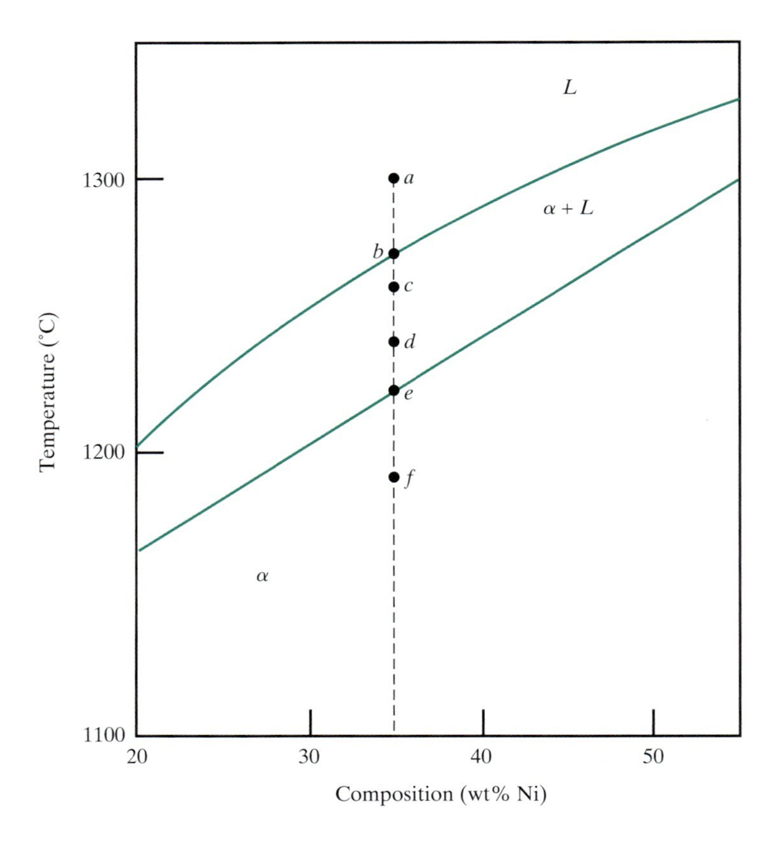

Figure **8.3.1**

Portion of the copper–nickel phase diagram.

forms at point b on Figure 8.3.1. Once that solid forms, it doesn't change. The material then cools to point c, and so the next solid that forms, which is the shell around the outside of the particle, has a composition given by the tieline at point c. This process continues until all of the liquid is gone.

Figure 8.3.2

Initial structure formed during the beginning of nonequilibrium cooling of a Cu–Ni mixture, based on the phase diagram in Figure 8.3.1.

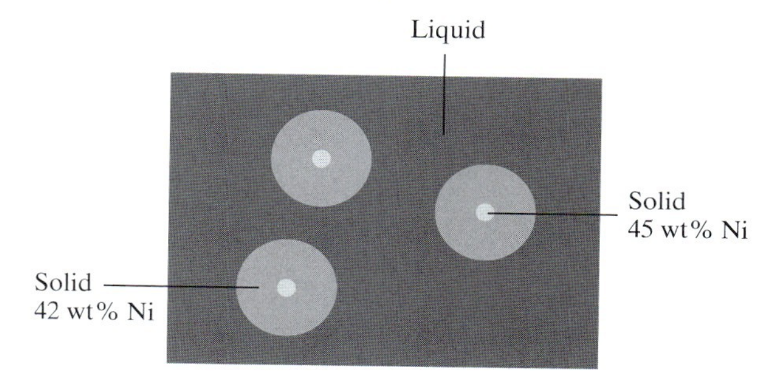

Guided Inquiry: Nonequilibrium Cooling

8.3.1 Draw the microstructure that forms on equilibrium cooling at points *a, b, c, d,* and *e* in Figure 8.3.1.
This is the same microstructure we discussed in Section 6.3.

8.3.2 In Figure 8.3.2, what is the composition of the first solid that forms?

8.3.3 In Figure 8.3.2, what is the composition of the second solid that forms?

8.3.4 The material shown in Figure 8.3.2 is cooled further to point *c* (Figure 8.3.1). What will be the composition of the solid that forms at that point?

8.3.5 Draw the microstructure once it has cooled to point *d*.
Use shading to illustrate the different compositions.

8.3.6 The material is cooled further until there is no liquid left. Draw the final microstructure that results.

8.3.7 How is the microstructure different between equilibrium cooling and non-equilibrium cooling?
Equilibrium cooling was discussed in Section 6.3.

Concept Check 8.3.1

- **How can you transform the microstructure obtained on nonequilibrium cooling to the structure obtained on equilibrium cooling?**

The microstructure that results from nonequilibrium cooling is called a "cored" structure. The important thing to remember about a cored structure is that the composition of the grain is not uniform; it changes from the center of the grain to the grain boundary. An important consequence of a cored structure is that the grain boundaries will melt before the center of the grains, and it will even start melting below the melting temperature shown on the phase diagram (point *d* on Figure 8.3.1). Clearly, this could cause problems if the material were used at high temperatures—the grain boundaries could start melting and the material would lose its mechanical integrity, even when you would think it should be all solid. We can fix the cored structure by heating the material to high temperatures and allowing the atoms to move around to form the equilibrium structure with uniform composition throughout the grains.

Application Spotlight **Scaffolds for Tissue Engineering**

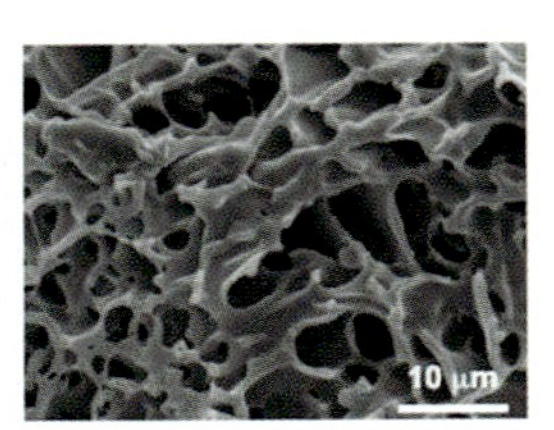

(Carl Simon/National Institute of Standards and Technology)

Traditional biomedical implants use materials that are placed in the body permanently (see, for example, Figure 2.1.1). But these materials all have disadvantages, so it would be desirable to create implants that slowly degrade away and allow the body's own tissue to grow back. This is the concept of tissue engineering, which typically involves creating a material that includes cells, proteins, or other chemicals that will promote tissue growth. These scaffolds must be porous to allow tissue to grow in and the growth factors to be easily released. While there are several approaches to making these scaffolds, one of the most effective is thermally induced phase separation of polymers, a process that relies on the transformation of a polymer system from one phase to two phases. By carefully controlling the various processing parameters, such as temperature, cooling and heating rates, and concentrations, the microstructure of the scaffold can be controlled, ranging from small, unconnected pores to large, open pores, as shown in the figure.

8.4 Isothermal Transformation Diagrams

LEARN TO: **Predict microstructures that arise from a given heat treatment.**
Design a heat treatment to produce a desired microstructure.

We will now apply the concepts of kinetics to the practical issue of microstructure in steel. Steel is an alloy consisting primarily of iron, plus carbon and sometimes other elements. Carbon increases the yield strength of the steel, while the other elements also change the strength as well as other properties. For example, stainless steel includes chromium, which provides protection against corrosion. For this section we are going to consider only a specific type of steel: plain carbon steel that consists of 99.24 wt% iron and 0.76 wt% carbon.

Before we can get to microstructure development, we need to know what the different microstructures of steel are. Figure 8.4.1 shows part of the iron–carbon phase diagram. This looks like a eutectic phase diagram but is actually called *eutectoid* because the high-temperature phase is solid instead of liquid. The key microstructures on this phase diagram are as follows:

- *Austenite*: The high-temperature γ phase that exists above the eutectoid temperature.
- *Ferrite*: The α phase.
- *Cementite*: A line compound with the chemical formula Fe_3C.
- *Pearlite*: The eutectoid microstructure of ferrite and cementite lamellae. It is important to remember that pearlite is not a phase; it is a microstructure.

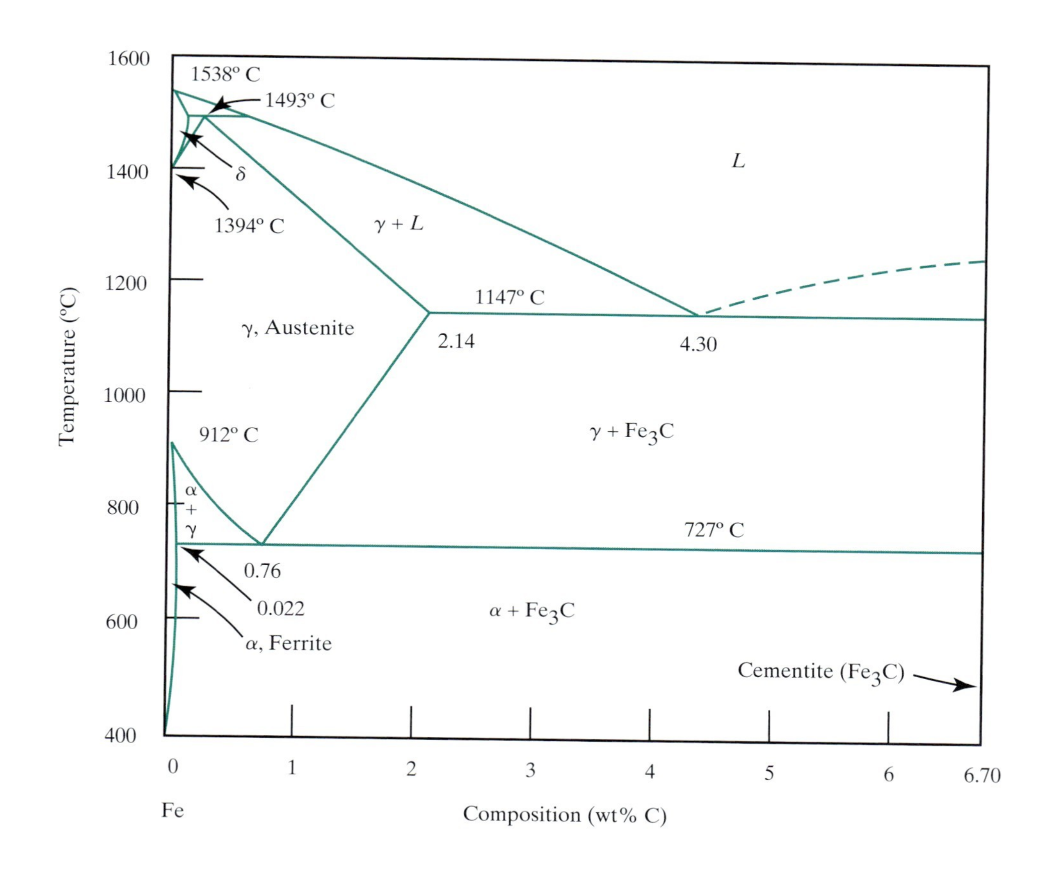

Figure 8.4.1
Iron–carbon phase diagram.

Figure 8.4.2 shows the *isothermal transformation diagram* for plain carbon steel of eutectoid composition. There is a lot of information on this figure that you have not learned yet; we will be going through it step-by-step. The first thing to notice is the overall U-shape of the curve. This is the typical phase-transformation curve that we saw in Figure 8.1.4. At high temperatures pearlite is formed, as would be expected based on the phase diagram. The thickness of the pearlite lamellae depends on the transformation temperature. Go back for a moment and look at Figure 6.3.3. This figure shows how the lamellae form by diffusion of the atoms. At higher temperatures diffusion is faster, so the atoms can travel further and the lamellae are thicker. At lower temperatures diffusion is slower, so the atoms travel less and the lamellae are thinner. This means that in the austenite-to-pearlite region we get coarse pearlite at high temperatures and fine pearlite at low temperatures.

Figure 8.4.2
Isothermal transformation diagram for eutectoid plain carbon steel.

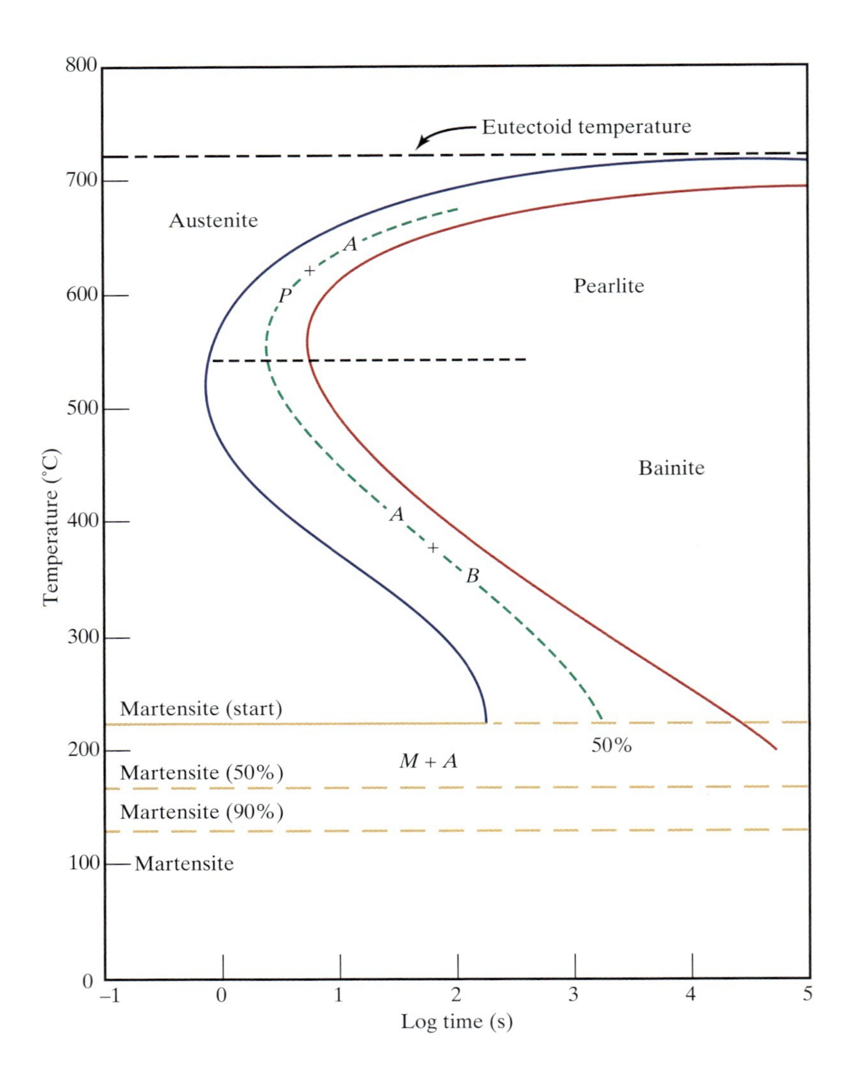

Below the "nose" of the diagram we no longer get pearlite. Instead we get something called *bainite*. Like pearlite, bainite consists of two phases; ferrite and cementite. However, bainite has a different microstructure than pearlite. At these low temperatures diffusion is so slow that even very thin lamellae cannot form. Instead, what we get are thin needles of cementite in ferrite (Figure 8.4.3).

If we cool austenite so fast that neither pearlite nor bainite has time to form, we get yet another form of steel called *martensite*. Martensite is unique because it is formed by a *diffusionless transformation*, in which the atoms do not diffuse to form a new phase. Instead, the crystal structure just shifts slightly so that a new phase is formed. Austenite is FCC and martensite is body-centered tetragonal (BCT). Figure 8.4.3 shows the BCT crystal structure as black lines, with the original face-centered cubic (FCC) (100) plane in green. You should be able to see how the BCT structure was formed by stretching the FCC unit cell slightly. Because the martensitic transition is diffusionless, the amount of martensite that forms does not depend on time; instead, it depends simply on the temperature we cool down to. Although the isothermal transformation diagram shows the temperatures at which we get different amounts of martensite, in practice we will cool all the way to room temperature, which means that any austenite that has not converted into pearlite or bainite will convert to martensite. Martensite is the strongest and most brittle form of steel.

There are two other microstructures that are not shown on the isothermal transformation diagram. *Spheroidite* is formed by heating pearlite or bainite to a high temperature below the eutectoid temperature and holding for a while. Since a sphere has the smallest surface area for a given volume, the free energy of the system is lowered if the cementite forms spherical particles in the ferrite. A typical heat treatment to form spheroidite is 700° C for 24 hours. Spheroidite is the most ductile form of steel.

Finally, if we take martensite and heat it up we can form *tempered martensite*. It is similar to spheroidite in that it consists of spherical particles of cementite in ferrite. The difference is that the particles in tempered martensite are much smaller, because the temperature is much lower and because we are starting with no cementite present. A typical heat treatment to form tempered martensite is 200° C for 1 hour. Tempered martensite has the best balance of strength and ductility for steel. In summary, the types of microstructures that can form are the following:

- *Pearlite:* Cooling austenite to a temperature close to the eutectoid temperature and holding. A higher temperature results in coarser pearlite.
- *Bainite:* Cooling austensite to a lower temperature and holding.
- *Martensite:* Rapidly cooling austenite so there is no time for pearlite or bainite to form.
- *Spheroidite:* Heating pearlite or bainite to 700° C and holding for 24 hours.
- *Tempered martensite:* Heating martensite to 200° C and holding for 1 hour.

The different microstructures are illustrated in Figure 8.4.3.

One more important point: When you change temperatures during a heat treatment, the nucleation and growth process starts all over again. This means that on the isothermal transformation diagram, every time you change temperatures you must begin again from zero. **The questions in this section will give you practice in using the isothermal transformation diagram to design heat treatments and predict the microstructures that develop.**

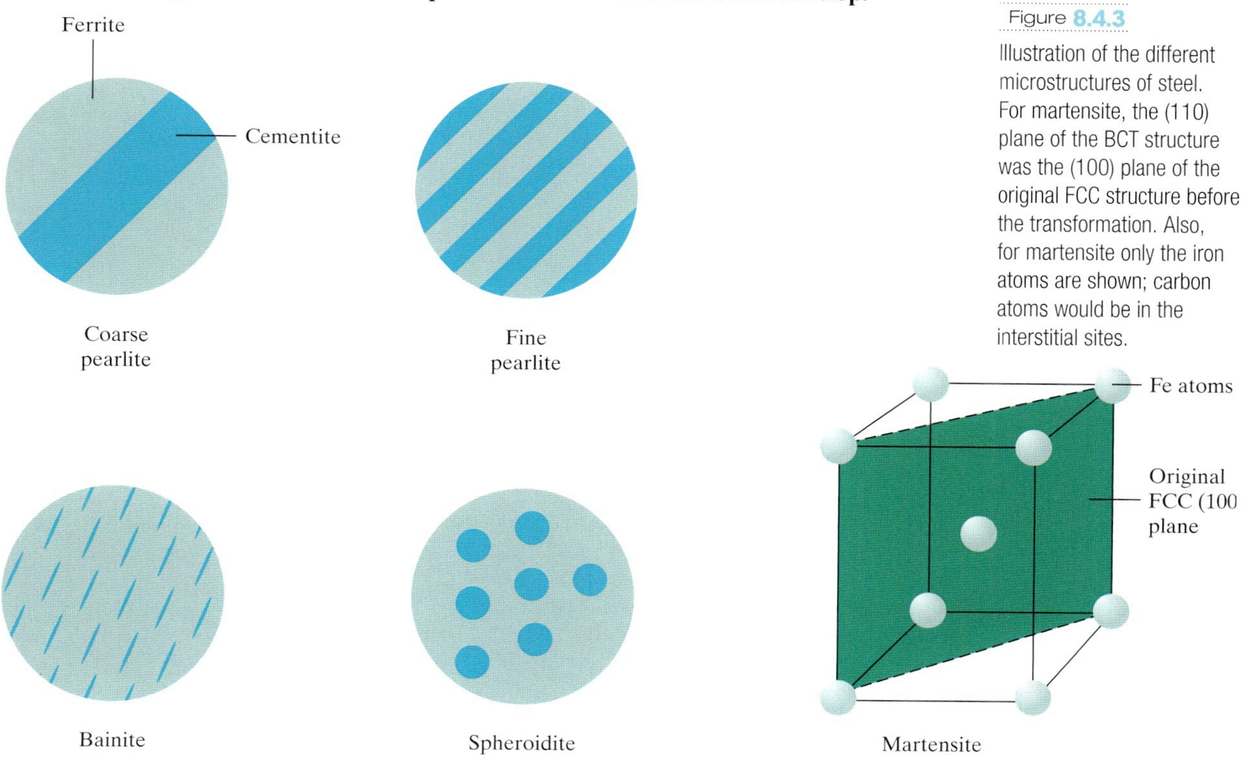

Figure 8.4.3

Illustration of the different microstructures of steel. For martensite, the (110) plane of the BCT structure was the (100) plane of the original FCC structure before the transformation. Also, for martensite only the iron atoms are shown; carbon atoms would be in the interstitial sites.

Guided Inquiry: Isothermal Transformations

8.4.1 What does "isothermal" mean?

8.4.2 If you are going to draw a heat treatment on an isothermal transformation diagram, what kinds of lines are you allowed to draw (horizontal, vertical, curved, etc.)?

Think about what "isothermal" means.

8.4.3 An iron–carbon alloy of 0.76 wt% carbon is heated to 750° C, rapidly cooled to 450° C, held for 10 seconds, then rapidly cooled to room temperature. What is the microstructure?

8.4.4 A eutectoid iron–carbon alloy is heated to 750° C, cooled to 650° C, and held for 10 seconds. What is the microstructure at this point? (The alloy is still at 650° C.)

8.4.5 After the heat treatment of question 8.4.4, the alloy is then cooled rapidly to 450° C and held for 10 seconds. How much of the remaining austenite is now bainite?

Remember, when you change temperature, time starts at zero again. So consider a heat-treatment time of only 10 seconds at 450° C.

8.4.6 Considering the total solid present (including the pearlite made in question 8.4.4), how much of the solid is bainite?

8.4.7 The alloy is now cooled rapidly to room temperature. What is the final microstructure?

Concept Check **8.4.1**

What microstructures will be formed for each of the following heat treatments:

- **Start with 50% bainite, 50% martensite at room temperature, heat rapidly to 750° C, cool rapidly to 600° C, hold for 3.2 seconds, cool rapidly to 500° C, hold for 3.2 seconds, cool rapidly to room temperature.**
- **Start with 50% bainite, 50% martensite at room temperature, heat rapidly to 750° C, cool rapidly to 600° C, hold for 100 seconds, cool rapidly to room temperature.**
- **Start with 50% bainite, 50% martensite at room temperature, heat to 200° C, hold for 60 minutes, cool rapidly to room temperature.**

EXAMPLE PROBLEM 8.4.1

What microstructure will be formed with the following heat treatment: Start with 50% bainite, 50% fine pearlite at room temperature, heat rapidly to 750° C, cool rapidly to 650° C, hold for 31.6 seconds, cool rapidly to 400° C, hold for 31.6 seconds, cool rapidly to room temperature, heat to 200° C, hold for 1 hour, cool to room temperature.

The heat treatment is shown on the figure below.

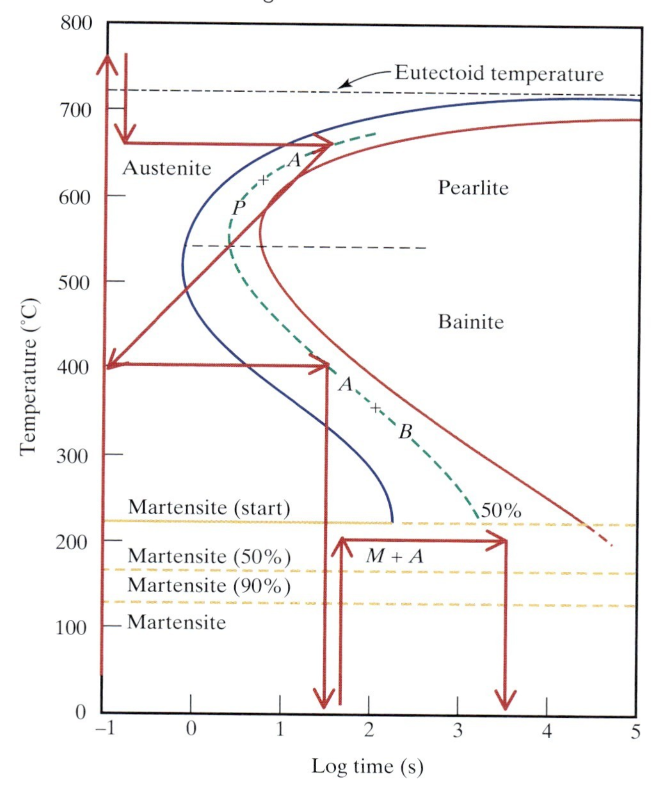

The steps in this process are as follows:

- *Heat to 750° C:* Erases the previous history and turns everything into austenite.
- *Cool rapidly to 650° C, hold for 31.6 seconds:* On the log scale, 31.6 seconds is $10^{1.5}$ seconds. Because this is in the upper half of the pearlite region, we get coarse pearlite. This results in 50% coarse pearlite and 50% austenite.
- *Cool rapidly to 400° C, hold for 31.6 seconds:* On the log scale, 31.6 seconds is $10^{1.5}$ seconds. This results in half of the remaining austenite turning into bainite, so in total we now have 50% coarse pearlite, 25% bainite, 25% austenite. Note that the line that goes diagonally to the left from 650° C to 400° C does not mean that time goes backwards. It indicates that when you change temperature the microstructural change restarts from time equal to zero.
- *Cool rapidly to room temperature:* Any remaining austenite becomes martensite, so in total we now have 50% coarse pearlite, 25% bainite, 25% martensite.
- *Heat to 200° C, hold for 1 hour:* This step is not part of the isothermal transformation diagram. As described above, this heat treatment converts martensite to tempered martensite. So in total we now have 50% coarse pearlite, 25% bainite, 25% tempered martensite.
- *Cool to room temperature:* Since no austenite is left, this doesn't do anything.

The final microstructure is 50% coarse pearlite, 25% bainite, 25% tempered martensite.

EXAMPLE PROBLEM 8.4.2

Start with 50% fine pearlite and 50% martensite, make 50% bainite and 50% tempered martensite.

For this problem it may look as if you could turn the martensite directly into tempered martensite. However, this won't work because you have to erase the pearlite, and when you do that you will also erase the tempered martensite. So take the following steps:

- Heat to 750° C to erase the previous microstructure and turn everything into austenite.
- Cool rapidly to 400° C, hold for 31.6 (= $10^{1.5}$) seconds to create 50% bainite and 50% austenite.
- Cool rapidly to room temperature to turn the remaining austenite into martensite. This results in 50% bainite and 50% martensite.
- Heat to 200° C for 1 hour to turn the martensite into tempered martensite. This results in 50% bainite and 50% tempered martensite.
- Cool to room temperature.

8.5 Continuous Cooling Transformation Diagrams

LEARN TO: **Predict microstructures that arise from a given heat treatment.**
Design a heat treatment to produce a desired microstructure.

In the previous section we looked at isothermal transformations. However, instead of holding a constant temperature it is often easier to just heat the steel up above the eutectoid temperature to "erase" the previous microstructure and then let it cool down at some constant rate. If we do this we cannot use the isothermal transformation diagram to predict the structure. Instead we need to use a *continuous cooling transformation diagram*, as shown for plain carbon eutectoid steel in Figure 8.5.1. You should immediately notice one big difference between this diagram and the isothermal transformation diagram: there is no bainite on the continuous cooling diagram. This is because of the kinetics of bainite formation. To get bainite we have to cool to below 550° C. But let's say we use a constant cooling rate fast enough to avoid forming pearlite. If we do that, it turns out the cooling rate is so fast that the material doesn't spend enough time below 550° C for bainite to form. The result is that neither pearlite or bainite can form, and we get only martensite.

Unfortunately it is not possible to use the continuous cooling diagram directly to get cooling rates. This is because the time axis is in a log scale, and even more importantly because the rate you would calculate off the graph depends on what temperature you start at. There are several critical cooling rates to keep in mind; these are illustrated in Figure 8.5.2.

- Anything less than about 10° C/s is slow, and will give coarse pearlite.
- Anything above about 25° C/s is fast, and will give fine pearlite.

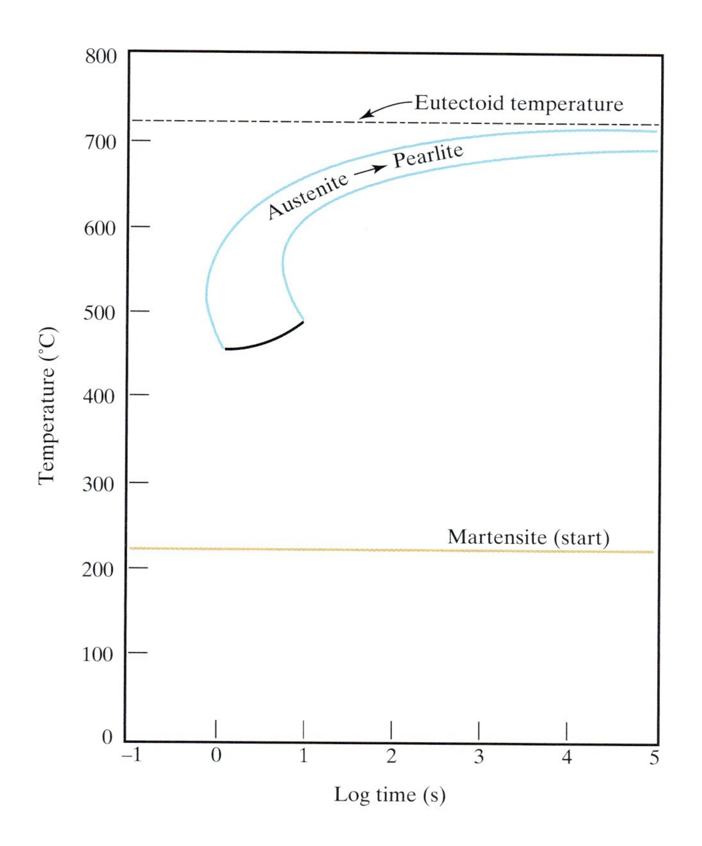

Figure 8.5.1
Continuous cooling transformation diagram for plain carbon eutectoid steel.

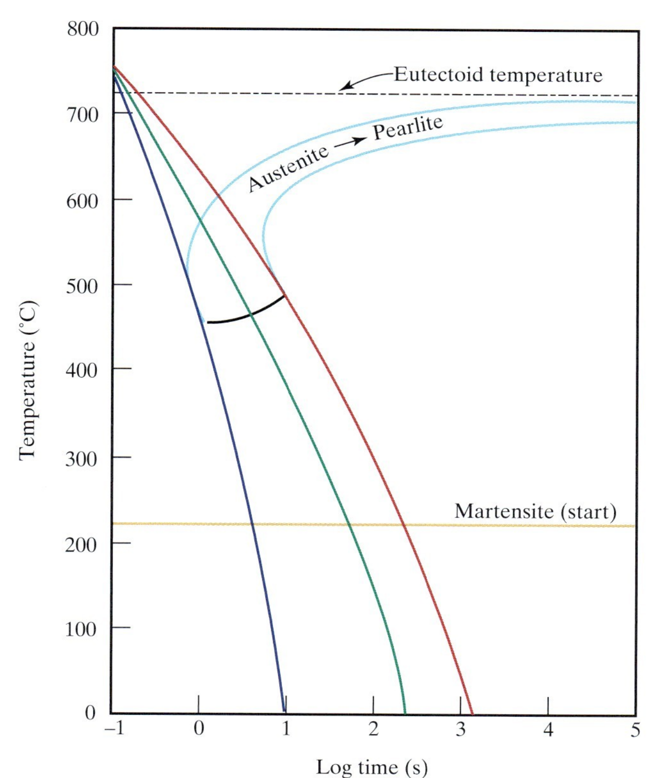

Figure 8.5.2
Continuous cooling transformation diagram for eutectoid steel with critical cooling rates for various microstructures: blue–100% martensite, 140° C/s; red–100% fine pearlite, 35° C/s; green–50% martensite, 50% fine pearlite.

- Anything slower than 35° C/s gives 100% pearlite.
- Anything faster than 140° C/s gives 100% martensite.

Also illustrated in Figure 8.5.2 is a cooling rate that gives 50% pearlite and 50% martensite. You can figure out the percentages by seeing where the cooling line cuts across the black line between the pearlite start and finish lines. However, because we can't get cooling rates directly from the graph, if you are given a cooling rate between 35° and 140° C/s you can't say what the percentages of pearlite and martensite are; you can only say there is a mixture of the two. You will now get some practice using the continuous cooling transformation diagram.

Guided Inquiry: Continuous Cooling Transformations

8.5.1 What microstructure will form from the following heat treatment: start with 50% bainite, 50% martensite at room temperature, heat rapidly to 750° C, cool to room temperature at 30° C/s?

8.5.2 What microstructure will form from the following heat treatment: start with 50% bainite, 50% martensite at room temperature, heat rapidly to 750° C, cool to room temperature at 150° C/s?

8.5.3 What microstructure will form from the following heat treatment: start with 50% bainite, 50% martensite at room temperature, heat rapidly to 750° C, cool to room temperature at 85° C/s?

Remember, when you are at a cooling rate that gives you a mixture of two phases you can't say what the percentages of the two phases are, only that the two phases are present.

Concept Check 8.5.1

- **What continuous cooling heat treatment can be used to obtain each of the following microstructures?**
  - **100% bainite.**
  - **Mixture of coarse pearlite and martensite.**
  - **Mixture of fine pearlite and martensite.**

Summary

This chapter first described the general aspects of phase transformations, then the application of these concepts to a specific material; plain carbon steel. The chapter focuses on steel because it provides a rich example of how microstructure can be changed with processing; pearlite, martensite, bainite, and combinations of all three. But you should be aware that all materials can be transformed through different processing schemes, and that both isothermal and continuous cooling diagrams can be created for all types of materials.

More broadly, in this chapter we have seen specific examples of one part of the MSE triangle; the structure–processing side. In the next chapter we connect these to the third aspect of the triangle; properties. By understanding the fundamental mechanisms involved in determining the strength of a metal, you will see how to control the properties through processing.

Key Terms

Activation energy
Austenite
Bainite
Cementite
Continuous cooling transformation diagram
Diffusionless transformation
Eutectoid reaction
Ferrite
Growth
Isothermal transformation diagram
Kinetics
Martensite
Nucleation
Nuclei
Pearlite
Phase transformations
Spheroidite
Tempered martensite

Problems

Skill Problems

8.1 Determine the microstructure for eutectoid steel that has undergone the following heat treatments. For each heat treatment you may assume that you start with austenite at 760° C.

a. Rapidly cool to 625° C, hold for 10 s, then quench to room temperature.
b. Rapidly cool to 600° C, hold for 4 s, rapidly cool to 450° C, hold for 10 s, then quench to room temperature.
c. Rapidly cool to 665° C, hold for 1,000 s, then quench to room temperature.
d. Rapidly cool to 350° C, hold for 150 s, then quench to room temperature.

8.2 On the isothermal transformation diagram for an iron–carbon alloy of eutectoid composition, sketch and label time–temperature paths to produce the following microstructures:

a. Start with 100% fine pearlite at room temperature, produce 100% coarse pearlite.
b. Start with austenite at 760° C, produce 100% tempered martensite.
c. Start with 50% bainite and 50% martensite, produce 25% coarse pearlite, 25% bainite, and 50% martensite.

8.3 What microstructure is produced by cooling an Fe–C alloy of eutectoid composition at the following cooling rates? You may assume in each case that you start with austenite at 760° C.

a. 1° C/s
b. 20° C/s
c. 50° C/s
d. 175° C/s

8.4 Briefly describe a heat treatment that could be used in converting an Fe–C alloy of eutectoid composition from one microstructure to the other as given below. You may use either isothermal or continuous cooling heat treatments.

a. Spheroidite to tempered martensite.
b. Tempered martensite to pearlite.
c. Bainite to martensite.
d. Martensite to pearlite.
e. Pearlite to tempered martensite.

Conceptual Problems

8.5 If the critical radius for nucleation is 10 nm and a nucleus with a radius of 5 nm forms, what will happen to it? Explain why.

8.6 Equation (8.1.2) indicates that the rate of solidification should increase as the temperature increases, but Figure 8.1.4 shows that at some temperatures the rate decreases as the temperature increases. Explain this apparent contradiction.

8.7 Explain why phase transformations occur slowly at low temperatures.

8.8 Imagine a strange world where diffusion is faster at lower temperatures. For this world draw the isothermal transformation diagram.

8.9 At low temperatures, why does bainite form instead of pearlite?

8.10 Why does spheroidite form when pearlite is heated?

8.11 What microstructures in plain carbon steel cannot be formed by continuous cooling? Explain why.

8.12 A sample of paraffin wax is extensively purified so that it contains no impurities, dust, dirt particles, etc. If this sample is melted in a container and then cooled and allowed to crystallize, what type of nucleation will occur; homogeneous or heterogeneous? Explain your answer.

8.13 If the contact angle between a metallic solid and a surface is 180°, which will be favored during solidification of that metal in the presence of the surface; homogeneous or heterogeneous nucleation? Explain your answer.

8.14 Under nonequilibrium cooling, do you expect the variation in composition across the grain in the final cored structure to be greater or less as the cooling rate decreases? Explain your answer.

Collapse of the I-35W bridge in Minneapolis in 2007 during rush hour. The bridge experienced sudden failure due to overloading of a buckled gusset plate. (Joe Ferrer/Shutterstock)

Mechanical Behavior

One of the most important aspects of materials is how they respond to forces. This response is determined by their stiffness, elasticity, resistance to deformation, resistance to breaking, and many other properties. Understanding these properties is a key aspect of being able to choose the correct material for a particular application and to use it effectively. By the end of this chapter you will:

Be able to calculate mechanical properties for materials.
Compare the properties for different materials.
Understand how structure affects mechanical properties.

9.1 Stress–Strain Curves

LEARN TO: Calculate mechanical properties.
Draw and identify stress–strain curves for different materials.

To begin, we need to understand stress and strain. You are probably used to thinking in terms of force and deformation. However, the response of a material to force depends on the size of the piece we have. For example, think about trying to bend a thin copper wire versus a copper pipe; it takes a lot more force to bend the pipe than to bend the wire. Instead of force and deformation, materials engineers use the terms *stress* and *strain*. We can have different kinds of stress and strain, but for now we will just talk about tensile stress and tensile strain, which result from a "pulling" force as shown in Figure 9.1.1. The tensile stress, which has the symbol σ (sigma), is given by the following:

$$\sigma = \frac{F}{A_0} \tag{9.1.1}$$

This is called the engineering stress, because it is calculated based on the initial area. True stress would be calculated based on the actual area, A. The tensile strain, which has the symbol ε (epsilon), is given by by the following:

$$\varepsilon = \frac{L - L_0}{L_0} = \frac{\Delta L}{L_0} \tag{9.1.2}$$

Figure 9.1.1

Deformation of a bar of a material due to application of a tensile force, F. The bar has an initial length L_0 and an initial cross-sectional area A_0. After application of the force, the bar has become longer and thinner, with a new length L and a new cross-sectional area A.

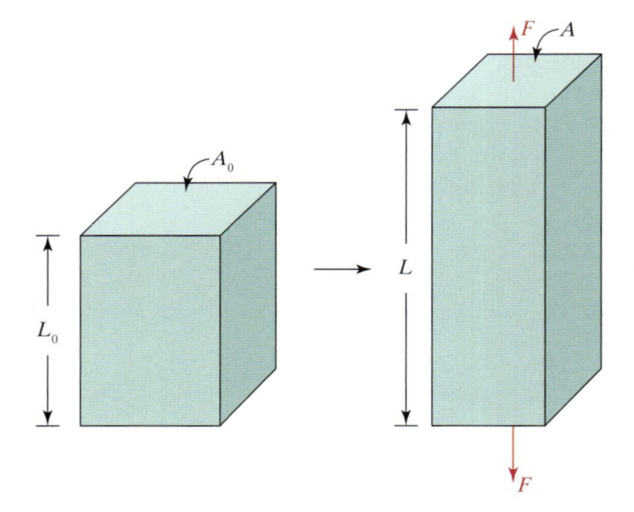

This is the engineering strain. As with stress, we can also define the true strain, although this is not used as often as engineering strain, so we won't consider it further.

In physics you learned about *Hooke's law* for a spring:

$$F = -kx \tag{9.1.3}$$

which says that the length of the spring is directly proportional to the force placed on it. The constant k is the spring constant, which depends on the spring. We can define a similar equation for a material subjected to a stress:

$$\sigma = E\varepsilon \tag{9.1.4}$$

In this equation the constant E is called the elastic modulus, Young's modulus, or simply *modulus*. Like the spring constant, the modulus defines the stiffness of the material—that is, how much force it takes to stretch it by a certain amount. A material with a higher modulus is stiffer and does not deform as much for a given stress.

If we do a tensile test by pulling on the material at a constant rate, we get a graph of stress versus strain, called a stress–strain curve. Figure 9.1.2 shows a generic stress–strain curve that we will use to understand what we can get from these curves. It is important to note that no material has a stress–strain curve like Figure 9.1.2.

Figure 9.1.2

Generic stress–strain curve. The important characteristics of a stress–strain curve are shown on this graph.

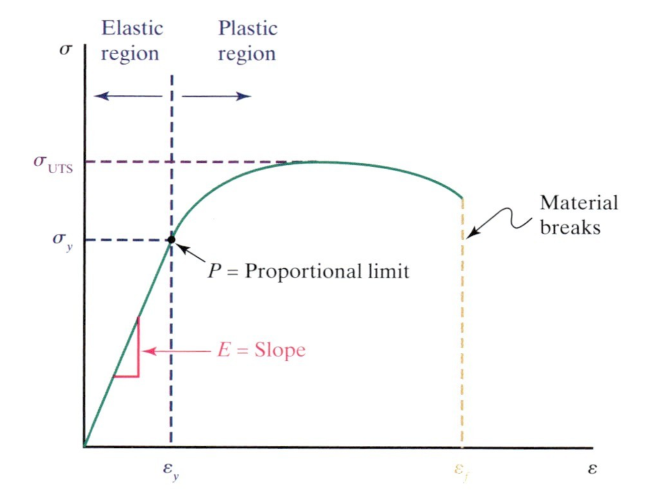

The first important thing to note is the difference between the *elastic* and *plastic* regions. In the elastic region, the stress–strain curve is linear, and Hooke's law applies. This means that the slope in the elastic region is the elastic modulus. In the elastic region the deformation of the material is completely recoverable; if the stress is removed, the material will return to its original shape. The proportional limit is the point at which the curve is no longer linear. Beyond the proportional limit the deformation is permanent and cannot be recovered. We call this type of deformation plastic, and so this part of the stress–strain curve is the plastic region. If you take a coat hanger and try to bend it, you can feel the proportional limit as the point at which you feel the hanger start to "give."

Figure 9.1.2 also shows the mechanical properties we can get from a stress–strain curve. These properties are as follows:

- *Yield stress* (σ_y): The stress at the proportional limit; it represents the stress at which the material begins to yield, or permanently deform.
- *Yield strain* (ε_y): The strain at the proportional limit; it represents the strain at which the material begins to yield, or permanently deform.
- *Ultimate tensile strength* (σ_{UTS}): The maximum stress on the stress–strain curve.
- *Fracture strain or ductility* (ε_F): The maximum strain on the stress–strain curve; it represents the strain at which the material breaks. It is also often given as a percentage, which is $\varepsilon_F \times 100$.
- *Toughness*: The area under the stress–strain curve; it represents the energy required to break the material. Note that this is the first of three different definitions of toughness that we will see.

The stress–strain curves for real materials are shown in Figure 9.1.3. We can see that they are very different from each other, reflecting the different properties of these types of materials. For example, we know that ceramics are very brittle. On the stress–strain curve this is represented by a low ductility and a lack of a plastic region. The biggest difference

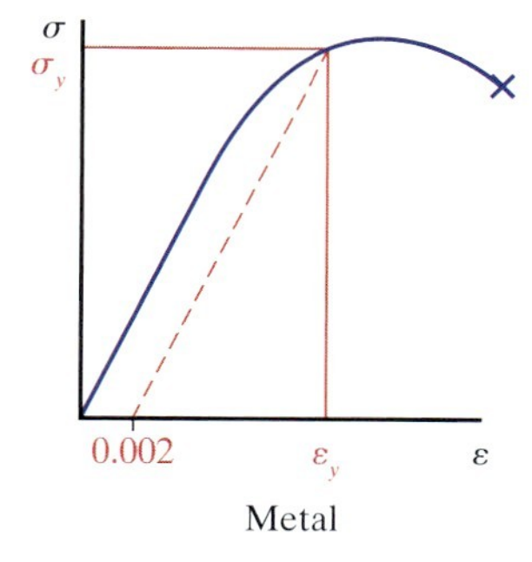

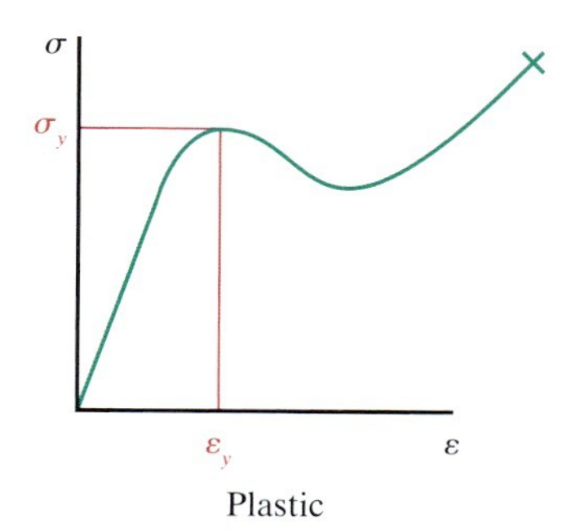

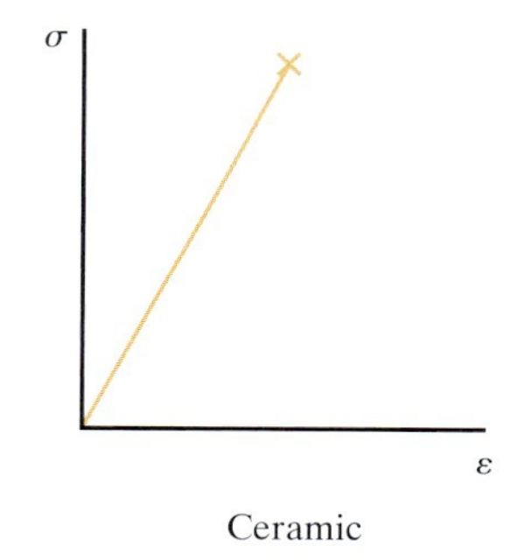

Figure 9.1.3

Typical stress–strain curves for metals, plastics, and ceramics. Determination of the yield point for metals and plastics is shown. Ceramics do not usually have a yield point. Other types of stress–strain curves are also possible for these materials.

between the real curves for metals and plastics in Figure 9.1.3 and the generic curve in Figure 9.1.2 is that real curves do not have a well-defined proportional limit. Therefore, we need alternate methods to determine the yield point. The definitions are different for metals and plastics, for no reason other than tradition. For a metal, we use the 0.2% offset yield. This is shown in Figure 9.1.3. The steps to determine the yield point are as follows:

- Draw a line parallel to the initial linear portion of the curve, but begin it at a strain of 0.2% (0.002 strain).
- Find where this line intersects the stress–strain curve. This intersection is the yield point.
- The yield stress and yield strain are the stress and strain at the yield point.

For polymers, finding the yield point is much easier; the yield point is simply the first maximum on the stress–strain curve.

If you wanted to obtain a stress–strain curve for a material, how would you do it? The American Society for Testing and Materials (ASTM) writes standards for how to conduct tests and specifications for materials. You can find almost any test you want—how to measure density, strength, refractive index, resistance to chemicals, and hundreds of others. These tests have been created by a group of experts, and are then checked by different labs to see if the instructions are clear and if they can get the correct answer. There are several standards for stress–strain measurements for different materials, but they all have the same basic requirements; the sample is machined or molded to a specified shape, then placed into a machine that pulls it at a constant rate. The force is measured, and then by knowing the initial length and cross-sectional area of the specimen, the force and length are converted to stress and strain.

Now that you know all the things you can find from a stress–strain curve, you will practice doing it using the curve shown in Figure 9.1.4.

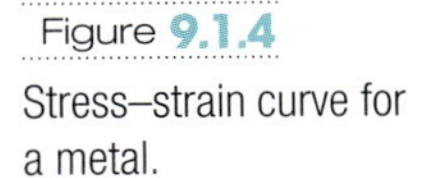
Figure 9.1.4

Stress–strain curve for a metal.

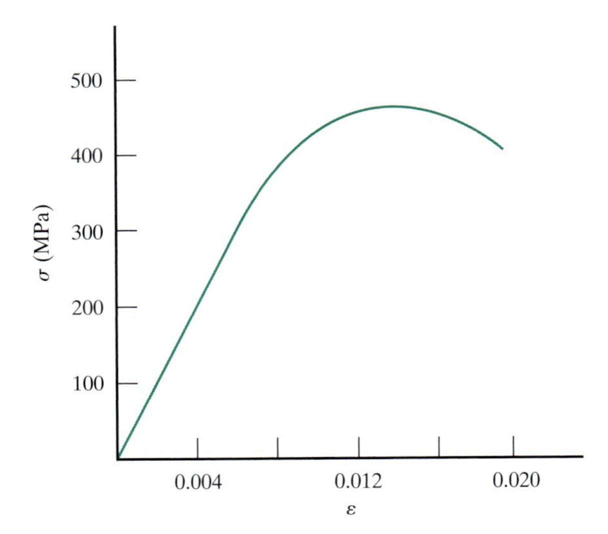

Guided Inquiry: Property Calculations

9.1.1 What is the elastic modulus for the stress–strain curve in Figure 9.1.4?

9.1.2 What is the yield strength for the stress–strain curve in Figure 9.1.4?
Don't use the end of the linear region; that is not a realistic approach. Use the method described above.

9.1.3 What is the yield strain for the stress–strain curve in Figure 9.1.4?

9.1.4 What is the tensile strength for the stress–strain curve in Figure 9.1.4?

9.1.5 What is the ductility for the stress–strain curve in Figure 9.1.4?

9.1.6 A beam is being designed with the metal with the stress–strain curve shown in Figure 9.1.4. The beam requires a safety factor of 2.5. What is the maximum stress you will allow for this beam?
Make sure to think logically about how to use a safety factor. Many students use an equation blindly without thinking about it, and they end up with a stress that is guaranteed to ensure the material will fail!

Concept Check **9.1.1**

- **A material has a yield stress of 350 MPa. A cylindrical rod of this material has a diameter of 5 cm. If the design requires a safety factor of 1.8, what is the maximum force you will allow this rod to experience?**

EXAMPLE PROBLEM 9.1.1

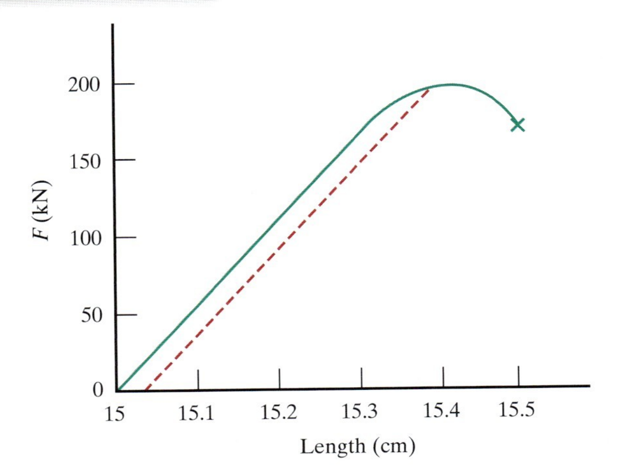

The force–elongation curve of a metal shown above was measured on a cylindrical rod that has an initial diameter of 2 cm and an initial length of 15 cm. Determine the modulus, yield strength, yield strain, tensile strength, and ductility of this material.

The graph above is in units of force and length, but all the properties are in units of stress and strain. The shape of the graph won't change when we convert the units, so the easiest way to do this is to get the values we need off the graph for each property and then do the conversion.

Modulus:

Two points in the linear region of the curve are the following:

$F = 0$ kN, $L = 15$ cm

$F = 165$ kN, $L = 15.3$ cm

Converting these values to stress and strain, we use the following equations:

$$\sigma = \frac{F}{A_0}$$

$$\varepsilon = \frac{L - L_0}{L_0} = \frac{\Delta L}{L_0}$$

Using these equations on the two points gives the following:

$$\sigma = \frac{0 \text{ kN}}{\pi(10^{-2}\text{ m})^2} = 0 \text{ MPa}, \ \varepsilon = \frac{15 \text{ cm} - 15 \text{ cm}}{15 \text{ cm}} = 0$$

$$\sigma = \frac{165 \text{ kN}}{\pi(10^{-2}\text{ m})^2} = 525.2 \text{ MPa}, \ \varepsilon = \frac{15.3 \text{ cm} - 15 \text{ cm}}{15 \text{ cm}} = 0.02$$

The modulus is

$$E = \frac{\sigma}{\varepsilon} = \frac{525.2 \text{ MPa}}{0.02} = 26.3 \text{ GPa}$$

Yield strength and strain:

Since this is a metal, the yield point is determined from the 0.2% offset. Use the following equation to determine the length corresponding to 0.002 strain:

$$\Delta L = \varepsilon L_0 = (0.002)(15\text{ cm}) = 0.03\text{ cm}$$

This means we have to draw our line from a length of 15.03 cm on the length axis. This is shown as the dotted line in the graph on the previous page. From where this dotted line crosses the curve we get the following:

$$\sigma_y = \frac{190{,}000\,\text{N}}{\pi(10^{-2}\,\text{m})^2} = 605\,\text{MPa}$$

$$\varepsilon_y = \frac{15.4\text{ cm} - 15\text{ cm}}{15\text{ cm}} = 0.027$$

For tensile strength, we take the maximum point on the curve:

$$\sigma_{\text{UTS}} = \frac{200{,}000\,\text{N}}{\pi(10^{-2}\,\text{m})^2} = 637\,\text{MPa}$$

For ductility, we take the strain at failure:

$$\varepsilon_f = \frac{15.5\text{ cm} - 15\text{ cm}}{15\text{ cm}} = 0.033$$

9.2 Bond-Force and Bond-Energy Curves

LEARN TO: **Draw and interpret bond-force and bond-energy curves.**
Describe the atomic basis for mechanical properties.

By understanding the relationship between modulus and bond strength, we can predict whether one material will be more rigid than another simply by knowing the types of bonds that each material has. Before we do that, however, we need to understand something about the strength of bonds and the forces between atoms. We can imagine these forces by using a simple analogy. Imagine that we have two balloons, one with a positive charge and one with a negative charge. When the balloons are far apart, there is an attractive force pulling them together, and this force increases as the balloons get closer. When the balloons touch, however, the force becomes repulsive, pushing them apart, and this repulsive force gets stronger the more you try to push the balloons into each other. We can imagine these balloons as atoms. When two atoms are far apart, there is an attractive force, which increases as they get closer. When the atoms get so close that their electron clouds start to overlap, there is a repulsive force that prevents them from getting any closer together. You will now examine how these opposite forces interact to define the length of a bond between two atoms.

Guided Inquiry: Bond Forces

9.2.1 When two atoms are separated, is the force between them attractive or repulsive?

Imagine you have a pair of oppositely charged atoms, such as a sodium ion and a chloride ion. The answer would be the same for any pair of atoms, but this is the easiest example to consider.

9.2.2 When two atoms get so close together that their electron clouds overlap, is the force attractive or repulsive?

9.2.3 Use your answers from questions 9.2.1 and 9.2.2 to sketch a graph of the force between two atoms versus the distance between them. On this graph, make attractive forces positive and repulsive forces negative.

9.2.4 On your graph, is there a distance at which the atoms are at an equilibrium distance from each other? What is the value of the force at that distance?

Figure 9.2.1(a) shows the actual bond-force curve for a pair of atoms. If you integrate the bond-force curve, you get the bond-energy curve, which is shown Figure 9.2.1(b). Use these graphs to answer the next set of questions.

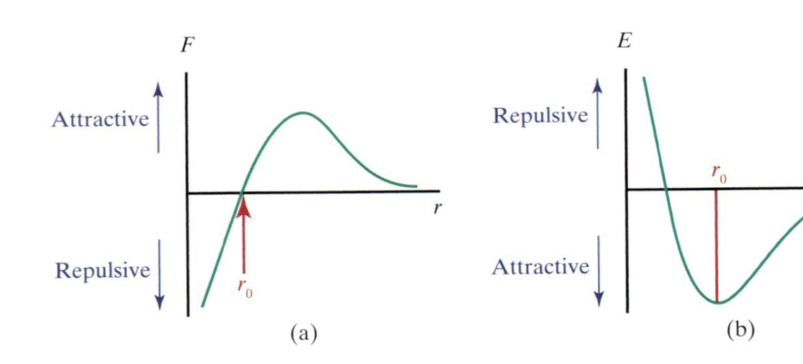

Figure 9.2.1
(a) Bond-force curve, showing the force between two atoms as a function of their distance from each other; (b) bond-energy curve, showing the energy in the bond between two atoms as a function of their distance from each other. Remember that lower energy is more favorable.

Guided Inquiry: Bond Length

9.2.5 At distances greater than r_0, is the net force between the atoms attractive or repulsive?

9.2.6 At distances less than r_0, is the net force between the atoms attractive or repulsive?

9.2.7 At what distance is the net force between the two atoms zero?

9.2.8 The length of a bond between two atoms is the distance at which the atoms are at an equilibrium separation. What would be the net force between atoms in a bond?

Think of statics or physics. When something is in equilibrium, what is the net force on it?

9.2.9 On the bond-force curve shown in Figure 9.2.1, what is the bond length?

9.2.10 The bond length is also the distance at which the atoms are at their lowest energy. On the bond-energy curve in Figure 9.2.1, what is the bond length?

Concept Check 9.2.1

- **Compare the bond-force curves shown in the figure below. Which one has the shorter bond length?**

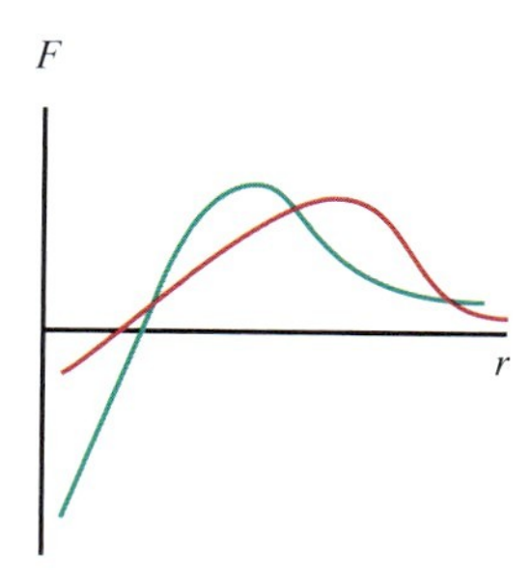

The bond-force and bond-energy curves change depending on the strength of the bonds between the atoms. Figure 9.2.2 shows how the curves change. These curves are directly related to properties of the material. Let's consider the modulus. From Section 9.2.1 we know that the modulus represents how much a material will stretch for a given amount of force. If the same force is applied, the material with the higher modulus will stretch less. When we stretch the material, we are really stretching the bonds between atoms, so the bonds stretch less in a material with a higher modulus. By examining bond-force curves, you will now see how the strength of a bond is related to modulus.

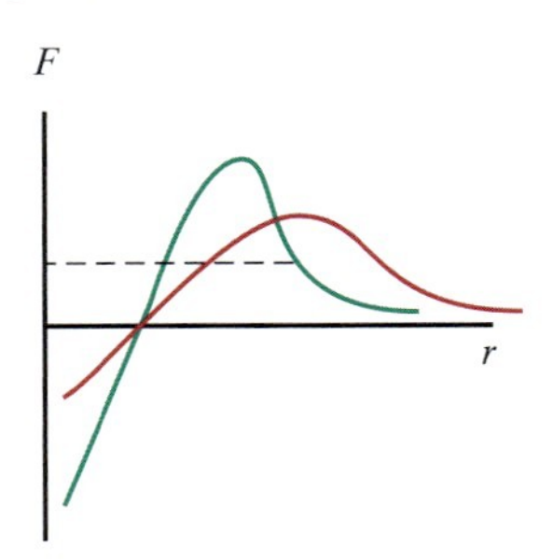

Figure 9.2.2

Comparison of the bond-force curves for two materials with different bond strengths. The material with the green curve has a higher bond strength than the one with the red curve.

Guided Inquiry: Modulus

9.2.11 On Figure 9.2.2, mark the equilibrium (unstretched) bond distance for each type of material.

Make sure you are keeping track of time in your group so you don't fall behind.

9.2.12 The force indicated with the dashed line in Figure 9.2.2 is applied to each of these materials. Which one has stretched more?

9.2.13 Which material has the higher modulus?

9.2.14 In general, how will the modulus of a material change as the bond strength increases?

9.2.15 What kinds of bonds are present in sodium chloride?

9.2.16 What kind of nonbonding interactions are present in polystyrene?

9.2.17 Compare the interactions (bonding and nonbonding) between atoms in sodium chloride and polystyrene. Which ones are weaker?

9.2.18 Comparing sodium chloride and polystyrene, in which one are you pulling on weaker bonds or interactions when you stretch it?

9.2.19 Based on your answer to question 9.2.17, which has a higher modulus; sodium chloride or polystyrene?

Concept Check 9.2.2

- **Rank the following materials from lowest to highest modulus based on their bond strengths: copper, potassium bromide, PVC.**

9.3 Strength of Metals

LEARN TO: **Describe the atomic basis for mechanical properties.**
Calculate resolved shear stress and yield strength.

As we learned in Section 9.1, the yield stress of a material is the stress at which permanent deformation begins. Below the yield stress, in the elastic region, deformation occurs by stretching of bonds. When the force is released, the bonds can return to their original length, and thus the deformation is recoverable. In contrast, above the yield stress, bonds are broken, and thus the deformation is not recoverable. In this section, we begin by figuring out what actually happens to a crystal when the bonds break. Then we will see how the yield stress can be calculated.

Guided Inquiry: Plastic Deformation

9.3.1 Describe the bond-breaking process that needs to occur for the deformation in Figure 9.3.1 to happen.

9.3.2 How many total bonds need to break in the process shown in Figure 9.3.1?

9.3.3 How many bonds need to break simultaneously in the process shown in Figure 9.3.1?

9.3.4 Describe the bond-breaking process that needs to occur for the deformation in Figure 9.3.2 to happen.

9.3.5 How many total bonds need to break in the process shown in Figure 9.3.2?

9.3.6 How many bonds need to break simultaneously in the process shown in Figure 9.3.2?

9.3.7 Compare the processes from Figures 9.3.1 and 9.3.2. Which requires less total energy? Which requires less energy per step in the process?

9.3.8 Compare a crystal with a dislocation and a crystal without a dislocation. Which one requires less force to deform?

Concept Check **9.3.1**

- **Compare a crystal with a dislocation and a crystal without a dislocation. Which one is stronger?**

The process of dislocation motion that leads to the deformation shown in Figure 9.3.2 is called *slip*. A more detailed diagram showing slip is given in Figure 9.3.3. The process shown in these figures results in plastic deformation—the material is permanently deformed as a result of the dislocation motion because the atoms have changed positions. The process shown in Figure 9.3.1 also leads to plastic deformation, but it takes a lot more force because many bonds have to be broken simultaneously. Because only one bond breaks at a time during slip, a crystal with dislocations has a lower yield strength than a perfect crystal with no dislocations.

Slip can occur only on certain directions and planes. First you will identify what the slip direction is.

Figure 9.3.1

Illustration of one possible bond-breaking mechanism that leads to plastic deformation.

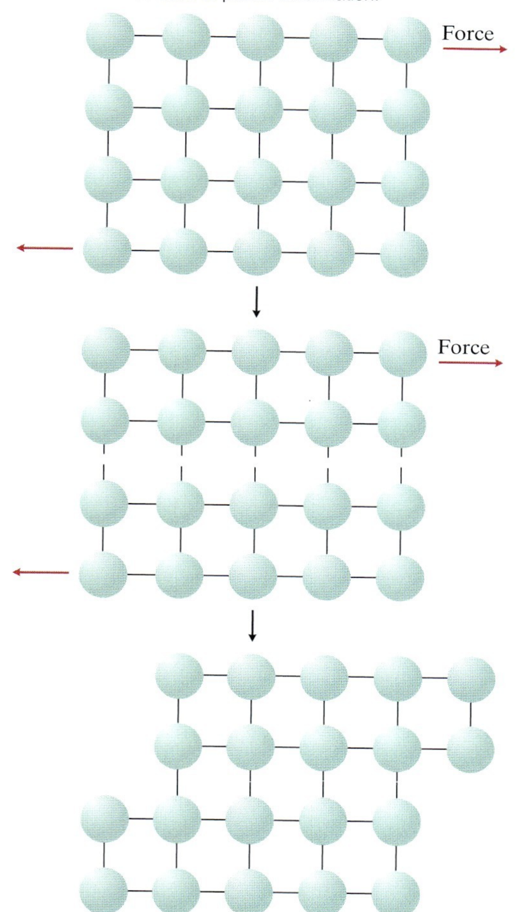

Figure 9.3.2

Bond-breaking process when a dislocation is present.

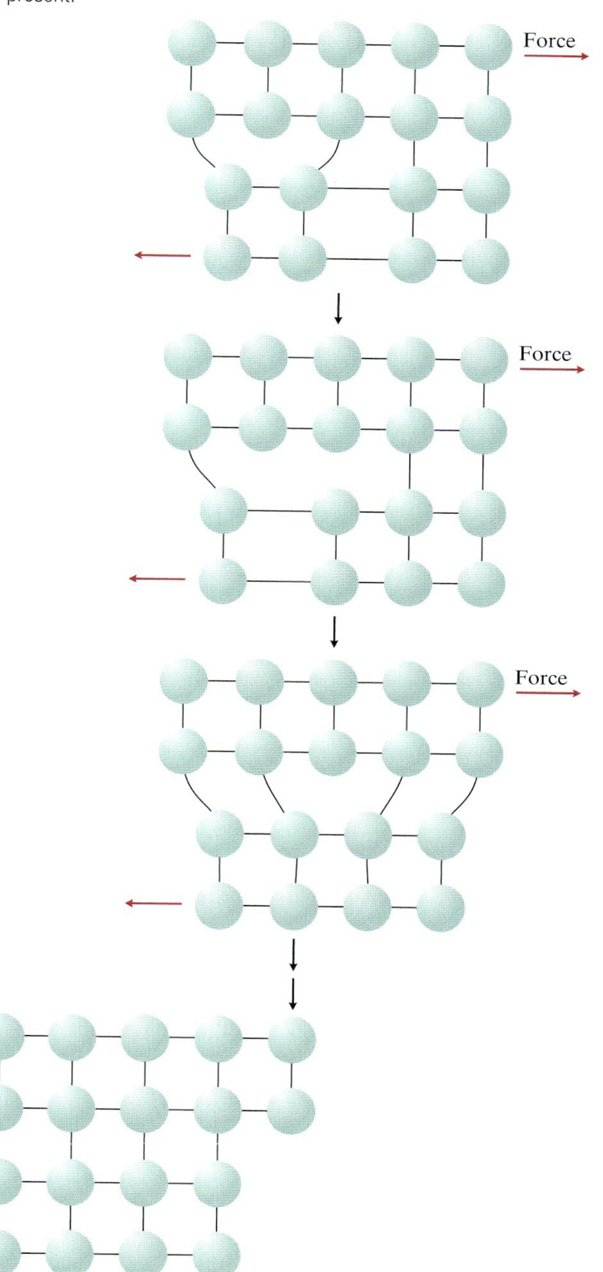

Figure 9.3.3

The process of slip leading to permanent deformation of a crystal.

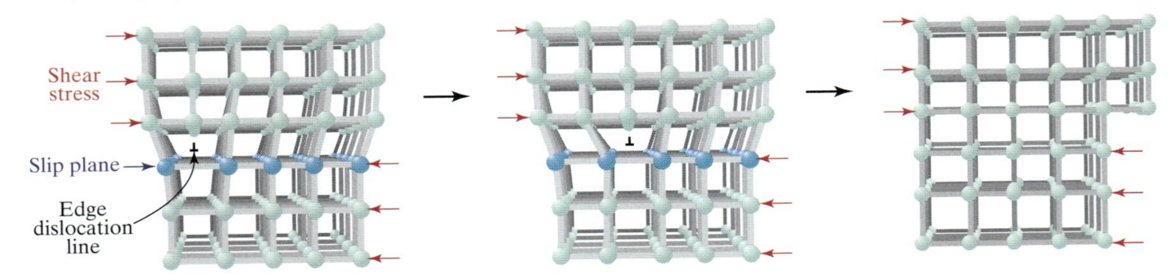

Guided Inquiry: Slip Direction and Plane

9.3.9 Based on Figures 9.3.2 and 9.3.3, what is the relationship between the slip direction and the direction of the Burgers vector?

9.3.10 For copper, what is one possible slip direction?
You may want to go back to Sections 4.4 and 4.5 for some help.

9.3.11 For copper, what is one possible slip plane?

Concept Check 9.3.2

- **What is a possible slip direction for iron?**
- **What is a possible slip plane for iron?**

The slip plane contains both the dislocation and the slip direction, and is always the plane of highest density in that crystal. A slip system is the combination of slip direction and slip plane. The notation for a slip system is $\{hkl\}\langle mnp\rangle$.

TABLE 9.3.1 Slip systems for various metals

Metals	Crystal structure	Number of slip systems
Cu, Al, Ni, Ag, Au	FCC	12
Fe	BCC	48
Cd, Zn	Hexagonal close-packed (HCP)	3

Guided Inquiry: Slip Systems

9.3.12 Table 9.3.1 shows the number of slip systems for different kinds of crystals. Which metal has more possible slip systems; Cu or Cd?

9.3.13 Which metal can more easily undergo slip; Cu or Cd?

9.3.14 Which metal is more ductile; Cu or Cd?

9.3.15 Which type of crystal is more ductile; FCC or hexagonal close-packed? Explain why.

9.3.16 If you had a piece of copper and wanted to make it stronger, using the concept of slip, what would you need to change about the dislocation motion?
Should it be harder or easier for dislocations to move?

Concept Check 9.3.3

- **If you reduce the ability of dislocations to move in iron, will the iron be stronger or more ductile (deformable)?**

We can calculate the yield strength of a material by considering how much force it takes to cause slip to occur. The force shown in Figure 9.3.3 is a shear (sliding) force causing the dislocation to move. However, we don't usually have a force applied along the slip direction. For the general case of a force applied in some arbitrary direction, we need to consider how much of the applied force is resolved into the slip direction. Figure 9.3.4 illustrates the geometry involved. We won't go through the trigonometry in any detail here.

Figure 9.3.4

The geometry of slip.

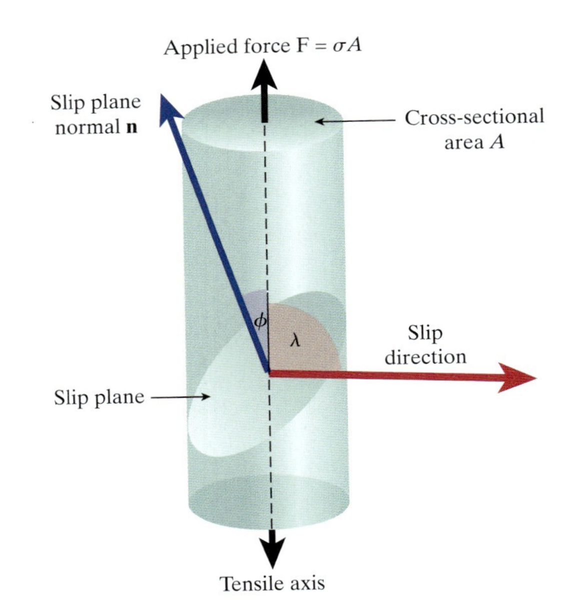

However, the result is as follows:

$$\tau = \sigma \cos\phi \cos\lambda \tag{9.3.1}$$

where ϕ is the angle between the slip plane normal and the direction of the applied stress, and λ is the angle between the slip direction and the direction of the applied stress. The shear stress τ is the resolved shear stress because it is the component of the applied stress that is resolved into the slip direction, and it is a shear stress.

To calculate the yield stress, we recognize that the yield stress is the stress at which slip just begins to occur. We can therefore rewrite the above equation as follows:

$$\tau_{CRSS} = \sigma_y \cos\phi \cos\lambda \tag{9.3.2}$$

In this equation, σ_y is the yield strength and τ_{CRSS} is the *critical resolved shear stress*; it is "critical" because it is the minimum stress needed for slip to occur. We can use this equation to calculate the yield strength of a crystal. However, before we can do that, we need to know one more important thing. We can use the definition of the dot product between two vectors to calculate the cosine of the angle between them:

$$\cos\theta = \frac{\vec{A} \cdot \vec{B}}{|\vec{A}||\vec{B}|} \tag{9.3.3}$$

For two directions in a crystal, $[hkl]$ and $[mnp]$, this equation becomes

$$\cos\theta = \frac{hm + kn + lp}{\sqrt{h^2 + k^2 + l^2}\sqrt{m^2 + n^2 + p^2}} \tag{9.3.4}$$

Note that because we can calculate the cosines directly, there is no need to calculate the actual angles ϕ and λ.

With these relationships, you will now see how to calculate the yield strength of a crystal.

The next questions are based on the following scenario: We want to calculate the yield strength along [010] for the (110)[$\bar{1}$11] slip system in a BCC crystal. The critical resolved shear stress is 30.0 MPa.

Guided Inquiry: Calculating Yield Strength

9.3.17 In the problem above, what variable are we solving for in the shear-strength equation?

9.3.18 What is the slip direction?
The biggest mistake in these calculations is identifying the directions incorrectly. Make sure everyone in your group agrees.

9.3.19 What is the slip plane?

9.3.20 What is the normal to the slip plane?
Go back and look at Concept Check 4.3.3.

9.3.21 What is the direction of the applied stress?

9.3.22 What is the yield strength?

Concept Check **9.3.4**

- **What is the yield strength you calculated in question 9.3.22?**

EXAMPLE PROBLEM 9.3.1

A single crystal of a BCC metal has a critical resolved shear stress of 18 MPa. What is the yield stress of this crystal in the [110] direction if slip occurs in the $(101)[11\bar{1}]$ slip system?

To do this, we first need to identify the various directions involved:

Stress direction: [110]

Slip direction: $[11\bar{1}]$

Slip plane normal: From Section 4.3, and specifically question 4.3.11, in cubic crystals, the normal to a plane has the same indices as the plane. So the normal to (101) is [101].

ϕ is the angle between the stress and the slip plane normal:

$$\cos\phi = \frac{1x1 + 1x0 + 0x1}{\sqrt{1^2 + 1^2 + 0}\sqrt{1^2 + 0^2 + 1^2}} = 0.5$$

λ is the angle between the stress and the slip direction:

$$\cos\lambda = \frac{1x1 + 1x0 + 0x(-1)}{\sqrt{1^2 + 1^2 + 0^2}\sqrt{1^2 + 1^2 + (-1)^2}} = 0.816$$

We can now calculate the yield stress from Equation 9.3.2:

$$\sigma_y = \frac{\tau_{CRSS}}{\cos\phi\cos\lambda} = \frac{18\,\text{MPa}}{(0.5)(0.816)} = 44.1\,\text{MPa}$$

The yield stress is higher than the critical resolved shear stress, as we would expect since only some of the applied yield stress is resolved into the slip direction.

9.4 Strengthening Mechanisms in Metals

LEARN TO: **Describe strengthening mechanisms in metals.**
Predict which of two given metals is stronger.

Dislocation motion is what causes plastic deformation in metals. One way to reduce plastic deformation might be to get rid of all dislocations in a crystal. While this would work, there is no way to accomplish it; there are always dislocations present. Therefore, to reduce plastic deformation and increase yield strength, we need to find ways to stop dislocation motion. In this section, we will describe some important characteristics of dislocations, and you will use this information to predict how to make a metal stronger.

In Figure 9.4.1 we see that around a dislocation, the bonds between atoms are stretched and compressed. If there are impurity atoms in the correct locations, the stretching and compressing is reduced, and the bonds are closer to their unstrained lengths. **The next questions will show you how this is related to making metals stronger.**

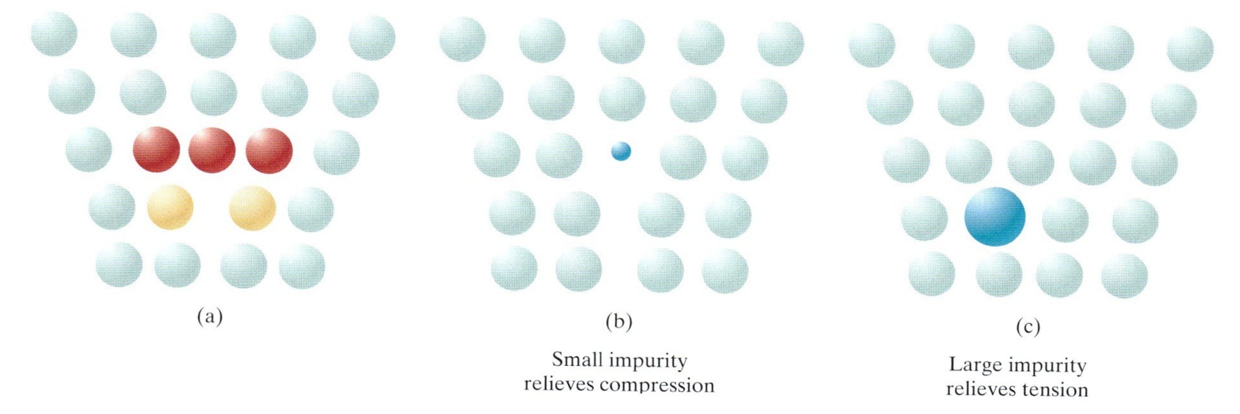

Figure 9.4.1
The effect of impurity atoms on the bond strain around a dislocation. Red atoms have bonds that are in compression, and yellow atoms have bonds that are in tension. Blue atoms are impurities.

Guided Inquiry: Solid Solution Strengthening

9.4.1 In Figure 9.4.1, which crystal or crystals have their bonds closer to the unstrained lengths?

9.4.2 In Figure 9.4.1, which crystal or crystals are in a lower energy state?

9.4.3 If a dislocation moves so that an impurity is in the positions shown in Figure 9.4.1(b) and (c), will it take more or less energy to move it from that position compared to there being no impurity present?

Remember, lower energy is more favorable, so if it is already low energy, it will take more energy to move it from that state.

9.4.4 Which is stronger; a crystal with impurities or without impurities?

In which case is it harder to move the dislocation? This is called solid solution strengthening.

Concept Check 9.4.1

- Which is stronger; pure aluminum or aluminum with 10 wt% copper?

Figure 9.4.2 shows a schematic diagram of how the crystal structure changes at a grain boundary and at the boundary of a *precipitate*. A precipitate is a second phase of a different material incorporated into the crystal. At these boundaries, the atoms are disordered because of the transition from one crystal structure to another. As a result, the slip direction and slip plane "disappear" at the boundary; they are no longer present for slip to occur. You will now see how this relates to the strength of metals.

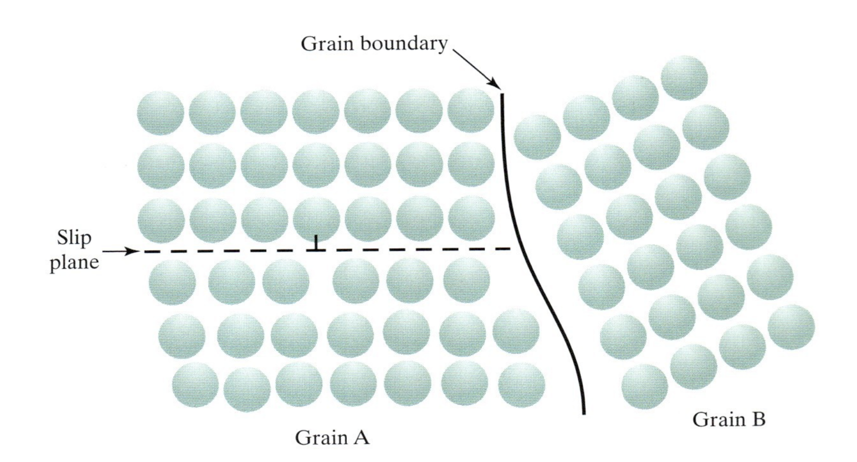

Figure **9.4.2**

Change in crystal structure at a grain boundary and the boundary of a precipitate.

Guided Inquiry: Grain-Size Reduction

9.4.5 When a dislocation reaches a grain boundary, is the slip system still present?

9.4.6 When a dislocation reaches a grain boundary, can the dislocation continue to move?

9.4.7 If the size of the grain is reduced, will the dislocation move a longer or shorter distance before reaching the grain boundary?

9.4.8 Does reducing the grain size make a material stronger, weaker, or have no effect?

9.4.9 Does adding precipitates make a material stronger, weaker, or have no effect?
A precipitate is a second phase. The interface at a precipitate is very similar to a grain boundary. This is called grain-size reduction (for grains) or precipitation strengthening (for precipitates).

Concept Check 9.4.2

- **Which is stronger; pure copper with 10-μm grains or pure copper with 100-μm grains?**

Application Spotlight Strength of Nanomaterials

As you have seen in this section, as the grain size gets smaller, metals will get stronger, because the grain boundaries block dislocation motion. The mechanism is actually a little more complicated than this, involving dislocations "piling up" at the grain boundary and "pushing" dislocations across that grain boundary. As grains get smaller, the amount of pile-up gets smaller, so it is less likely a dislocation will cross the grain boundary. The resulting relationship between grain size and strength is called the Hall–Petch relation. However, there is a limit to this effect; in nanomaterials with a grain size less than 100 nm, the material actually becomes more ductile (less strong) as grain size gets smaller. This is called the reverse Hall–Petch relation. It is believed that this occurs because the grains are so small there is only one dislocation in each grain, so pile-up cannot occur. Instead, other deformation mechanisms occur, such as sliding of grain boundaries past each other. Taking advantage of the reverse Hall–Petch effect could result in new materials with greatly increased fracture toughness and formability at low temperatures.

Figure 9.4.3 shows two dislocations near each other. If two dislocations approach each other like this, the regions of tension and compression begin to overlap. This means that the bonds become even more strained than when only one dislocation is present. One way to create more dislocations is to deform the material—when a crystal is plastically deformed, dislocations are created. You will now see how the number of dislocations in a crystal affects its strength.

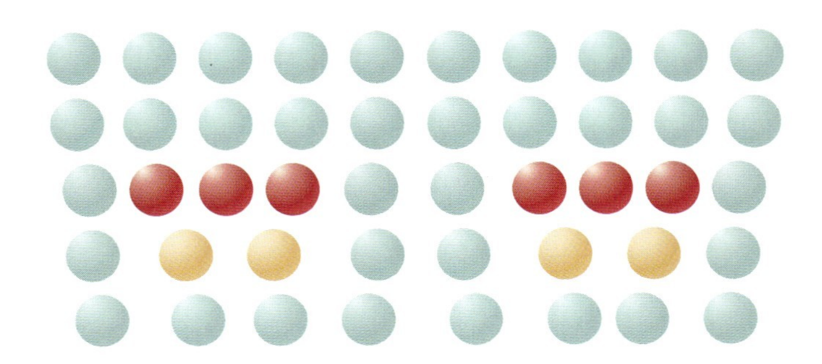

Figure **9.4.3**

Two dislocations near each other. Red atoms have bonds that are in compression, and yellow atoms have bonds that are in tension.

Guided Inquiry: Cold-Working

9.4.10 When two dislocations approach each other, is there more or less bond strain?

9.4.11 When two dislocations approach each other, does the energy become higher or lower?

9.4.12 When two dislocations approach each other, is it harder or easier for them to keep moving closer together?

9.4.13 Which is stronger; a crystal with many dislocations or a crystal with a few dislocations?

This is called strain hardening or cold-working.

Concept Check **9.4.3**

- **Which is stronger; undeformed copper or copper that has been plastically deformed?**

Annealing is the process of heating up a metal to some temperature below its melting temperature for the purpose of changing its properties. The change in the structure of the metal during annealing is shown in Figure 9.4.4. During annealing three changes occur, in this order:

- *Recovery:* The number of dislocations is reduced.
- *Recrystallization:* The grains change from being elongated to being equiaxed (more spherical).
- *Grain growth:* The grains get bigger.

How does annealing change the strength of a metal? Think about everything else you have seen so far in this section.

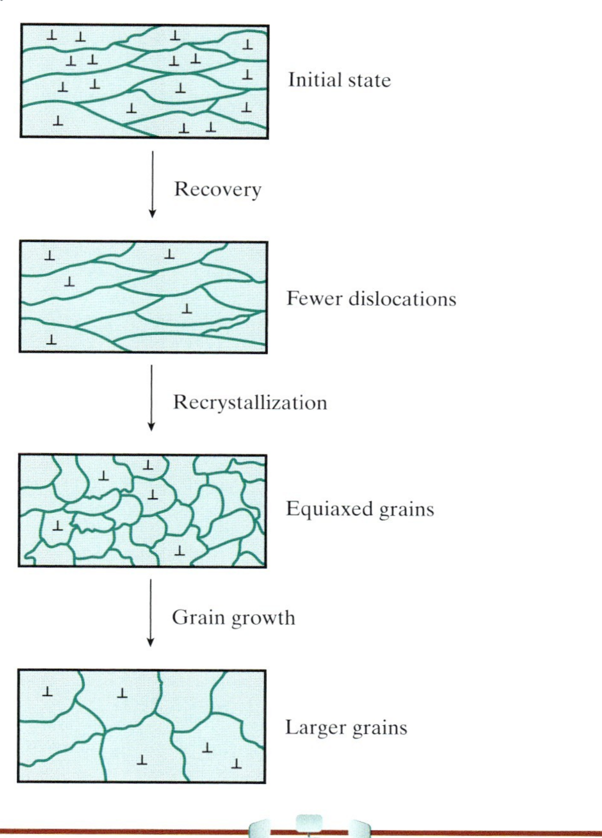

Figure 9.4.4
Changes in structure of a metal during annealing.

Guided Inquiry: Annealing

9.4.14 Which is stronger; a material that has been annealed or one that has not been annealed?

Concept Check 9.4.4

- **Which is stronger; unannealed copper or copper annealed at 750° C?**

9.5 Structure–Property–Processing Relationships in Steel

LEARN TO: **Predict mechanical properties for a given microstructure.**

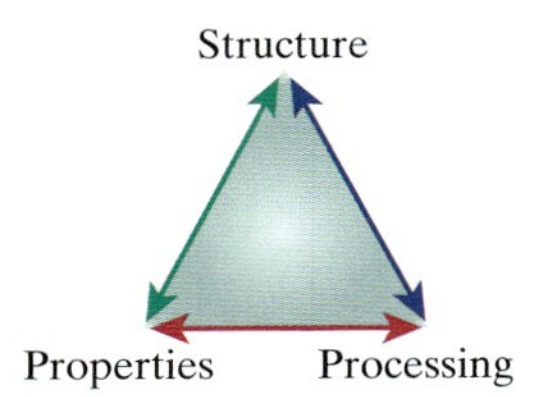

Figure 9.5.1
The MSE triangle.

At the heart of materials science and engineering are structure–property relationships; look at Figure 9.5.1, which is the same as Figure 2.2.1. The MSE triangle illustrates what materials science and engineering is all about: using processing to tailor the structure of a material, which in turn affects its properties. At this point, you have learned enough so that you can understand these relationships and make predictions about properties just by knowing how the material was processed. We will use steel to illustrate this point, by bringing together what you already know about strengthening mechanisms and the microstructures of steel.

Guided Inquiry: Properties of Steel

9.5.1 Compare the single phases of cementite and ferrite. Which has a higher concentration of carbon?
You may want to go back to Chapter 8 to review the phases and microstructures of steel.

9.5.2 Which do you expect to be stronger; cementite or ferrite? Explain why.
Think about which strengthening mechanism is relevant for this comparison. Are these single phases for which solid solution strengthening is relevant? Are these two-phase, for which precipitation hardening is relevant? Has there been plastic deformation or annealing?

9.5.3 How many phases are present in pearlite?

9.5.4 Which is stronger; aluminum with a few copper precipitates or aluminum with many copper precipitates? Explain why.

9.5.5 Based on your answer to question 9.5.4, which is stronger; coarse pearlite or fine pearlite? Explain why.

9.5.6 Which is stronger; coarse pearlite or spheroidite? Explain why.

9.5.7 Martensite is a single phase with a body-centered tetragonal crystal structure. It is the hardest and strongest form of steel. Propose an explanation as to why it is the strongest.

The following table shows the hardness measured for different heat treatments of a eutectoid carbon–iron alloy. Hardness is related to strength—a material with a higher hardness also has a higher strength.

Heat treatment	Brinell hardness	Microstructure
Cool at 30° C/s	270	
Cool at 5° C/s	210	
Cool at 5° C/s, then reheat to 700° C and hold for 24 hours	180	
Quench (cool as fast as possible)	680	

9.5.8 Fill in the empty cells of the table.

Concept Check 9.5.1

- **Rank coarse pearlite, fine pearlite, and spheroidite in order from least ductile to most ductile.**

9.6 Polymer Properties

LEARN TO: **Predict properties of polymers.**

One difference between polymers and other materials is the range of properties that they can have. Most metals are somewhat similar to each other: moduli for metals are around 100–200 GPa, yield strengths are generally a few hundred MPa, and percent elongations are a few tens percent. In contrast, polymers have a wide range of properties. Table 9.6.1 shows typical values for three different polymers. The biggest range is seen in the modulus, which can vary by 3 orders of magnitude!

TABLE 9.6.1 Typical mechanical properties for polystyrene, low-density polyethylene, and elastomers

Material	Modulus (GPa)	Yield strength (MPa)	Elongation (%)
Polystyrene	3	25–70	1–3
LDPE	0.2	10–15	100–500
Elastomer	0.003	No yield	100–800

Polymer properties depend on the degree of crystallinity or crosslinking, and on what the T_g and T_m of the polymer are. In order to predict the properties, we need to think about the state of matter for each phase that is present in the polymer. We will begin by looking at the stress–strain curves in Figure 9.6.1. This figure shows three curves that are typical for different types of polymers. To begin, you will relate these curves to the mechanical behavior of the materials.

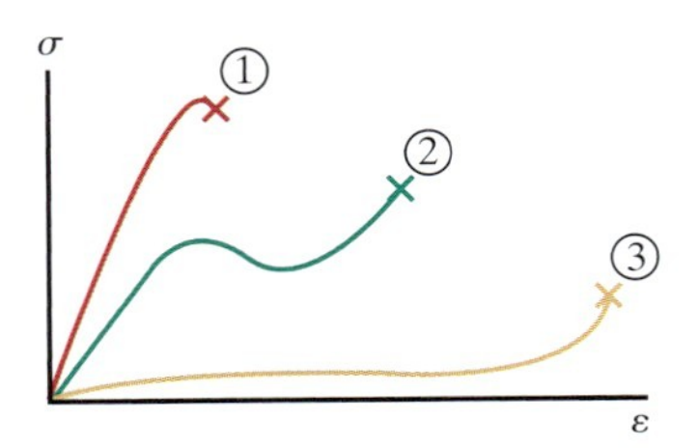

Figure 9.6.1

Three typical stress–strain curves for different types of polymers.

Guided Inquiry: Polymer Properties

9.6.1 Is a material with a high modulus hard to bend or easy to bend?

9.6.2 Is a material with a high modulus rigid or flexible?

9.6.3 Is a material with a high percent elongation easy to stretch or hard to stretch?

9.6.4 Is a material with a high percent elongation ductile or brittle?

9.6.5 Classify polymers 1, 2, and 3 in Figure 9.6.1 as being either rigid or flexible, and as being either ductile or brittle.

Concept Check 9.6.1

- **You identify a material that is more rigid than polymer 1. How is its stress–strain curve different from the one for polymer 1?**
- **You identify a material that is more ductile than polymer 2. How is its stress–strain curve different from the one for polymer 2?**

The glass transition and melting temperatures are the points at which there is a change in the state of matter of the glassy and crystalline phases of a polymer. At temperatures below the T_m the crystalline phase is solid, while above T_m it is liquid. At temperatures below the T_g, the amorphous phase is solid, while above T_g it is liquid. If the amorphous phase is liquid but the crystalline phase is solid, the crystalline phase acts to hold the liquid regions together, preventing flow. Crosslinks will do the same thing.

The crystalline phase will scatter light, making a semicrystalline polymer appear opaque. If no crystals are present, there is nothing to scatter the light, so the material is transparent. Figure 9.6.2 shows an example of the difference in transparency between a semicrystalline and an amorphous polymer.

With this information we can predict the properties of a polymer by knowing its T_g, T_m, and whether or not it is crystalline. You will now see how to do that.

Figure **9.6.2**

Polystyrene cup (left) and high density polyethylene milk bottle (right). Polystyrene is amorphous and HDPE is semicrystalline. Note the difference in transparency.
(Photos courtesy of Lifeprints Photography)

Application Spotlight **Challenger Accident**

In January 1986, I was on a trip with my college concert band from Boston to Florida. On January 28, we were driving from Orlando to Titusville, where many Kennedy Space Center workers live, to play a concert that night. While on the drive we saw the Challenger launch, but it looked different from what I had seen on TV. I quickly learned from the radio that the shuttle had exploded shortly after lift-off. While many technical and managerial issues contributed to the disaster, the fundamental cause was failure of a rubber O-ring in the solid rocket boosters. I remember it was very cold that day, below freezing. I had never been to Florida before, and it surprised me that Florida could be that cold. As you have seen in this section, above T_g, a polymer is flexible, but below T_g, it becomes hard and brittle. The cold temperature that day reduced the flexibility of the rubber O-ring so it could not deform and seal the opening; hot gases escaped from the booster and ultimately caused the explosion. In the resulting investigation, the most dramatic moment came when physicist Richard Feynman quietly asked for a glass of ice water, put the O-ring in it, and then demonstrated what happened.

(Photo courtesy of NASA)

Guided Inquiry: Predicting Polymer Properties

9.6.6 Does an amorphous polymer have a T_g, T_m, or both?

9.6.7 Does an amorphous polymer have a crystalline phase?

9.6.8 Is an amorphous polymer opaque or transparent?

9.6.9 What is the state of matter of an amorphous polymer above T_g?
Make sure to look at the text preceeding. Sometimes the information you need is there.

9.6.10 What is the state of matter of an amorphous polymer below T_g?

9.6.11 Is an amorphous polymer below T_g rigid or flexible? Is it brittle or ductile?

9.6.12 What states of matter are present for a semicrystalline polymer below T_g and below T_m?

9.6.13 What states of matter are present for a semicrystalline polymer above T_g but below T_m?

9.6.14 What states of matter are present for a semicrystalline polymer above T_g and above T_m?

9.6.15 For each situation in questions 9.6.12–9.6.13, classify whether the polymer at that temperature is rigid or flexible, brittle or ductile, and opaque or transparent.

9.6.16 Consider a polymer that is amorphous and crosslinked. At a temperature below T_g, classify whether it is rigid or flexible, brittle or ductile, and opaque or transparent. Do the same if the polymer is above T_g.

9.6.17 If you now think about the amount of crystallinity, how do you think your answers to question 9.6.15 might change if I asked you to compare a polymer that was 90% crystalline and a polymer that was 10% crystalline?

Concept Check **9.6.2**

Isotactic polystyrene has a glass transition temperature of 100° C and a melting temperature of 240° C:

- **What are the properties of isotactic polystyrene at 150° C?**
- **What are the properties of isotactic polystyrene at room temperature?**
- **High density polyethylene is 90% crystalline while low density polyethylene is 40–60% crystalline. What would be the difference in properties between these two polymers?**

9.7 Properties of Ceramics

LEARN TO: **Describe types of ceramic materials and their properties.**

What are ceramics? For many people, their first thought will be things like pottery and tile, and most dictionary definitions refer to clay. But ceramics are much more than that. A more general definition is provided by the Ceramic Tile Institute, which defines ceramic as follows:

> *Inorganic, nonmetallic materials, the article or coating being made permanent and suitable for utilitarian and decorative purposes by the action of heat at temperatures sufficient to cause sintering, solid-state reactions, bonding, or conversion, partially or wholly to the glassy state.*

Although this definition is focused on decorative materials like tile, it covers all aspects of ceramics. A more pragmatic definition is the following: We can all identify whether something is metal or plastic fairly easily; anything else is usually a ceramic. This definition may seem silly, but it works fairly well. So, for example, ceramics include diamond, sand, window glass, and silicon for computer chips.

Ceramics have a wide range of properties that make them useful for many applications. Table 9.7.1 provides a list of properties and the applications for ceramics that result from those properties.

TABLE 9.7.1 Properties and applications of ceramics

Property	Application
High strength and stiffness.	Glass and carbon fibers for composites.
Low thermal conductivity.	Thermal insulators, such as space shuttle tiles.
Low electrical conductivity.	High-voltage insulators. Capacitors.
Ionic conductivity.	High-temperature fuel cells.
High electrical conductivity.	High-temperature superconductors.
High temperature stability.	Refractory crucibles for melting metals.
Optical transparency.	Windows. Fiber optic cables.

As we have discussed in previous chapters, materials can be either crystalline or amorphous. Unlike polycrystals, glasses are transparent because there are no grain boundaries to scatter light. Unlike single crystals, glasses have isotropic properties because the atoms are arranged randomly. Figure 9.7.1 shows the structures of crystalline and amorphous silica (silicon dioxide). In both cases the structure is made up of repeating SiO_4 tetrahedra that are linked together.

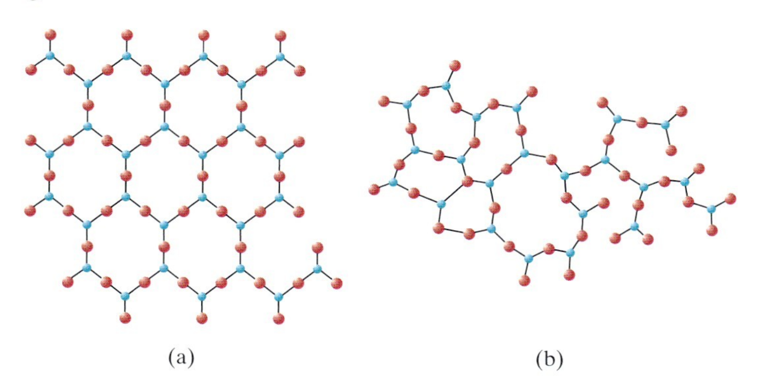

Figure 9.7.1 (a) Structures of crystalline and (b) amorphous silica. The blue atoms are silicon and the red atoms are oxygen. These are two-dimensional views of a three-dimensional structure. In reality each silicon atom is connected to four oxygen atoms, forming a tetrahedron.

To make glass, additives are added to change the properties. The primary additives are the following:

- *Network formers:* These compounds also form tetrahedra and are added to change the chemical properties. Some examples are B_2O_3 and GeO_2. For example, borosilicate glass contains B_2O_3 to increase chemical resistance, and is used for chemistry glassware.
- *Network modifiers:* These compounds break up the silica tetrahedra and lower the softening point to make the glass easier to form. Some examples are Na_2O (lime) and CaO (soda). Common window glass is soda–lime–silicate glass.

The composition of a glass affects various thermal transition temperatures that are important for fabrication processes. The transitions that are important for glass are the following:

- *Melting point:* The material can flow easily enough to be a liquid.
- *Working point:* The material can be easily deformed.
- *Softening point:* The maximum temperature at which the material will not lose its shape. Below the softening point you can handle the material without deforming it, while above the softening point it will deform easily.
- *Annealing point:* Diffusion is fast enough that residual stresses can be easily removed by holding at that temperature.
- *Strain point:* The maximum temperature for plastic deformation. Below the strain point the material behaves as a completely brittle material with no plastic deformation.

Figure 9.7.2 shows viscosity versus temperature for various glasses, along with the viscosity corresponding to each of the thermal transitions. The next questions will show you how to use this graph.

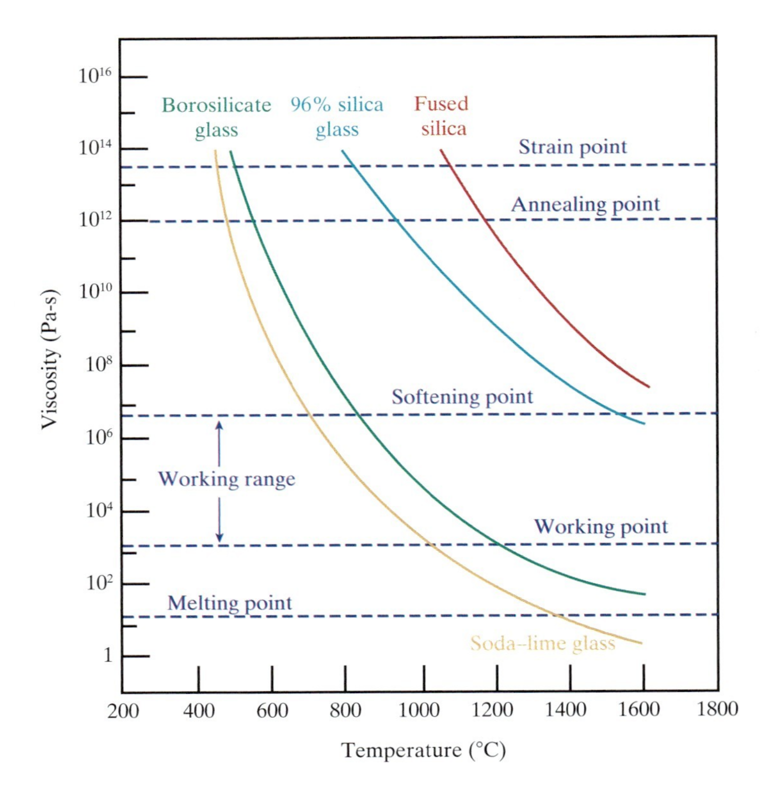

Figure 9.7.2

Thermal transitions and corresponding viscosities for several types of silica-based glass.

Guided Inquiry: Glass Structure and Properties

9.7.1 What is the viscosity at the softening point for soda–lime glass?

9.7.2 What is the viscosity at the softening point for borosilicate glass?

9.7.3 What is the viscosity at the softening point for 96% silica glass?

9.7.4 For any glass, what is the viscosity at the softening point?

9.7.5 For any glass, what is the viscosity at the working point?

9.7.6 At what temperature is the softening point for soda–lime glass?

9.7.7 At what temperature is the softening point for 96% silica glass?

9.7.8 If your answers for questions 9.7.6 and 9.7.7 are different, explain how the structure of the glass causes this difference.

Concept Check 9.7.1

- **Of the glasses shown in Figure 9.7.2, which has the highest viscosity at its softening point?**
- **Of the glasses shown in Figure 9.7.2, which has the highest temperature for its softening point?**

Because of the strong bonding present in ceramics (ionic and covalent) the properties of ceramics are different from metals. Figure 9.7.3 shows typical stress–strain curves for a ceramic and a metal. We will now use these curves to understand how the mechanical properties of ceramics differ from those of metals.

Figure 9.7.3

Typical stress–strain curves for a ceramic and a metal.

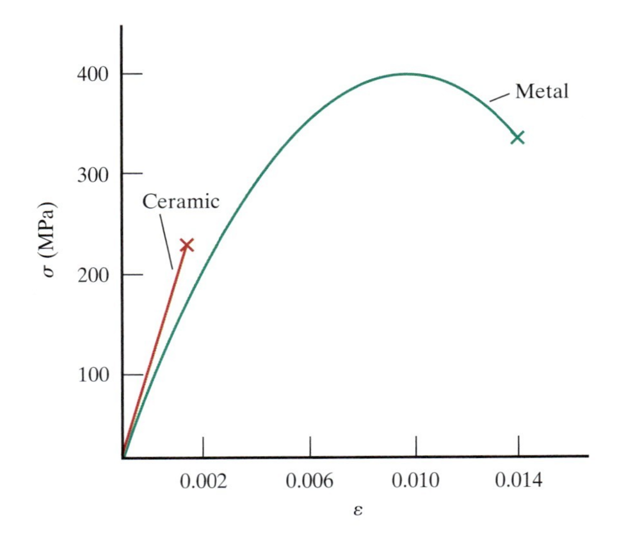

Guided Inquiry: Ceramic Properties

9.7.9 Which material has a higher modulus; ceramic or metal?

9.7.10 Which material has the higher ductility; ceramic or metal?

9.7.11 Which material has the higher tensile strength; ceramic or metal?

9.7.12 Identify the approximate yield point of the metal.

9.7.13 Does the ceramic have a yield point? Explain why or why not.

9.7.14 Describe the properties of a ceramic. Compared to a metal is it stiff or flexible? Brittle or ductile?

Concept Check 9.7.2

- **A polymer has a lower stiffness and strength but a higher ductility than either a ceramic or a metal. Add its stress–strain curve to Figure 9.7.3.**

9.8 Fracture

LEARN TO: Describe types of fracture.
Calculate critical stress and crack length for failure.

Cracks are almost always present in materials, yet quite often the material can remain in service for years without breaking. On the other hand, materials that appear to have no cracks may suddenly fail with no warning, often with catastrophic consequences. Behavior like this can make failure seem like a random event, but in fact, we can predict the point at which failure occurs. The best example of the difference is described by J. E. Gordon in his book *Structures, or Why Things Don't Fall Down* (New York: Plenum, 2003). He tells the story of a cook who noticed a crack in the floor of his ship's galley. The crack was so small that it was ignored, but the cook decided to mark how long the crack was each day. The crack steadily grew larger, until one day the ship suddenly broke in two without warning. We know about this story because the piece of deck with the cook's markings was recovered.

This story illustrates the important difference between *stable crack growth* and *unstable crack growth*. We can summarize these two types of crack growth as follows:

- *Stable crack growth:* When the crack is below a certain critical length, it will grow only if the force applied increases. This means that if the force is constant, the crack will just stay the same length.
- *Unstable crack growth:* When the crack gets to a certain critical length, it will grow spontaneously, at supersonic speeds, leading to failure.

Obviously, it is important to be able to determine when the transition from stable to unstable growth occurs. This is the realm of fracture mechanics. There are many different ways to approach fracture mechanics. We are going to stick with a simple approach, called the Griffith approach. In 1920, A. A. Griffith developed a simple approach to fracture mechanics based on energy considerations. In Figure 9.8.1, we see an elliptical crack in a material. As the crack grows, two things happen:

1. New surface is created.
2. Bond strain is released.

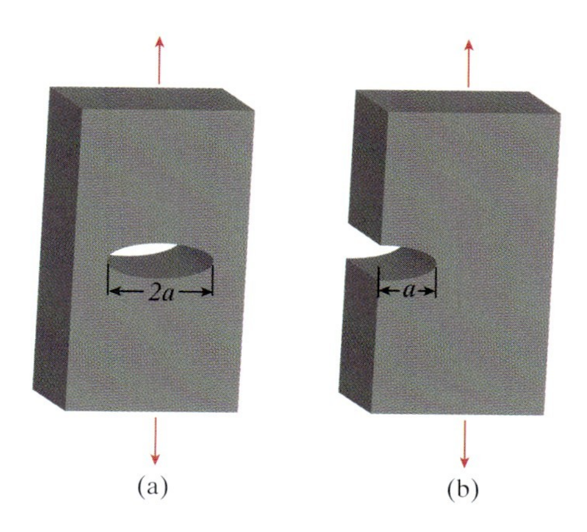

Figure 9.8.1
Elliptical cracks in materials, showing both (a) an internal crack and (b) an external crack.

The idea of bond strain needs some discussion. Figure 9.8.2 is an expanded view of the region around a crack tip. First, the presence of the sharp crack tip causes a

Figure 9.8.2

Close-up view of atoms near a crack tip. The applied stress, shown by red arrows, causes the bonds away from the crack tip to be stretched. There is no force applied at the free surface of the crack, so the bonds just above and below the crack are not stretched.

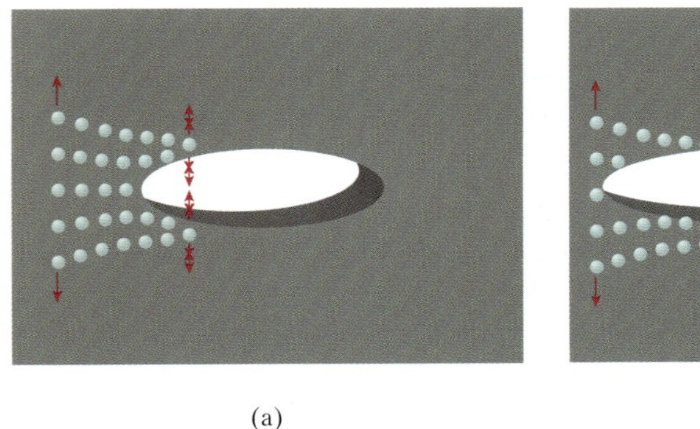

(a)

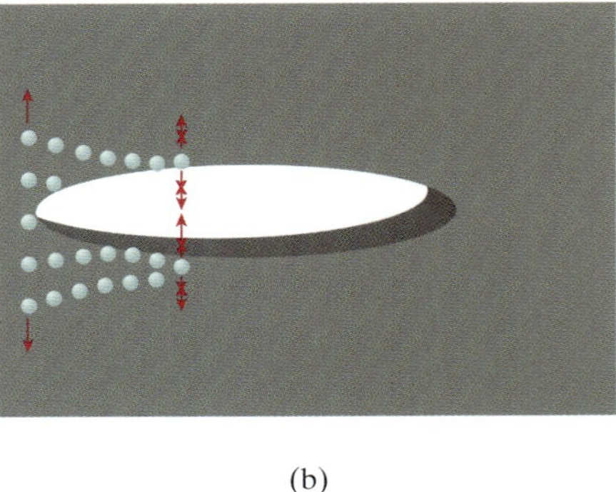

(b)

stress concentration, which is an increase in the stress around the crack tip higher than the applied stress. This stress concentration will occur anytime there is a discontinuity in the material: a crack tip, a corner, a piece of dust, etc. Figure 9.8.3 shows how the stress varies around a crack tip.

If we now look at the bonds close to the crack tip, we will see that they are stretched due to the application of the force. However, right at the surface of the crack they cannot be stretched because simple statics indicates there cannot be any force on those atoms (see Figure 9.8.2); no force is being applied on the free surface of the crack, so there can't be any force acting on those atoms. As a result, atomic bonds near the crack surface aren't stretched, and as the crack grows, more bonds become unstretched. Putting all of this information together, you will now see how it relates to whether a crack will be stable or will be unstable and cause catastrophic failure.

Figure 9.8.3

Stress concentration near a crack tip. The applied stress is σ_0. The stress felt in the material increases to a maximum of σ_m at the crack tip.

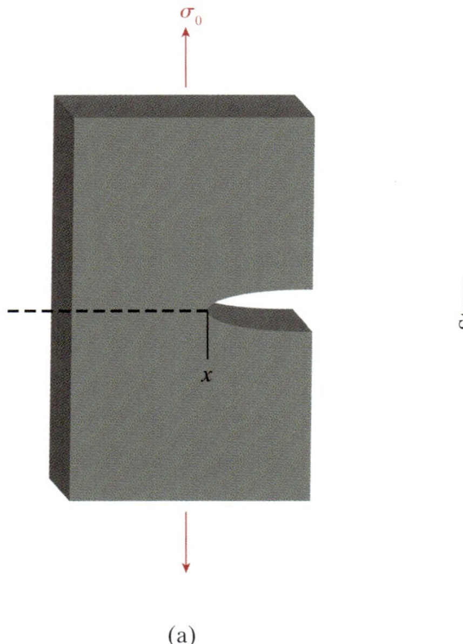

(a)

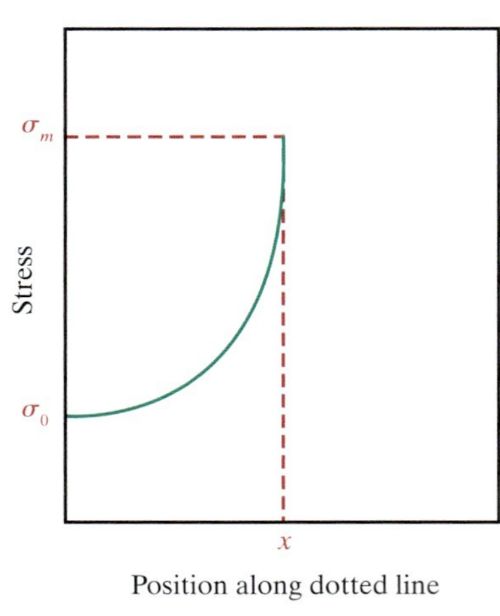

(b)

Guided Inquiry: Energy of Crack Growth

9.8.1 When a bond is stretched, is the energy of the bond increased or decreased?
Higher energy corresponds to a less favorable situation, lower energy to a more favorable situation.

9.8.2 As a crack grows, are more or fewer bonds stretched?

9.8.3 As a crack grows, does the energy from bond stretching increase or decrease?

9.8.4 When surface is created, does the energy of the material increase or decrease?
Surfaces and interfaces are always unfavorable because the atoms at the surface or interface are not bonded the way they want to be.

9.8.5 As a crack grows, is more or less surface created?

9.8.6 As the crack grows, does the energy from surface formation increase or decrease?

9.8.7 As a crack grows, if the magnitude of the energy change from creating surfaces is greater than the magnitude of the energy change from releasing bond strain, will the energy go up or down?
Think carefully about this one! Remember, increasing surface is unfavorable, while releasing bond strain is favorable.

9.8.8 For the situation in question 9.8.7, will the crack grow spontaneously? Explain your answer.

9.8.9 As a crack grows, if the magnitude of the energy change from creating surfaces is less than the magnitude of the energy change from releasing bond strain, will the energy go up or down?

9.8.10 For the situation in question 9.8.9, will the crack grow spontaneously? Explain your answer.

9.8.11 Based on your answers to questions 9.8.8 and 9.8.10, explain the difference between stable and unstable crack growth.

Concept Check 9.8.1

- **When stable crack growth occurs, is the energy of the material increasing or decreasing?**
- **When unstable crack growth occurs, is the energy of the material increasing or decreasing?**

Before we can go any further we need to define the length of a crack. In fracture mechanics we use the symbol a in the equations for failure. The variable a is related to the crack length, but the definition of a depends on whether the crack is internal or external. This is because a is the length of half of an ellipse. If the crack is external, a is equal to the crack length. If the crack is internal, a is equal to half the crack length (see Figure 9.8.1). So overall, if the measured crack length is L, then we can say

$$a = L \text{ (external crack)}$$

$$a = \frac{L}{2} \text{ (internal crack)} \tag{9.8.1}$$

It is important to keep these definitions in mind when doing calculations.

There are several ways to use fracture mechanics to determine whether a material will fail. The most straightforward would be to calculate the fracture stress. This can be done with the following equation:

$$\sigma_f = \sqrt{\frac{2E\gamma}{\pi a}} \tag{9.8.2}$$

where E is the modulus and γ is the surface energy plus any energy absorbed due to plastic deformation. Although this might seem like a simple way to determine if a material will fail, it is actually not very useful because we don't have a good way of figuring out values for γ. The one good thing about this equation is that it shows you the relationship between the crack size and the failure stress. As the crack gets bigger, the failure stress gets smaller.

Another way to determine if a material will fail is with the strain energy release rate, G_c. The strain energy release rate is defined as 2γ, and has units of J/m^2. We can use G_c to determine failure according to

$$\frac{\pi\sigma^2 a}{E} \geq G_c \rightarrow \text{unstable growth} \tag{9.8.3}$$

where σ is the applied stress, a is related to the crack length as defined above, and E is the modulus.

We can also use the *critical stress intensity parameter*, K_{1c}, also called the fracture toughness.

The first subscript on K is the Roman numeral one, not the letter "ell."

K_{1c} is defined as $2E\gamma_{eff}$, where γ_{eff} includes both surface energy and plastic deformation. K_{1c} has units of $MPa\text{-}m^{1/2}$. We can use K_{1c} to determine failure according to

$$Y\sigma\sqrt{\pi a} \geq K_{1c} \rightarrow \text{unstable growth} \tag{9.8.4}$$

where Y is a geometric parameter, σ is the applied stress, and a is related to the crack length as defined above. You will now see how to use this equation.

Guided Inquiry: Fracture Calculations

9.8.12 Without doing any calculations, describe an overall strategy for solving the following problem:

A piece of 4340 steel is subjected to a stress of 1,000 MPa and has a critical stress intensity parameter of 45 MPa-$m^{1/2}$. If the largest crack present is 1.0 mm, will the steel fail? You may assume that the geometric parameter Y is equal to 1.0.

I want you to understand how you would approach this problem before you just jump in and start calculating numbers.

9.8.13 Without doing any calculations, describe an overall strategy for solving the following problem:

A piece of 4340 steel is subjected to a stress of 1,500 MPa and has a critical stress intensity parameter of 45 MPa-$m^{1/2}$. What is the largest internal crack that can be present without the material failing? You may assume that the geometric parameter Y is equal to 1.0.

9.8.14 Without doing any calculations, describe an overall strategy for solving the following problem:

The largest internal cracks present in a piece of 4340 steel are 0.5 mm long; you may assume there are no external cracks. The critical stress intensity parameter for this material is 45 MPa-$m^{1/2}$. What is the maximum stress that can be applied without the material failing? You may assume that the geometric parameter Y is equal to 1.0.

9.8.15 Solve the problem of question 9.8.13.

Concept Check 9.8.2

- **Two pieces of material are identical, except that one has 0.5-mm internal cracks and one has 1.0-mm internal cracks. Which fails more easily?**
- **What is the answer for question 9.8.15?**

EXAMPLE PROBLEM 9.8.1

A plate of a polymer fails at a stress of 150 MPa. What is the largest internal crack length present on this plate? You can assume there are no surface cracks. The fracture toughness is 5.0 MPa-m$^{1/2}$ and $Y = 1.2$.

$$K_{Ic} = Y\sigma_f\sqrt{\pi a}$$

Values for the variables are as follows:

$K_{1c} = 5.0$ MPa-m$^{1/2}$

$Y = 1.2$

$\sigma_f = 150$ MPa

$a = ?$

We can rearrange the equation to solve for a:

$$a = \frac{1}{\pi}\left(\frac{K_{1c}}{Y\sigma_f}\right)^2 = \frac{1}{\pi}\left(\frac{5\ \text{MPa} - \text{m}^{1/2}}{(1.2)\,(150\,\text{MPa})}\right)^2 = 0.00025\,\text{m} = 0.25\,\text{mm}$$

Since this is an internal crack, the crack length is $2a$, so the largest internal crack is 0.5 mm long.

9.9 Fatigue

LEARN TO: Calculate time and stress amplitude for failure.

Have you ever taken a paperclip and bent it back and forth until it broke? That is an example of fatigue. Let's look at what is happening more closely. If you just bend the paperclip, it won't break. But as you bend it back and forth, it becomes weaker and weaker, until it just falls apart. In some ways this failure process is like the failure you learned about in Section 9.8. Cracks form at some flaw where there is a stress concentration, and the crack grows slowly until it reaches a critical size. There are, however, some important differences. The stresses that cause fatigue failure are much lower than what you would calculate using the approach in Section 9.8. This is because the crack grows a little bit each time the paperclip is bent. Fatigue failure is also brittle, even in materials that show plastic deformation under constant-elongation-rate tests (such as for a stress–strain curve). This means that fatigue failures often occur without warning.

Fatigue occurs whenever there is a cyclic load. Obvious places for it to occur would be wherever there is vibration. Another example is a bicycle pedal shaft; the stress on the shaft changes direction as you pedal the bicycle and can ultimately lead to fatigue failure. The same type of failure occurs in car axles.

Figures 9.9.1 and 9.9.2 show different aspects of fatigue. Figure 9.9.1 shows how the stress changes as a function of time for the simple case of a sinusoidally changing stress. Figure 9.9.2 is called an *S–N* curve, which can be used to predict fatigue failure. The rest of this section will give you practice interpreting these different graphs.

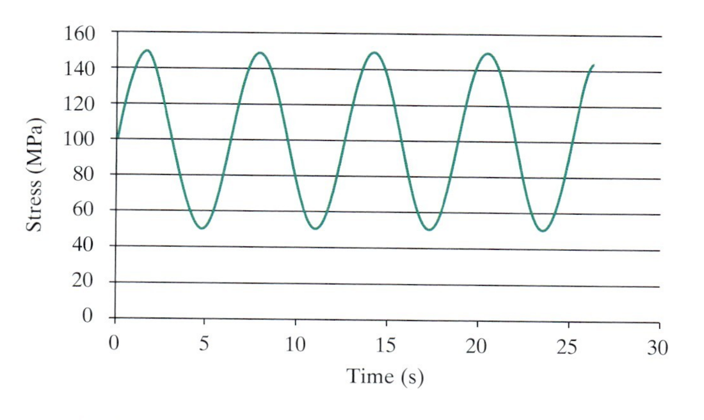

Figure 9.9.1

Stress versus time for a cyclic applied stress.

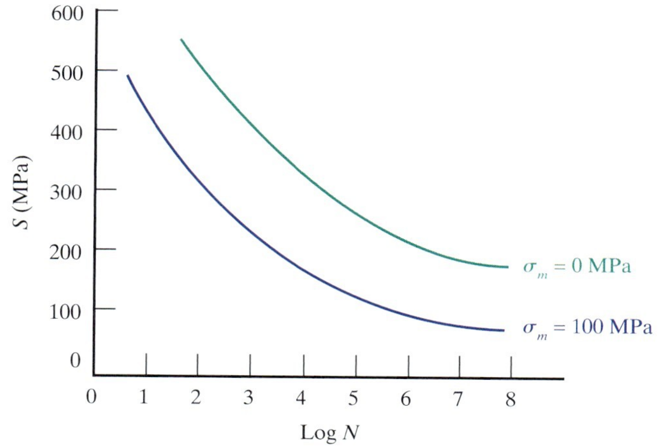

Figure 9.9.2

S–N curve for two different maximum stress levels applied to the same material.

Guided Inquiry: Cyclic Stress

9.9.1 What is the average stress for the cyclic stress shown in Figure 9.9.1?

9.9.2 What is the maximum stress for the cyclic stress shown in Figure 9.9.1?

9.9.3 What is the amplitude of the stress for the cyclic stress shown in Figure 9.9.1?

9.9.4 How long is one cycle for the cyclic stress shown in Figure 9.9.1?

9.9.5 How many cycles occur in 18 seconds for the cyclic stress shown in Figure 9.9.1?

To answer the next questions you need the following definitions:

- *Average stress* = σ_m
- *Stress amplitude* = S
- *Number of cycles* = N
- *Fatigue strength* = stress amplitude that causes failure
- *Fatigue life* = number of cycles that causes failure

On an *S–N* curve, the line indicates the combinations of fatigue strength and fatigue life that cause failure.

Guided Inquiry: Fatigue Curves

9.9.6 In Figure 9.9.2, for a mean stress of 0 MPa and a fatigue life of 100,000 cycles, what is the fatigue strength?

9.9.7 In Figure 9.9.2, for a mean stress of 0 MPa and a fatigue strength of 300 MPa, what is the fatigue life?

9.9.8 In Figure 9.9.2, if the stress is applied at a frequency of 10 Hz, how long will it take for the material to fail at a mean stress of 0 MPa and a fatigue strength of 300 MPa?

9.9.9 For a given fatigue life, which mean stress level has the higher fatigue strength; 0 MPa or 100 MPa?

9.9.10 For a given fatigue strength, which mean stress level has the higher fatigue life; 0 MPa or 100 MPa?

9.9.11 Is a material more susceptible to fatigue at higher or lower mean stress levels?

Concept Check 9.9.1

- **In Figure 9.9.2, if the stress is applied at a frequency of 10 Hz, the material fails in 16 minutes and 40 seconds. What is the stress amplitude that caused that failure?**
- **A material is subjected to fatigue. In which case will it last longer; a mean stress of 50 MPa or a mean stress of 80 MPa?**

EXAMPLE PROBLEM 9.9.1

A part made of 2024-T6 aluminum is subject to a cyclic stress at a frequency of 10 Hz. If the part must last at least 280 hours before failing, what is the maximum stress amplitude allowed for the cyclic stress? See data in Figure 9.9.3.

First we need to figure out how many cycles this amount of time is:

$$10 \text{ Hz} = 10 \text{ cycles/s}$$

$$280 \text{ hours} = 1{,}008{,}000 \text{ s}$$

$$1{,}008{,}000 \text{ s} \times 10 \text{ cycles/s} = 10{,}080{,}000 \text{ cycles}$$

From Figure 9.9.3, for 10^7 cycles, failure occurs at a stress amplitude of 175 MPa. Therefore, the maximum stress amplitude allowed is 175 MPa.

There is one more point to make about fatigue. Ferrous alloys (metal alloys based on iron, like steel) have something called an endurance limit. Figure 9.9.3 shows a comparison between a ferrous alloy with an endurance limit and a nonferrous alloy without an endurance limit. An endurance limit is a limiting fatigue strength for failure. As long as the stress amplitude stays below the endurance limit, the material will never fail by fatigue. In contrast, nonferrous alloys don't have an endurance limit. They will always fail in fatigue, although if the stress amplitude is low enough, the fatigue lifetime may be longer than you ever need to worry about.

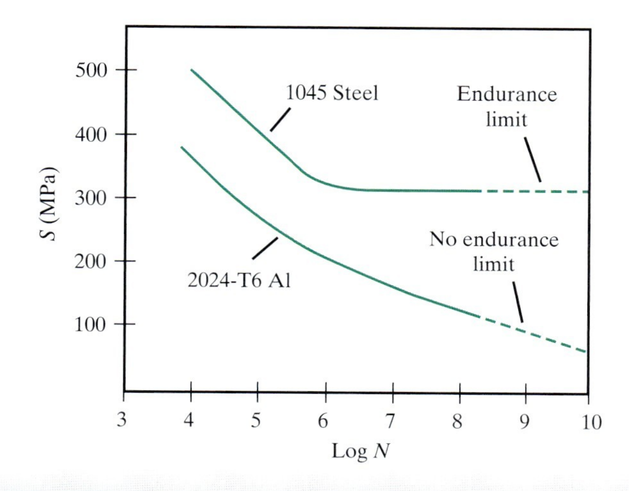

Figure 9.9.3

Comparison of S–N curves for ferrous and nonferrous alloys.

9.10 Hardness

LEARN TO: Describe hardness tests.
Calculate hardness from experimental data.
Correlate hardness to tensile strength.

The measurement of mechanical properties using a stress–strain test as described in Section 9.1 is a destructive test, meaning that the sample is changed or destroyed during the measurement. Sometimes, however, we would like to have a nondestructive test available to determine the strength of a material. For example, quality assurance in a manufacturing plant relies on being able to measure the properties of a product using nondestructive tests. One type of measurement that can serve this purpose is the *hardness* test. A schematic of various types of hardness tests are shown in Table 9.10.1. For example, in the Brinell test a steel ball is pushed into the surface of the material, using a fixed force. The resulting indentation has a diameter which depends on how far the steel ball penetrated into the surface. The hardness can then be calculated from the load, the diameter of the steel ball, and the diameter of the indentation.

TABLE 9.10.1 Types of hardness tests

Test	Indenter	Shape of indentation: Side view	Shape of indentation: Top view	Load	Formula for hardness number
Brinell	10 mm sphere of steel or tungsten carbide	D, d	d	P	$\text{BHN} = \dfrac{2P}{\pi D\left[D - \sqrt{D^2 - d^2}\right]}$
Vickers	Diamond pyramid	136	d_1, d_1	P	$\text{VHN} = 1.72P/d_1^2$
Knoop microhardness	Diamond pyramid	t; $l/b = 7.11$; $b/t = 4.00$	b, l	P	$\text{KHN} = 14.2P/l^2$
Rockwell					
A, C, D	Diamond cone	120; t		60 kg, 150 kg, 100 kg	$R_A =$, $R_C =$, $R_D =$ } $100 - 500t$
B, F, G	1/16 in. diameter steel sphere	t		100 kg, 60 kg, 150 kg	$R_B =$, $R_F =$, $R_G =$ } $130 - 500t$
E, H	1/8 in. diameter steel sphere			100 kg, 60 kg	$R_E =$, $R_H =$ } $130 - 500t$

Units: *Brinell:* P (kg), D (mm), d (mm); *Vickers:* P (kg), d (mm); *Knoop:* P (g), l (mm); *Rockwell:* t (measured in units of 0.002 mm); t is the difference between the depth of the indentation from an initial minor load, usually 10 kg, and the major load indicated in the table.
Source: H. W. Hayden, W. G. Moffatt, and J. Wulff, *The Structure and Properties of Materials*, Vol. 3: *Mechanical Behavior*, John Wiley & Sons, Inc., NY, 1965.

Brinell hardness is just one type of hardness. There are many others that use different geometries and hardness calculations. Table 9.10.1 summarizes the different types of hardness. As you will see in the questions that follow, hardness is a very useful way for comparing the properties of different materials.

TABLE 9.10.2 Hardness values of some carbon steels

Material	BHN	VHN	KHN	Tensile strength (MPa)
1040 steel, rolled	201	211	223	620
1040 steel, annealed	149	155	169	515
1040 steel, quenched	217	228	240	722

Guided Inquiry: Hardness

9.10.1 Rank the materials in Table 9.10.2 from lowest to highest Brinell hardness.

9.10.2 Rank the materials in Table 9.10.2 from lowest to highest Vickers hardness.

9.10.3 Rank the materials in Table 9.10.2 from lowest to highest Knoop hardness.

9.10.4 Rank the materials in Table 9.10.2 from lowest to highest tensile strength.

9.10.5 What is the relationship between hardness and tensile strength?

9.10.6 If you were asked to compare the tensile strengths of two materials using a nondestructive test, how would you do it?

9.10.7 Explain on the basis of microstructure, your rankings in questions 9.10.1–9.10.3. *Remember the different microstructures of steel and how those relate to strength.*

Concept Check 9.10.1

- **Aluminum alloy 2024-T4 has a Brinell hardness of 120, while aluminum alloy 2117-T4 has a Brinell hardness of 70. Which alloy is more ductile?**

As you have just seen, a higher hardness means the material has a higher tensile strength. Various mathematical equations have been developed which allow you to calculate the tensile strength if you know the hardness. These equations are empirical, which means they are developed based on doing the measurements and aren't derived from any underlying theory. This means that the equations are different for different hardness scales, and even for different classes of alloys: there is one equation for Brinell hardness for steels, another for Brinell hardness for aluminum, another for Vickers hardness for aluminum, etc. But hardness can still be a very useful measurement and is used commonly.

9.11 Viscoelasticity

LEARN TO: **Predict the properties of viscoelastic materials.**

So far we have considered only the elastic and plastic behavior of materials. But polymers can also behave in a unique way that is called *viscoelastic*. To understand this behavior we first need to understand what we mean when we say something is *elastic* or *viscous*. You may be familiar with these terms but not in the way they are meant here. When you say that something is elastic you probably mean that it is stretchy, like a rubber band. And when you say something is viscous you probably mean that it flows very slowly, like molasses. Put those definitions aside, because that is not what we mean here. There are specific definitions of elastic and viscous related to how materials respond to stress and strain. Under these definitions window glass is elastic and water is viscous. The behaviors of these types of materials are shown in Figure 9.11.1. The next set of questions will help you understand what we mean when we say a material is elastic.

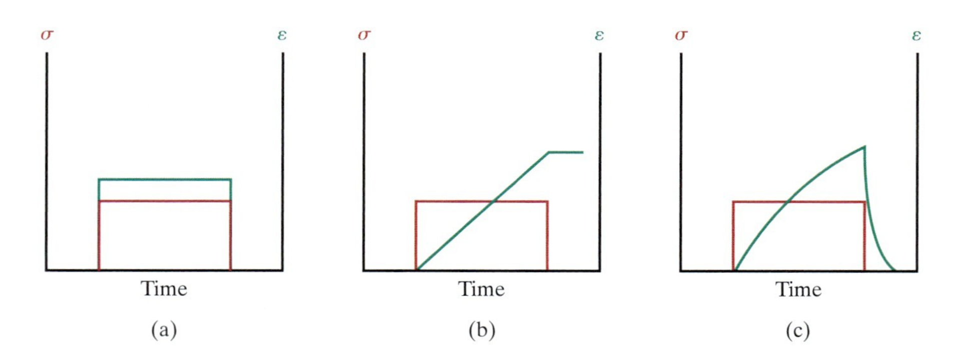

Figure 9.11.1
Strain versus time under constant applied stress for different types of materials: (a) elastic; (b) viscous; (c) viscoelastic.

Guided Inquiry: Elastic Materials

9.11.1 What is Hooke's law for stress and strain?
See Section 9.1.

9.11.2 For a material with a modulus of 1.3 GPa, if a stress of 50 MPa is applied, what will be the strain?

9.11.3 Based on Hooke's law, if the stress does not change, will the strain change?

9.11.4 Based on Hooke's law, if the stress doubles, what will happen to the strain?

9.11.5 Based on Hooke's law, if stress applied to an elastic material is constant, will the strain change over time?

9.11.6 Based on Hooke's law, if I want to increase the strain in an elastic material, what has to happen to the applied stress?

Concept Check 9.11.1

- **A stress is applied to an elastic material, held for 5 minutes, and then the applied stress is cut in half. What happens to the strain when the stress is reduced?**

The behavior of an elastic material, as you just figured out, is shown in Figure 9.11.1(a). In addition to a strain that doesn't change with time, if the stress is released in an elastic material, the strain goes back to zero, and it does this instantly. We can think of an elastic material as behaving like a spring.

Now let's consider a viscous material. In this situation, viscous does not mean it is resistant to flowing, as we normally think about. For materials, viscous means that it behaves as an ideal fluid. Water is an example of a viscous material. If you apply a stress to water, what happens to it? It flows. If we think about this flow as a deformation, what we see is that even under constant stress, the strain continues to increase over time. Further, if we remove the stress, the water doesn't flow back to where it was. This means the strain that occurred during flow does not recover. This behavior is shown in Figure 9.11.1(b).

A viscoelastic polymer has behavior that is partially elastic and partially viscous. This means that the strain will increase over time, even under constant stress, but if the stress is removed, the strain will go back to zero. However, it doesn't go back to zero instantly. This behavior is shown in Figure 9.11.1(c). You will now compare the behavior of these different types of materials.

Guided Inquiry: Elastic vs. Viscoelastic Materials

9.11.7 How does the strain change with time under constant applied stress for an elastic material?

9.11.8 How does the strain change with time under constant applied stress for a viscous material?

9.11.9 How does the strain change with time under constant applied stress for a viscoelastic material?

9.11.10 For an elastic material with a modulus of 1.3 GPa, if a stress of 50 MPa is applied, what will be the strain immediately afterward?
Remember Hooke's Law.

9.11.11 For the elastic material in question 9.11.10, what will be the strain if you apply the stress and wait for 5 minutes?

9.11.12 If the material in question 9.11.10 were viscoelastic, how would the strain right after the stress was applied compare to the strain 5 minutes later?

9.11.13 Based on your answers to questions 9.11.11 and 9.11.12, what happens to the modulus over time for elastic materials? For viscoelastic materials?

Concept Check 9.11.2

- **Draw a plot of modulus versus time for a viscoelastic material under constant applied stress.**

Polymers are viscoelastic because of chain motion. Long polymer molecules take time to move, so they can't react instantly as the individual atoms in a metal would. If the chains can move more, the polymer behaves more like a fluid and less like a solid. In the next questions you will explore these structure–property relationships.

Guided Inquiry: Mechanism of Viscoelasticity

9.11.14 Are polymer chains more mobile above T_g or below T_g?

9.11.15 Will a polymer be more viscous or more elastic above T_g? What about below T_g?
Think about the mobility of the chains. Does greater mobility mean it acts more like a liquid or more like a solid?

9.11.16 At constant stress, will the strain of a polymer increase more quickly above T_g or below T_g?

9.11.17 Does crosslinking a polymer make the chains more mobile or less mobile?
If you don't remember what a crosslink is, go back to Section 5.1.

9.11.18 Will a crosslinked polymer be more elastic or more viscous than a linear polymer?

9.11.19 At constant stress, will the strain of a crosslinked polymer increase more or less quickly than a linear polymer?

Concept Check 9.11.3

- **A constant stress is applied to a viscoelastic material at a temperature above T_g and a temperature below T_g. At which temperature will the strain be greater 10 minutes after applying the stress?**

9.12 Composites

LEARN TO: List types of composites.
Calculate composite properties.

It is very rare for a pure material to be used in an engineering application. In practice, no single material has all the properties we might need, so we use combinations of materials that combine their desired properties. The most common type of combination is a *composite*. A composite is simply a combination of two or more different materials. It is different from a mixture in that mixtures obey the rules for phase diagrams and kinetics that we saw in Chapters 6 and 8. Instead, composites are usually just physical mixtures in which each component might have dimensions of a few microns to millimeters. You may also have heard of nanocomposites. These are composites in which at least one component has dimensions of nanometers. Nanocomposites are interesting because they have unusual properties that often disobey the rules of conventional composites.

We usually distinguish between two components of a composite, the *reinforcement* and the *matrix*. The reinforcement is generally a fiber or a particle, which is embedded in the matrix. Usually the reinforcement adds strength or stiffness to the matrix, while the matrix acts as the glue that holds the reinforcement together. There are many different kinds of composites. Some examples are as follows:

- *Fiber-reinforced polymer composites:* This is what most people think of when they hear the word "composite." They consist of some kind of fiber reinforcement embedded in a polymer matrix. They are often used in aerospace, marine, and sports applications. Composite bicycle frames are carbon fibers embedded in an epoxy matrix.

Figure 9.12.1
Types of fiber-reinforced composites.

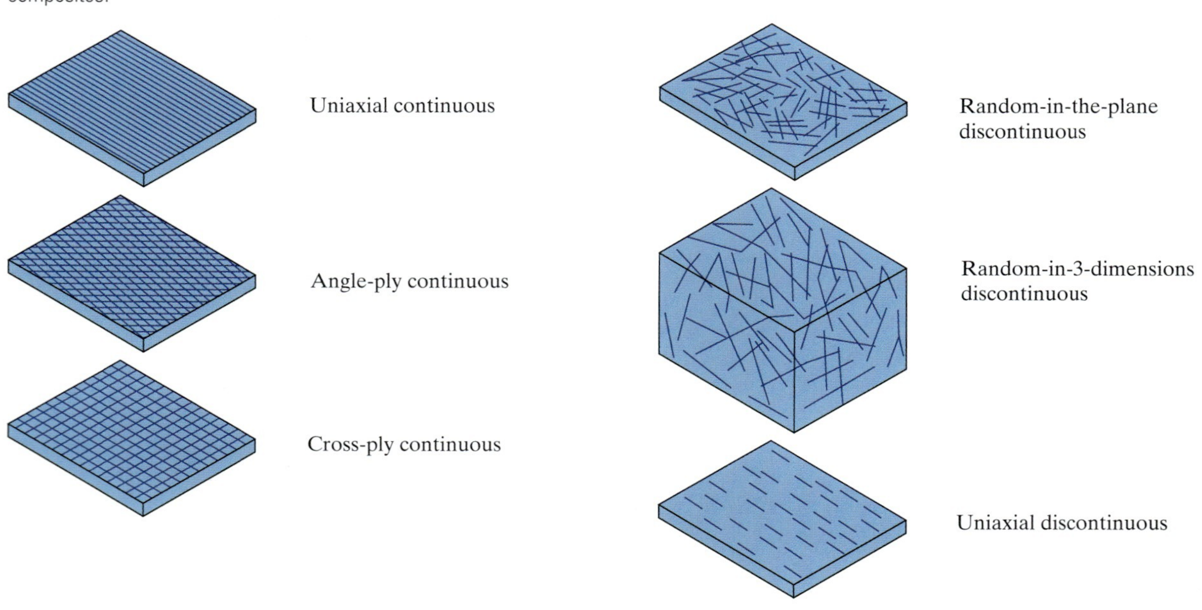

- *Rubber:* Most rubbers have some kind of reinforcement added to improve stiffness and durability. For example, car tires are made out of rubber with carbon black particulate reinforcement. Without the carbon black, rubber has almost no abrasion resistance and it could not be used for tires. Adding carbon black increases abrasion resistance and stiffness.
- *Cermets:* Cermets are composites of ceramics and metals. The ceramic provides high strength and stiffness, while the metal provides some ductility. Their main application is in cutting tools, such as drill bits for oil well drills.
- *Concrete:* Concrete is a composite of cement and an aggregate, such as rocks and sand. Cement alone has poor mechanical properties. Adding the aggregate makes concrete very strong in compression, although it is still weak in tension.

Figure 9.12.1 illustrates possible arrangements for fiber-reinforced composites. In particulate composites, the particles are just randomly arranged within the matrix. Tables 9.12.1 and 9.12.2 list many of the common matrix and reinforcement materials.

TABLE 9.12.1 Composite matrix materials

Material	Uses	Advantages	Disadvantages
Thermoset polymers	Continuous fiber composites Epoxies: aerospace Polyesters and vinyl esters: automotive, marine, household Polyimides: high-temperature aerospace	Low viscosity allows for easy infiltration of fibers	Limited shelf life Longer processing times to fully cure the resin
Thermoplastic polymers	Discontinuous fiber composites	Unlimited shelf life Shorter processing times	High viscosity
Metals	High-temperature composites	Higher use temperature than polymers	Expensive High-temperature processing
Ceramics	High-temperature and corrosive environments	High-temperature stability Good chemical stability	More brittle; fiber is added to enhance toughness

TABLE 9.12.2 Composite fiber materials

Material	Uses	Advantages	Disadvantages
Glass	All	Low cost Good insulator	Low modulus High density Susceptible to water and bases
Carbon	Aerospace	High strength and modulus Low density Electrically conductive	Expensive
Aramid	Aerospace Marine	High strength-to-weight ratio	Poor compressive strength Difficult to cut
Polyethylene	Armor and helmets Marine	Low density High strength-to-weight ratio	Low use temperature Poor adhesion to matrix Difficult to cut
Boron	Aerospace	High modulus Composites with high compressive strength	Expensive
Silicon carbide and alumina	Used with metal and ceramic matrices	High-temperature use	Expensive

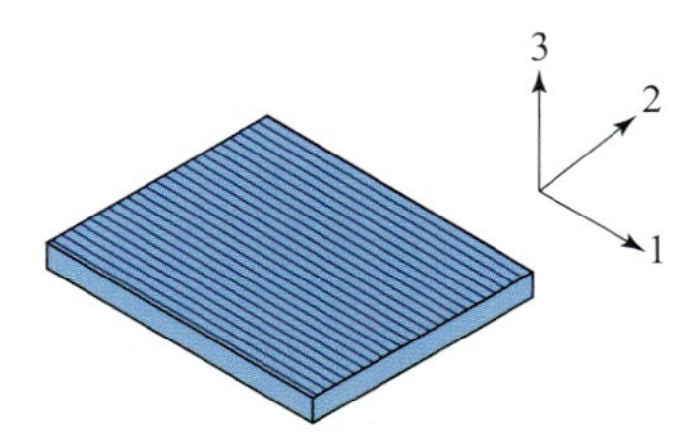

Figure 9.12.2
Definitions of the directions in an aligned continuous fiber composite.

The properties of a composite are a combination of the properties of the reinforcement and the matrix. We will limit ourselves to calculating the modulus and start with an aligned continuous fiber composite, as shown in Figure 9.12.2. The modulus in the 1-direction, along the fiber axis, is given by

$$E_1 = \phi_m E_m + \phi_f E_f = E_{\text{max}} \quad (9.12.1)$$

where ϕ is the volume fraction, E is the modulus, and the subscripts m and f refer to the matrix and the fiber. E_1 is often called the *longitudinal modulus*. As shown in the equation, this is also the maximum possible modulus we can get for a combination of two materials; we are going to use E_{max} in some equations that are coming up.

The modulus in the 2-direction, perpendicular to the fibers, is given by

$$\frac{1}{E_2} = \frac{\phi_m}{E_m} + \frac{\phi_f}{E_f} \quad (9.12.2)$$

E_2 is often called the *transverse modulus*. E_1 and E_2 also represent the upper and lower limits of modulus for particulate composites. Measured values for particulate composites usually fall in between these values.

For a discontinuous aligned composite, E_1 is calculated according to

$$E_1 = \phi_m E_m + \eta_1 \phi_f E_f \quad (9.12.3)$$

where η_1 is the fiber length correction factor, which accounts for having a short fiber. Values for this correction factor can be calculated, but for our purposes, we will note that a fiber with a larger aspect ratio—that is, length-to-diameter ratio—has a larger correction factor and thus provides better reinforcement.

We can use E_{max} to approximate the modulus of other fiber-reinforced composites shown in Figure 9.12.1. For a random-in-the-plane composite

$$E_1 = E_2 \approx \frac{3}{8} E_{\text{max}} \quad (9.12.4)$$

and for a random-in-3-dimensions composite

$$E_1 = E_2 = E_3 \approx \frac{1}{5} E_{\text{max}} \quad (9.12.5)$$

You will now use these equations to compare the properties of different types of composites.

Guided Inquiry: Composite Properties

9.12.1 For an aligned continuous fiber composite, which is higher; the longitudinal modulus or the transverse modulus?

9.12.2 Which has a higher value of E_2; an aligned continuous fiber composite or a cross-ply continuous fiber composite?

Look at the arrangement of the fibers. Composites are always stiffer when you pull in the direction of fibers.

9.12.3 Which has a higher modulus; a cross-ply continuous fiber composite or a random-in-3-dimensions composite?

9.12.4 In general, which has a higher modulus; a continuous fiber or discontinuous fiber composite?

9.12.5 What is the advantage of a discontinuous fiber composite compared to a continuous fiber composite?

9.12.6 You are asked to create a uniaxial continuous fiber composite with epoxy matrix and Kevlar fiber. If the longitudinal modulus must be at least 100 GPa, what volume fraction of fiber is needed?

Epoxy modulus = 2.41 GPa

Kevlar modulus = 131 GPa

What is the relationship between ϕ_m and ϕ_f?

Concept Check 9.12.1

- **Rank the following composites in order from highest E_1 to lowest E_1; uniaxial continuous fiber, cross-ply continuous fiber, random-in-3-dimensions discontinuous fiber.**
- **What is the answer for question 9.12.6?**

EXAMPLE PROBLEM 9.12.1

A continuous carbon fiber/epoxy composite is required to have a longitudinal modulus of 150 GPa.

a. What volume fraction of glass fiber must the composite have?
b. What will be the transverse modulus of the resulting composite?

The modulus of epoxy is 2.41 GPa and the modulus of carbon fiber is 230 GPa.

a. We start with Equation (9.12.1):

$$E_1 = \phi_f E_f + \phi_m E_m$$

Since the only things present are the epoxy and the carbon fiber, the sum of their volume fractions equals one, and this equation can be modified to

$$E_1 = \phi_f E_f + (1 - \phi_f) E_m$$

The values for the variable in this equation are as follows:

$E_1 = 150$ GPa

$E_f = 230$ GPa

$E_m = 2.41$ GPa

$\phi_f = ?$

Rearranging the above equation to solve for ϕ_f and plugging in the known values gives the following:

$$\phi_f = \frac{E_1 - E_m}{E_f - E_m} = \frac{150 \text{ GPa} - 2.41 \text{ GPa}}{230 \text{ GPa} - 2.41 \text{ GPa}} = 0.65$$

b. To calculate the transverse modulus we use Equation (9.12.2):

$$\frac{1}{E_2} = \frac{\phi_f}{E_f} + \frac{\phi_m}{E_m}$$

The values of the variables are as follows:

$E_f = 230$ GPa

$E_m = 2.41$ GPa

$\phi_f = 0.65$

$\phi_m = 0.35$

$E_2 = ?$

Rearranging Equation (9.12.2), and plugging in the known values gives the following:

$$E_2 = \left(\frac{\phi_f}{E_f} + \frac{\phi_m}{E_m}\right)^{-1} = \left(\frac{0.65}{230 \text{ GPa}} + \frac{0.35}{2.41 \text{ GPa}}\right)^{-1} = 6.75 \text{GPa}$$

Summary

Mechanical properties are probably the most common properties engineers deal with because so much of what engineers do is structural. As you have seen in this chapter, many different types of properties can be considered: tensile properties (as shown on stress–strain curves), fracture behavior, fatigue, hardness, and viscoelasticity. That's not all; even though this chapter is one of the longest ones in this book, we have barely scratched the surface of the types of properties that can be measured. The American Society for Testing and Materials (ASTM) has many volumes of standard test techniques for mechanical properties. Also, quite often there are specialized tests for particular products. For example, a test for garbage cans includes dropping one from a height of 6 feet to simulate being dropped off a porch, squeezing it 520 times to simulate being grabbed by a garbage truck once a week for 10 years, and similar tests.

While many other types of engineers deal with mechanical properties of materials, this chapter also illustrates what is different about materials engineering. Rather than looking up properties in a book and accepting those values, materials engineers control the properties of materials through the structure–property–processing relationships. We know how to adjust the strength of a metal by changing the grain size or annealing; we know how to adjust the stiffness of a polymer by changing the crystallinity. Through the MSE triangle, materials engineers can create new materials that meet the needs of other engineers.

Key Terms

Annealing
Composite
Continuous fiber composite
Critical resolved shear stress
Critical stress intensity parameter
Discontinuous fiber composite
Ductility
Elastic
Fatigue life
Fatigue strength
Fracture toughness
Hardness
Hooke's law
Longitudinal modulus
Matrix
Modulus
Network former
Network modifier
Plastic
Reinforcement
Slip
Stable crack growth
Strain
Stress
Transverse modulus
Ultimate tensile strength
Unstable crack growth
Viscoelasticity
Viscous
Yield strain
Yield stress

Problems

Skill Problems

9.1 A cylinder of an aluminum alloy with a radius of 2 cm and a length of 10 cm is subjected to a force of 40 kN. After this force is applied, the length of the cylinder becomes 10.005 cm.

a. Calculate the stress and strain on this cylinder.

b. Assuming that the deformation is purely elastic, calculate the modulus of aluminum.

9.2 A cylinder of PET with a radius of 0.5 cm and a length of 10 cm is subjected to a force of 4 kN. After this force is applied, the length of the cylinder becomes 10.2 cm.

a. Calculate the stress and strain on this cylinder.

b. Assuming that the deformation is purely elastic, calculate the modulus of PET.

9.3 The modulus of high-density polyethylene is 1.1 GPa. A cylinder of this material, with a radius of 1 cm and a length of 8 cm, is subjected to a force of 6 kN. Assuming that the deformation is purely elastic, what is the length of the cylinder after the force is applied?

9.4 The graph below gives a plot of force versus total sample length in a tensile test of a polymer. The sample is cylindrical, with an initial diameter of 1 cm and an initial length of 10 cm. From the data below, calculate the modulus, yield stress, yield strain, tensile strength, and % elongation.

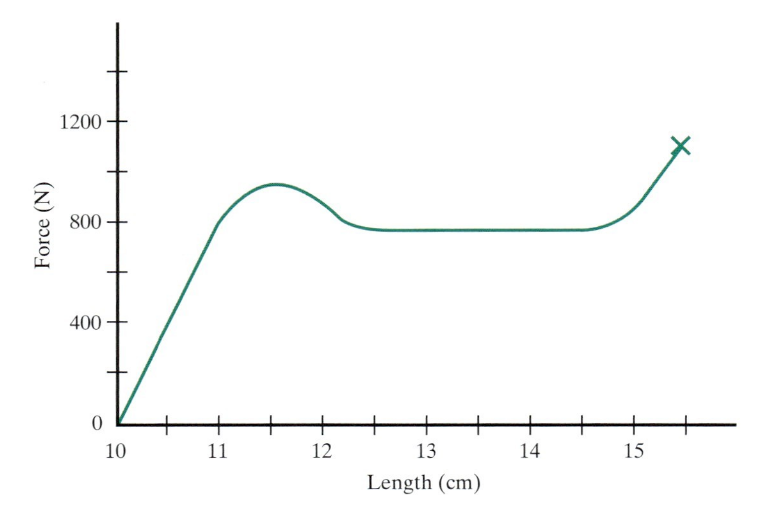

9.5 You are designing a crane with a steel cable that will be used to pick up objects. The radius of the cable is 10 cm, and the yield strength of the steel is 1070 MPa. Assuming a design factor of 2, what is the maximum force that can be sustained by this cable?

9.6 A steel cable with circular cross-section is determined to have a maximum applied load of 425 kN. The yield strength of the steel is 1070 MPa. Given a design factor of 1.8, what should the radius of the cable be?

9.7 A steel cable with a cross-sectional radius of 1.5 cm is being used as a support cable for a bridge. Using a safety factor of 1.5, what is the maximum force you will certify for this cable? Properties of the steel are as follows:

Modulus = 207 GPa
Yield strength = 220 MPa
Tensile strength = 400 MPa
Percent elongation = 23

9.8 On the bond-force curve below, indicate the equilibrium distance between two atoms.

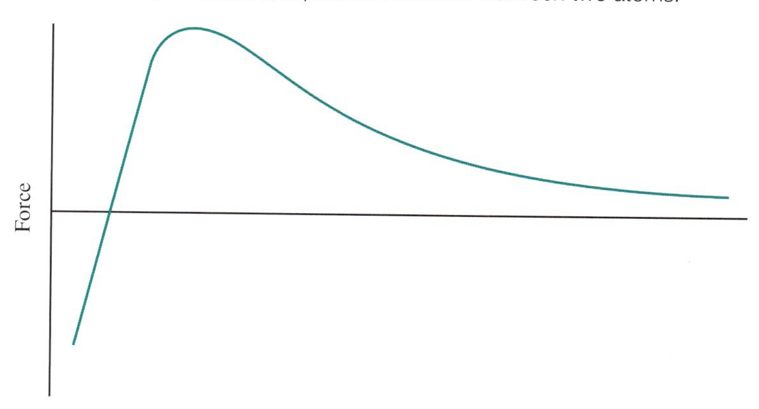

9.9 Stainless steel has an elastic modulus of approximately 200 GPa, while aluminum has an elastic modulus of approximately 70 GPa. Sketch the bond-force curves for these two materials, clearly showing the difference between the two.

9.10 Identify the possible slip planes and directions for aluminum.

9.11 In an FCC single crystal, slip is found to occur on the (111) plane in the [011] direction. A stress of 12 MPa applied along the [121] direction is just sufficient to cause slip to occur. What is the critical resolved shear stress?

9.12 A single crystal of a BCC metal has a critical resolved shear stress of 5.4 MPa. Slip occurs on the (011) plane in the $[\bar{1}\bar{1}1]$ direction. Calculate the yield stress in the [001] direction.

9.13 For each of the following pairs, indicate which will be more ductile, and provide a brief (one sentence) explanation as to why:

a. Copper with a grain size of 10 μm, or copper with a grain size of 1 μm.
b. A one-phase alloy of silver and copper, or pure silver.
c. Pure aluminum, or aluminum with Al_2O_3 particles.
d. Semicrystalline PET at room temperature, or semicrystalline PET at 100° C; for PET $T_g = 80°$ C, $T_m = 280°$ C.
e. Undeformed copper, or copper that has been plastically deformed.
f. Coarse pearlite or fine pearlite.
g. Fine pearlite or spheroidite.
h. Pure copper solidified at 900° C, or pure copper solidified at 600° C.

9.14 A particular processing application requires that the softening point of a glass be no higher than 1,000° C. Which of the glasses in Figure 9.7.2 meet this requirement?

9.15 Polycarbonate has a fracture toughness of 2.2 MPa-m$^{1/2}$. If the largest internal crack in a polycarbonate sheet is 0.2 mm long and there are no surface cracks, what is the failure stress? Take $Y = 1.2$.

9.16 A plate of 1040 steel fails at a stress of 350 MPa. What is the length of the largest internal crack present on this plate? You can assume there are no surface cracks. The fracture toughness is 54.0 MPa-$m^{1/2}$ and $Y = 1.8$.

9.17 A part made of 2024-T6 aluminum is subject to a cyclic stress at a frequency of 10 Hz. If the part must last at least 280 hours before failing, what is the maximum stress amplitude allowed for the cyclic stress? See data in Figure 9.9.3.

9.18 A part made of 1045 steel is subject to a cyclic stress at a frequency of 10 Hz. If the stress amplitude is 60 ksi, how many hours will the part last before failing due to fatigue? See data in Figure 9.9.3.

9.19 Consider a unidirectional continuous fiber-reinforced composite with epoxy as the matrix with 55% by volume fiber.

a. Calculate the longitudinal and transverse moduli when E-glass fiber is used.

b. Calculate the percentage increase in the longitudinal and transverse moduli if carbon fibers are used instead of E-glass.

Epoxy modulus = 2.41 GPa
E-glass fiber modulus = 72.5 GPa
Carbon fiber modulus = 230 GPa

9.20 What is the modulus of a random-in-the-plane composite containing 60 vol% carbon fiber in an epoxy matrix?

Epoxy modulus = 2.41 GPa
Carbon fiber modulus = 230 GPa

9.21 A continuous glass fiber/epoxy composite is required to have a longitudinal modulus of 40 GPa.

a. What volume fraction of glass fiber must the composite have?

b. What will be the transverse modulus of the resulting composite?

The modulus of epoxy is 2.41 GPa and the modulus of glass fiber is 72.5 GPa.

9.22 You are asked to create a uniaxial continuous fiber composite with polyester as the matrix and carbon fiber. If the longitudinal modulus must be at least 155 GPa, what volume fraction of fiber is needed?

Polyester modulus = 3.05 GPa
Carbon fiber modulus = 230 GPa

Conceptual Problems

9.23 For very low-temperature applications, polyethylene is preferred over polypropylene. Explain why.

9.24 Polypropylene bowls are microwave safe, while polyethylene bowls are not. Explain why.

9.25 You need a glass with a softening point of less than 1,200° C. Will 96% silica glass work for this application? See data in Figure 9.7.2. If 96% silica will not work, describe what you would do to create a new glass composition that could meet this requirement, and why that new composition results in a lower softening point.

9.26 A particular polymer has a critical stress intensity parameter of 5 MPa-$m^{1/2}$ and a strength of 150 MPa. It is being used as a plate that experiences tension. The maximum stress it is designed to experience is 80% of its strength. You are inspecting this plate, and the minimum crack length you can detect is 1.5 mm. Is the design stress appropriate to ensure that failure will not occur after your inspection? If not, calculate the maximum stress you will certify this plate to withstand. For this calculation use $Y = 1$.

9.27 Compare two pieces of silica glass; one with a 1 mm surface crack and one with a 2 mm surface crack. Which will fracture at the lower applied stress? Explain your answer.

9.28 Most people consider strength to be an intrinsic property of a material. Explain why this is actually not true. Use equations as appropriate.

9.29 When we discussed diffusion, we used the example of diffusing carbon into the surface of steel to make the surface harder. Explain why, on an atomic level, adding carbon to the surface of steel makes it harder.

9.30 Your friend shows you a bar of copper that has been annealed so there are very few defects in it. He bends the bar in his bare hands, hands it to you, and asks you to straighten it out. You know that you are stronger than your friend, but you cannot bend the bar. Why?

9.31 A steel manufacturer is making steel I-beams from a plain carbon steel of eutectoid composition. The original manufacturing process called for casting the molten steel, cooling to 750° C and holding there for 30 minutes, then cooling to 600° C and holding for 20 minutes, then cooling rapidly to room temperature. In an effort to speed up the production process, the hold at 600° C was eliminated. Mechanical tests using the new process show that the I-beams are too brittle for the application intended. Describe the microstructure for each of these two processes, and explain why the new process results in increased brittleness.

9.32 Nylon 66 has a fatigue life of 10,000 cycles at a mean stress of 10 MPa. If the mean stress increases, how will the fatigue life change?

9.33 You have two samples of copper; one that was solidified at 900° C, and one that was solidified at 700° C. Which has the higher Brinell hardness? Explain your answer.

9.34 You are working for a manufacturer of steel plates for biomedical implants. You are tasked with developing a non-destructive quality assurance test to make sure every plate your company makes has undergone the proper heat treatment. Describe the test you would use and why it is appropriate.

9.35 Draw a graph showing strain versus time at constant applied stress for polystyrene at 50° C and at 120° C. Put both curves on the same graph, clearly showing the differences, if any, between the two. Explain the reason for any differences between the two temperatures. The T_g for polystyrene is 100° C.

9.36 A polymer undergoes a creep experiment at a temperature below its T_g. Draw a graph showing strain versus time for this experiment.

Corrosion is a significant safety and economic issue. For example, the American Society of Civil Engineers Report Card for America consistently finds significant deficiencies in the nation's bridges, often due to corrosion. (Sofilou/Shutterstock)

Materials in the Environment

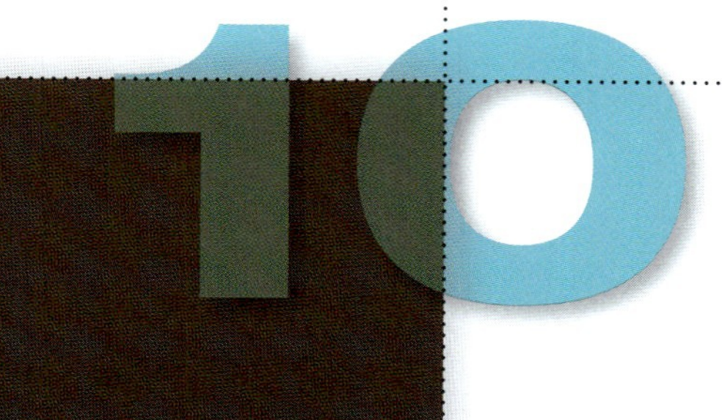

Because all materials are used in some type of environment, how the material interacts with that environment can play a large role in the kinds of properties it has, and particularly how those properties change with time. Think about the different kinds of environments a material might experience; a bridge will be exposed to water from rain or rivers, oils and other chemicals from cars, and sunlight. Materials used for a biomedical implant are exposed to a very salty solution with varying pH, and materials used in a computer must experience high temperatures due to the power generated in the computer chips. As engineers we cannot think just about the properties of materials as described in Chapter 9, we need to think about how the environment changes materials over time and what that can do to the properties. While this chapter does not focus much on changes in properties, it does describe what happens to the structure of various materials in the environments where they are used. By the end of this chapter you will:

Understand the mechanisms of corrosion.
Be able to determine if a metal has a natural protective coating.
Be able to describe how polymers behave in the environment.

10.1 Electrochemistry: How Does a Battery Work?

LEARN TO: Write corrosion reactions.
Calculate electrochemical cell potential.
Predict which metal in a galvanic couple will corrode.

Corrosion of metals is an electrochemical reaction. In order to understand how corrosion works and how to prevent it, we need to understand the electrochemical reactions that occur. We begin by considering the following reaction:

$$Mg\ (s) + Ni^{2+}\ (aq) \rightarrow Mg^{2+}\ (aq) + Ni\ (s)$$

This reaction says that if you place magnesium solid in a solution that contains Ni^{2+} ions (like a solution of $NiCl_2$), you will end up dissolving some of the magnesium to form magnesium ions, and depositing nickel metal on the magnesium. This is a reduction–oxidation reaction, or *redox reaction*, and it involves the transfer of electrons from one metal to the other. In order to see how the electrons are transferred, it is convenient to divide the redox reaction into its separate half-reactions. One half-reaction is a *reduction reaction*, in which a metal gains electrons. The other is an *oxidation reaction*, in which a metal loses electrons. For the redox reaction shown above, the half-reactions are as follows:

$$\text{Reduction: } Ni^{2+}\,(aq) + 2e^- \rightarrow Ni\,(s)$$

$$\text{Oxidation: } Mg\,(s) \rightarrow Mg^{2+}\,(aq) + 2e^-$$

If we add up these two half-reactions we get the overall redox reaction. We can remember which is reduction and which is oxidation with some memory devices:

OIL RIG = Oxidation Is Loss, Reduction Is Gain

LEO the lion says GER = Loss of Electrons is Oxidation, Gain of Electrons is Reduction

The metal that is oxidized is also called the *reducing agent* (because it reduces the other metal), while the metal that is reduced is called the *oxidizing agent* (because it oxidizes the other metal). Now you will practice figuring out the redox reaction between two metals.

Guided Inquiry: Redox Reactions

10.1.1 How many electrons are transferred between magnesium and nickel in the previous reaction?

You may have learned about electrochemical reactions in your chemistry class. If you have, make sure to help your other group members understand these concepts.

10.1.2 In the reaction between magnesium and nickel; which metal is oxidized and which metal is reduced?

10.1.3 In the reaction between magnesium and nickel; which metal is the oxidizing agent and which metal is the reducing agent?

10.1.4 When chromium is placed into a solution of $CoCl_2$ cobalt metal is deposited on the chromium. Which substance is oxidized and which is reduced? Which is the oxidizing agent and which is the reducing agent?

10.1.5 For the situation in question 10.1.4 chromium ions have a valence of +3 and cobalt ions have a valence of +2. Write the oxidation and reduction reactions for chromium metal placed in a solution of $CoCl_2$.

The number of electrons must be the same in both reactions.

10.1.6 Write the overall redox reaction for chromium metal placed in a solution of $CoCl_2$. Remember, this reaction should not have any electrons in it.

Concept Check 10.1.1

A piece of aluminum is placed in a solution of $NiCl_2$. Nickel metal is observed to deposit on the aluminum.

- Which chemical species is being reduced?
- Which chemical species is the reducing agent?
- What is the correct oxidation reaction that is occurring?
  See Table 4.2.1 for common valences of ions.
- What is the correct balanced redox reaction that is occurring?

It is possible to separate oxidation and reduction reactions by creating a *galvanic cell*, as shown in Figure 10.1.1. In this cell, electrons can flow through the wire, while ions can migrate through the salt bridge to maintain electrical neutrality. As a result of the electron flow, a voltage is created that can be measured. For the galvanic cell shown in Figure 10.1.1 the voltage is 1.10 V. In a galvanic cell, oxidation occurs at the anode and reduction occurs at the cathode. We can remember this by noting that the words "oxidation" and "anode" both begin with vowels, while the words "reduction" and "cathode" both begin with consonants. As you will see in the next questions, the redox reactions for a galvanic cell are the same as what you did previously.

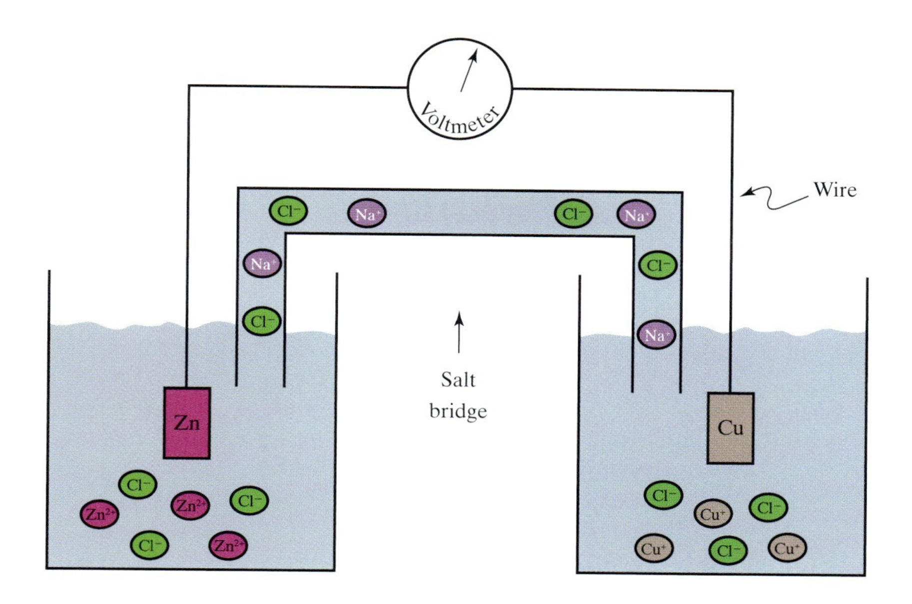

Figure 10.1.1

A zinc–copper galvanic cell. It is observed that the mass of the copper electrode increases while the mass of the zinc electrode decreases.

Guided Inquiry: Galvanic Cells

10.1.7 If a metal is being oxidized, will its mass increase or decrease?

Ions will dissolve in water. Neutral metal atoms are deposited on electrodes as a solid.

10.1.8 In Figure 10.1.1, which chemical species is oxidized and which is reduced?

10.1.9 Which electrode is the anode and which is the cathode?

10.1.10 Write the half-reactions and total balanced redox reaction for the galvanic cell of Figure 10.1.1.

10.1.11 Which way do the electrons flow in the wire?

Concept Check 10.1.2

- **The galvanic cell in Figure 10.1.1 is changed so that the zinc is replaced with nickel and the Zn^{2+} ions are replaced with Ni^{2+} ions. Write the balanced redox reaction for this new galvanic cell.**

For Figure 10.1.1 you were told that the zinc loses mass, and so it is being oxidized. But if you were not given that information, how would you know? The cell voltage tells you how strongly the electrons are being "pulled" by each metal, so if there was a way to compare the strength of that pulling you could figure it out. The way to do that is with the standard galvanic cell, which is shown in Figure 10.1.2 for zinc. This cell uses the standard hydrogen electrode as one of the electrodes. If we make a series of standard cells with every possible metal, we have a way to compare each metal to every other. Table 10.1.1 shows a list of standard potentials. You will notice that these are all listed as reduction reactions, so they are called *standard reduction potentials*. A larger voltage means that the metal pulls the electrons more strongly. To get the actual voltage created by a galvanic cell, you can use the standard potentials as follows:

$$E^0_{cell} = E^0_{ox} + E^0_{red} \tag{10.1.1}$$

where E^0 means the potential under standard conditions. Since Table 10.1.1 shows only reduction reactions, to get the potential for the oxidation reaction you just have to switch the sign on the reduction potential shown in the table. You will now use the standard reduction potentials to determine which metal in a galvanic cell is oxidized.

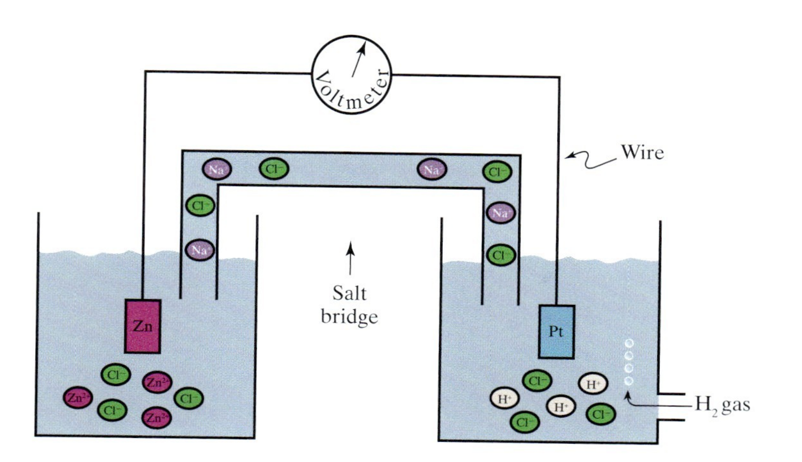

Figure 10.1.2 Standard galvanic cell for zinc. In a standard cell all concentrations are 1 M, all gas pressures are 1 atm, and the temperature is 25° C.

TABLE 10.1.1 Standard reduction potentials

Reaction	E^0 (Volts)
F_2 (g) + 2 e^- ⇌ 2 F^-	+2.87
Au^+ + e^- ⇌ Au (s)	+1.83
O_2 (g) + 4 H^+ + 4 e^- ⇌ 2 H_2O	+1.23
Pt^{2+} + 2 e^- ⇌ Pt (s)	+1.188
Pd^{2+} + 2 e^- ⇌ Pd (s)	+0.915
Hg^{2+} + 2 e^- ⇌ Hg (l)	+0.85
Ag^+ + e^- ⇌ Ag (s)	+0.7996
Tl^{3+} + 3 e^- ⇌ Tl (s)	+0.72
O_2 (g) + 2 H^+ + 2 e^- ⇌ H_2O_2 (aq)	+0.70
O_2 (g) + 2 H_2O + 4 e^- ⇌ 4 OH^- (aq)	+0.40
Cu^{2+} + 2 e^- ⇌ Cu (s)	+0.340
Bi^{3+} + 3 e^- ⇌ Bi (s)	+0.308
2 H^+ + 2 e^- ⇌ H_2 (g)	0.0000
Pb^{2+} + 2 e^- ⇌ Pb (s)	−0.13
Sn^{2+} + 2 e^- ⇌ Sn (s)	−0.13
Ni^{2+} + 2 e^- ⇌ Ni (s)	−0.25
Co^{2+} + 2 e^- ⇌ Co (s)	−0.28
Cd^{2+} + 2 e^- ⇌ Cd (s)	−0.40
Fe^{2+} + 2 e^- ⇌ Fe (s)	−0.44
Cr^{3+} + 3 e^- ⇌ Cr (s)	−0.74
Zn^{2+} + 2 e^- ⇌ Zn (s)	−0.7618
2 H_2O + 2 e^- ⇌ H_2 (g) + 2 OH^-	−0.8277
Mn^{2+} + 2 e^- ⇌ Mn (s)	−1.185
Ti^{2+} + 2 e^- ⇌ Ti (s)	−1.63
Al^{3+} + 3 e^- ⇌ Al (s)	−1.66
Mg^{2+} + 2 e^- ⇌ Mg (s)	−2.372
Na^+ + e^- ⇌ Na (s)	−2.71
Ca^{2+} + 2 e^- ⇌ Ca (s)	−2.868
K^+ + e^- ⇌ K (s)	−2.931
Li^+ + e^- ⇌ Li (s)	−3.0401

Note: (aq) = aqueous, (g) = gas, (l) = liquid, (s) = solid.

Guided Inquiry: Standard Reduction Potentials

10.1.12 What is the standard reduction potential for aluminum?

10.1.13 What is the standard reduction potential for lead?

10.1.14 Which metal has the stronger pull on electrons; aluminum or lead?

10.1.15 In an aluminum/lead galvanic cell, which metal is oxidized and which is reduced? Which is the cathode and which is the anode? Which metal is being corroded?
Use the concept of electrons being "pulled" by one of the atoms.

10.1.16 If you had two metals in a galvanic cell and were not sure which one was oxidized and which one was reduced, how could you decide?

Concept Check 10.1.3

For these questions, consider a standard galvanic cell with titanium and cadmium electrodes.

- **Which chemical species is being oxidized?**
- **Which electrode is the anode?**
- **What is the potential of this cell?**

The standard reduction potentials in Table 10.1.1 are taken under ideal conditions of 1 M concentrations, 1 atm of pressure, and 25° C. Clearly this is not realistic for actual applications. A more realistic approach is the *galvanic series*, which is shown in Table 10.1.2 (you will notice that Table 10.1.2 does not have any voltages listed). The galvanic series is measured in seawater and is just a relative ranking of which metal is a stronger reducing agent than another; metals lower in the table are stronger reducing agents (that is, they are more easily oxidized). As you will see in the next questions, we can use the galvanic series to predict which metal in a cell will corrode, but we cannot use it to calculate voltages.

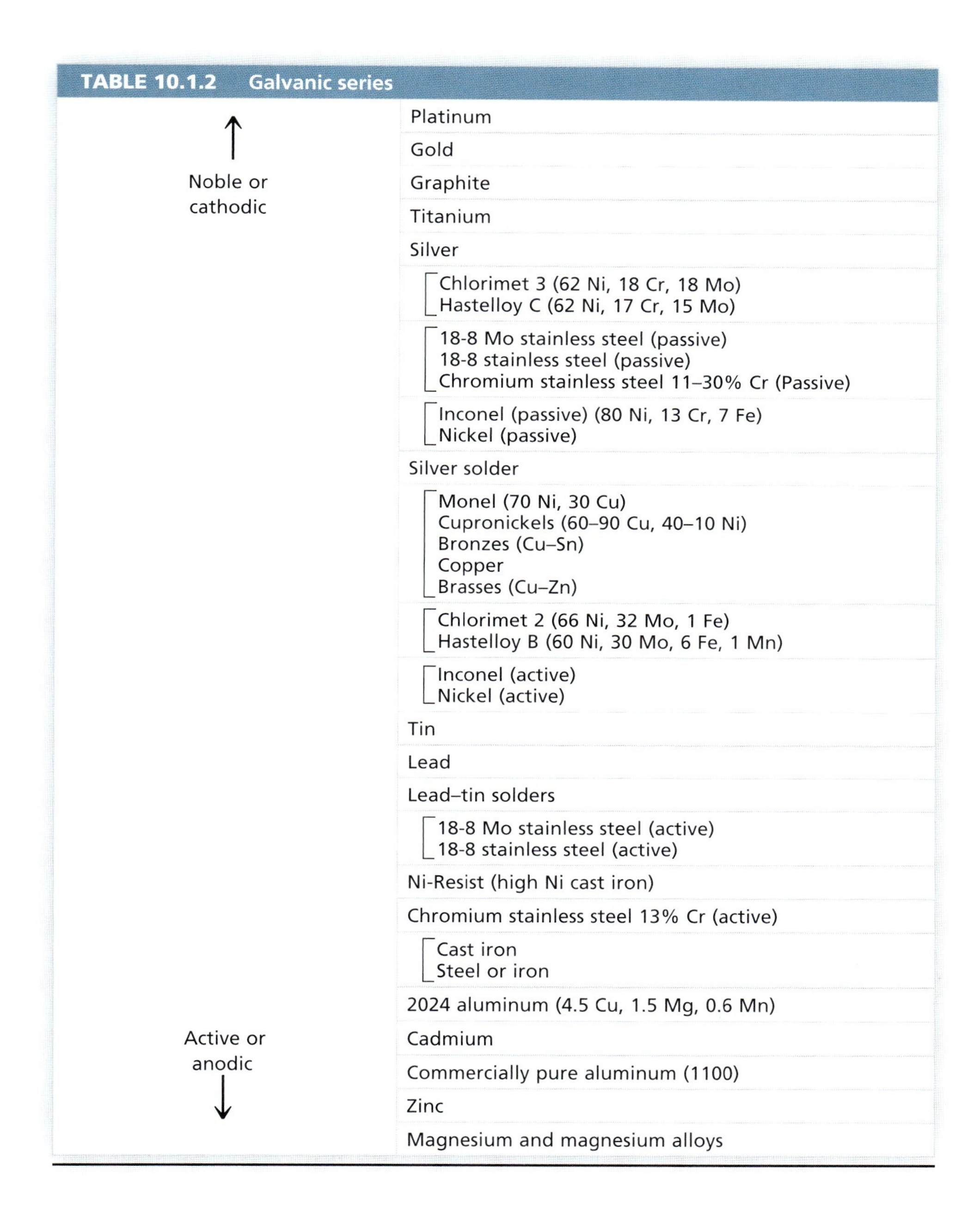

TABLE 10.1.2 Galvanic series

↑ Noble or cathodic	Platinum
	Gold
	Graphite
	Titanium
	Silver
	Chlorimet 3 (62 Ni, 18 Cr, 18 Mo) Hastelloy C (62 Ni, 17 Cr, 15 Mo)
	18-8 Mo stainless steel (passive) 18-8 stainless steel (passive) Chromium stainless steel 11–30% Cr (Passive)
	Inconel (passive) (80 Ni, 13 Cr, 7 Fe) Nickel (passive)
	Silver solder
	Monel (70 Ni, 30 Cu) Cupronickels (60–90 Cu, 40–10 Ni) Bronzes (Cu–Sn) Copper Brasses (Cu–Zn)
	Chlorimet 2 (66 Ni, 32 Mo, 1 Fe) Hastelloy B (60 Ni, 30 Mo, 6 Fe, 1 Mn)
	Inconel (active) Nickel (active)
	Tin
	Lead
	Lead–tin solders
	18-8 Mo stainless steel (active) 18-8 stainless steel (active)
	Ni-Resist (high Ni cast iron)
	Chromium stainless steel 13% Cr (active)
	Cast iron Steel or iron
	2024 aluminum (4.5 Cu, 1.5 Mg, 0.6 Mn)
Active or anodic ↓	Cadmium
	Commercially pure aluminum (1100)
	Zinc
	Magnesium and magnesium alloys

Guided Inquiry: Galvanic Series

10.1.17 According to standard reduction potentials, if aluminum is placed into contact with zinc, which will corrode?

10.1.18 According to the galvanic series, if aluminum is placed into contact with zinc, which will corrode?

10.1.19 According to both standard reduction potentials and the galvanic series, aluminum is one of the most easily oxidized metals. How is it possible to make aluminum boats?

Concept Check 10.1.4

- **Which is more likely to corrode; bronze or brass?**

10.2 Corrosion of Metals

LEARN TO: **Calculate corrosion rate.**
Describe factors affecting corrosion rate.
Describe types of corrosion.
Describe ways to prevent corrosion.

In the previous section you learned about the basic electrochemistry that underlies corrosion. In this section you will look at some of the practical aspects of how corrosion occurs, as well as ways to prevent it. The simplest case is shown in Figure 10.2.1, in which corrosion occurs uniformly over the surface of a metal. The metal "dissolves" away, and the piece of metal becomes thinner.

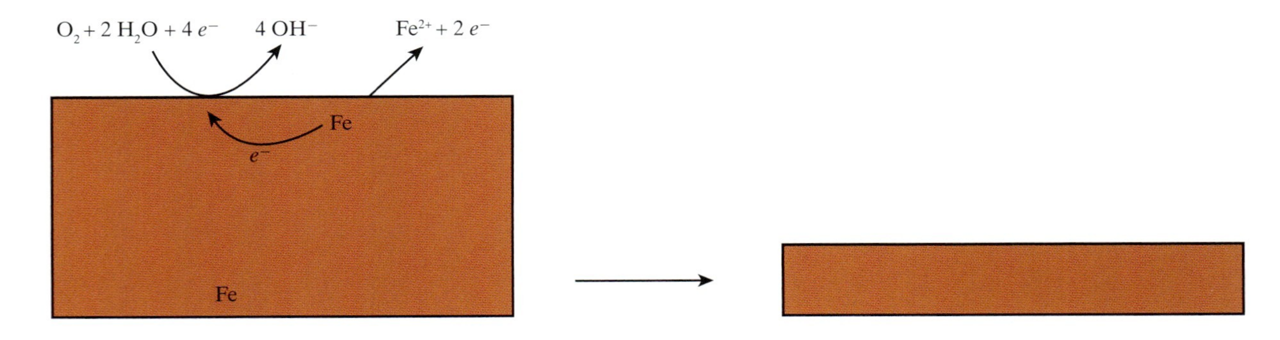

Figure 10.2.1
Illustration of uniform corrosion.

For the case of *uniform corrosion*, the rate can be determined from the following equation:

$$CPR = \frac{I}{A}\frac{M}{n\rho\mathcal{F}} \tag{10.2.1}$$

The variables in this equation are as follows:

- CPR = *corrosion penetration rate* = thickness lost/time
- I/A = current density over the area which is losing the thickness
- I = corrosion current
- A = area over which the material is being lost
- M = atomic mass of the metal
- n = number of electrons involved in the **unbalanced** oxidation half-reaction; it is **not** the number of electrons in the balanced reaction
- ρ = density of the metal
- $\mathcal{F}$ = Faraday's constant = 96,500 C/mol

An acceptable rate of corrosion is on the order of 0.5 mm per year.

The next questions are based on the following problem scenario: The corrosion current measured on a 3.0 cm × 3.0 cm plate of iron is 3.8×10^{-5} A. The corrosion product is FeO. What is the corrosion penetration rate in mm/year?

Guided Inquiry: Uniform Collosion

10.2.1 Identify the value and units for all of the quantities in Equation (10.2.1) for the above problem.

10.2.2 Using a unit analysis, determine the units for CPR based on your answer to question 10.2.1.

10.2.3 Calculate the CPR for the problem above in units of mm/year.

Concept Check 10.2.1

- What is the CPR for question 10.2.3?

EXAMPLE PROBLEM 10.2.1

The corrosion current measured on a 4.0 cm * 4.0 cm plate of copper is 4.2×10^{-5} A. The corrosion product is CuO. What should the initial thickness of the plate be if the thickness after 5 years must be at least 0.50 cm? You may assume that the corrosion occurs on only one side of the plate.

$$CPR = \frac{I}{A}\frac{M}{n\rho\mathcal{F}}$$

$I = 4.2 \times 10^{-5}\text{ A} = 4.2 \times 10^{-5}\text{ C/s}$

$A = 16\text{ cm}^2$

$M = 63.55\text{ g/mol}$

$n = 2$; since the problem says that CuO is formed, the valence of the copper ion is +2, so 2 electrons are involved in the oxidation reaction.

$\rho = 8.94\text{ g/cm}^3$

$\mathcal{F} = 96{,}500\text{ C/mol}$

$$\text{CPR} = \frac{4.2 \times 10^{-5}\,\text{C/s}}{16\text{ cm}^2}\;\frac{63.55\text{ g/mol}}{2(8.94\text{ g/cm}^3)(96{,}500\text{ C/mol})} = 9.7 \times 10^{-11}\text{cm/s}$$

There are 365 days/yr $\times$ 24 hrs/day $\times$ 3600 s/hr = 31,536,000 s/yr. Thickness lost in 5 years is as follows:

$$t = (9.7 \times 10^{-11}\text{ cm/s})(31{,}536{,}000\text{ s/yr})(5\text{yr}) = 0.015\text{ cm}$$

Since the final thickness must be 0.50 cm, the initial thickness must be 0.515 cm.

Uniform corrosion is easy to design around because you can determine exactly how much material is being lost. Unfortunately it is very rare, and we usually won't encounter it. We will now take a look at more common corrosion situations. One of these is *galvanic corrosion*, in which two metals are in contact with each other. This often occurs in boats, where you have many different metals in contact in a salt-water environment. A famous example of galvanic corrosion is the Statue of Liberty, in which the protective layer between the iron support structure and the copper shell began to fail, causing corrosion of the iron. A restoration project in the 1980s replaced the iron supports with stainless steel, and a new insulating layer of poly(tetrafluoroethylene) (Teflon®) was placed between the copper shell and the steel supports. Figure 10.2.2 illustrates a case in which a piece of copper is in contact with a piece of iron. You will use this figure to understand how galvanic corrosion works.

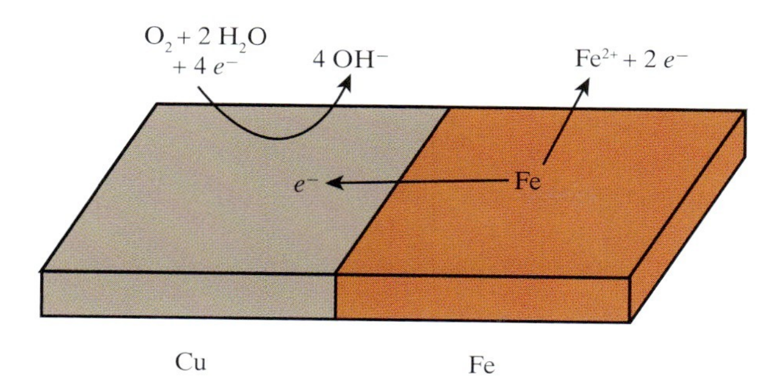

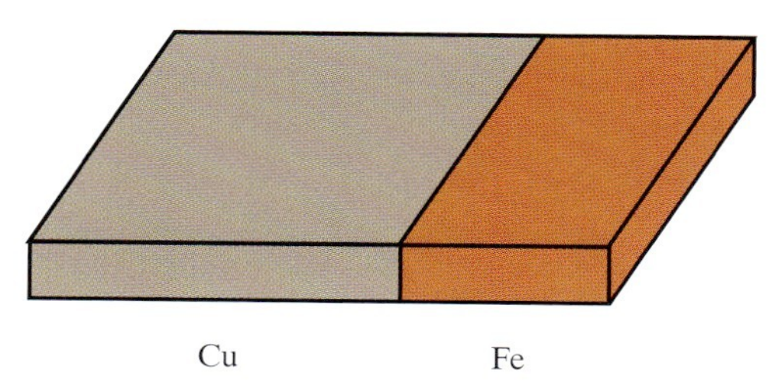

Figure **10.2.2**
Illustration of galvanic corrosion. In the bottom image the reactions are the same and the piece of copper is the same size, but the piece of iron is smaller.

Guided Inquiry: Galvanic Corrosion

10.2.4 In the situation in Figure 10.2.2, which metal will corrode?

10.2.5 In the situation in Figure 10.2.2, which metal is the anode?

10.2.6 If the corrosion occurs over the area shown in Figure 10.2.2, which has the larger anode; the top figure or the bottom figure?

10.2.7 Based on your answer to question 10.2.6 and Equation (10.2.1), which will corrode faster; the bottom figure or the top figure?

10.2.8 What is a general rule for how the corrosion rate changes as the ratio of anode area to cathode area changes?

Figure 10.2.2 keeps one electrode the same size so you can work out what happens. But the trend you have identified is true even if both electrodes change size. What counts is not the actual size but the ratio of the areas.

Concept Check 10.2.2

- **When two different metals are in contact and the ratio of the anode area to the cathode area increases, what will happen to the corrosion rate?**
- **A copper wire is connected to an aluminum junction box. Which metal will corrode?**

Another common corrosion situation is called *crevice corrosion*; this is illustrated in Figure 10.2.3. You may have noticed that if two pieces of metal are held together by a bolt, the area underneath the bolt head is the most corroded. This is an example of crevice corrosion, in which corrosion preferentially occurs in cracks or crevices between two pieces of metal, even if they are the same kind of metal. The next set of questions will guide you to understanding how crevice corrosion works.

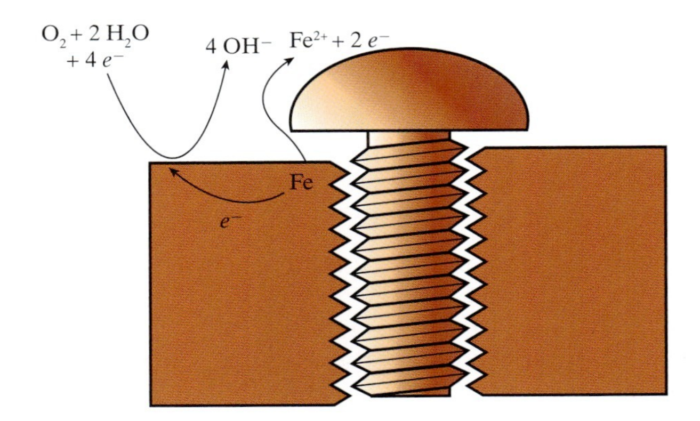

Figure **10.2.3**

Example of crevice corrosion. Because the water is stagnant, the oxygen underneath the bolt cannot be replenished, and so the water underneath the bolt has no oxygen present.

Guided Inquiry: Crevice Corrosion

10.2.9 What is the oxidation reaction that occurs in Figure 10.2.3? Where does it occur?

You just need to use the information in Figure 10.2.3.

10.2.10 What is the reduction reaction that occurs in Figure 10.2.3? Where does it occur?

10.2.11 What reactants are needed for the reduction reaction to take place?

10.2.12 Based on your answer to question 10.2.11, why can't the reduction reaction take place under the bolt?

Look at the reactants that are available. What reactants are needed for the reduction reaction?

10.2.13 Explain why corrosion occurs mostly under the bolt.

Concept Check **10.2.3**

- **List two other situations where crevice corrosion could occur.**

A variety of other types of corrosion can also occur. Here are a few examples:

- *Stress corrosion*: Corrosion that occurs in combination with stress. The presence of the stress can cause the corrosion to occur faster than it would without the stress. Some materials are more susceptible to stress corrosion than others.
- *Erosion corrosion*: In cases where there is fluid flow, the fluid can damage any protective layer on the metal surface and cause corrosion to occur more quickly. A related type of corrosion is cavitation damage, which occurs on boat propellers. The bubbles formed as the propeller spins damage the protective layer and cause corrosion.
- *Intergranular corrosion*: This is corrosion that occurs at grain boundaries. This is a particular problem in welding of steels, because the welding process removes chromium from the grain boundaries, which normally acts to reduce corrosion in stainless steels.

Corrosion prevention is a huge cost to society. For example, the Golden Gate Bridge in San Francisco has an extensive maintenance program to continuously touch up paint and replace corroding pieces. The more we can think about ways to reduce corrosion as structures are designed and built, the more money we can save on maintenance. Most corrosion-prevention approaches are common sense ideas that you can come up with yourself.

Guided Inquiry: Corrosion Prevention

10.2.14 List three ways to prevent or reduce the occurrence of uniform corrosion. *After you make your list, compare it with another group to see if they came up with anything different.*

10.2.15 What is one way to prevent crevice corrosion from occurring?

10.2.16 In Figure 10.2.2, which material is corroding?

10.2.17 In Figure 10.2.2, which material is being protected from corrosion?

10.2.18 Cathodic protection is the use of one metal as a sacrificial anode to prevent the corrosion of another metal. Based on your answer to question 10.2.17, list three other metals you could put into contact with copper to prevent the copper from corroding.

Concept Check 10.2.4

- **What metals could be used to protect iron from corroding?**

10.3 Oxide Formation

LEARN TO: **Determine whether an oxide is protective.**

Molecular oxygen (O_2) is a very strong oxidizing agent. As a result, most metals will oxidize in the presence of oxygen. For example, the electrochemical reactions for aluminum exposed to oxygen are as follows:

$$4\ Al \rightarrow 4\ Al^{3+} + 12\ e^-$$
$$3\ O_2 + 12\ e^- \rightarrow 6\ O^{2-}$$

Putting these two half-reactions together, the redox reaction is as follows:

$$4\ Al + 3\ O_2 \rightarrow 4\ Al^{3+} + 6\ O^{2-}$$

or

$$4\ Al + 3\ O_2 \rightarrow 2\ Al_2O_3$$

In other words, aluminum metal will react with oxygen to form aluminum oxide. The importance of this reaction is that the aluminum oxide forms on the surface of the aluminum metal. If this layer is continuous, it acts as a protective barrier, preventing any more oxygen or other oxidizing agents (such as water) from reaching the metal surface and causing further corrosion. Figure 10.3.1 shows the oxide layer for three different metals. As you will see in the next questions, some of these oxide layers protect the underlying metal and some don't.

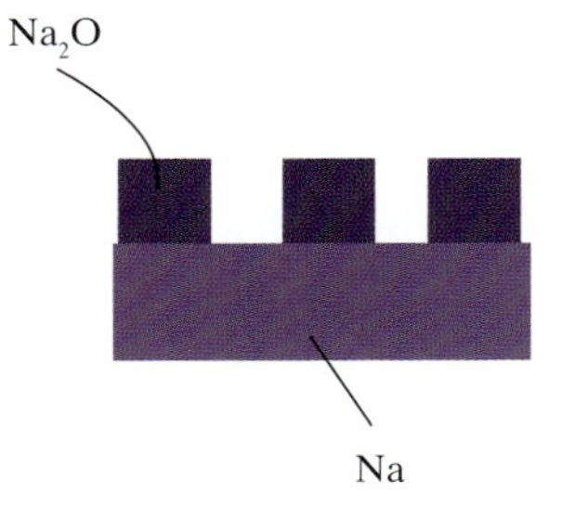

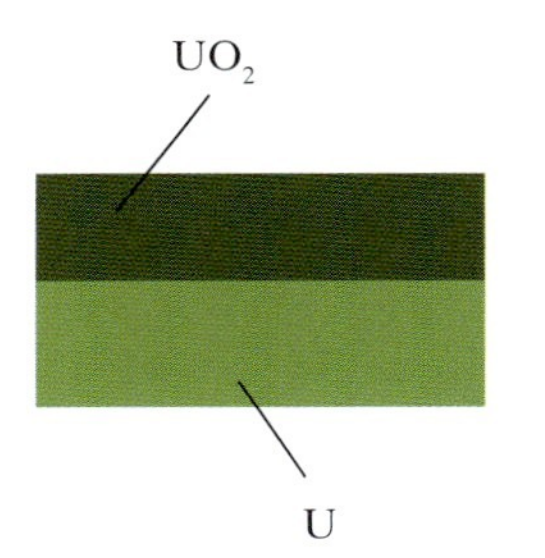

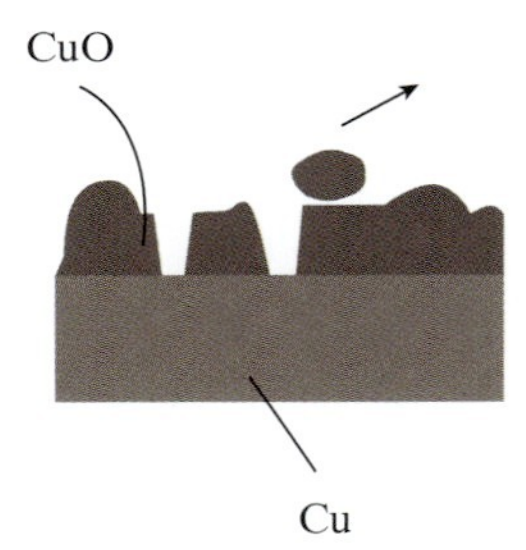

Figure 10.3.1 Schematic of oxide layer on sodium, uranium, and copper.

Guided Inquiry: Oxide Layers

10.3.1 For which metals is the oxide layer continuous?

10.3.2 For which metals is the oxide layer discontinuous?

10.3.3 Which metals will be stable in air? Which will continue to corrode in air?

Whether or not an oxide is protective can be calculated using the *Pilling–Bedworth ratio*. The P–B ratio is defined as follows:

$$P\text{–}B = \frac{\text{volume of oxide}}{\text{volume of metal used to create the oxide}} \tag{10.3.1}$$

Guided Inquiry: P-B Ratio

10.3.4 If P–B is less than 1.0, which has a greater volume; the oxide or the metal used to create it? Will the oxide take up less space or more space than the metal it replaces?

10.3.5 If P–B is less than 1.0, will the oxide completely cover the surface?

10.3.6 If P–B is equal to 1.0, will the oxide completely cover the surface?

10.3.7 If P–B is greater than 1.0, which has a greater volume; the oxide or the metal used to create it? Will the oxide take up less space or more space than the metal it replaces?

10.3.8 Based on your answers to questions 10.3.4–10.3.7, for each of the metals shown in Figure 10.3.1 state whether or not the P–B ratio is less than 1.0, equal to 1.0, or greater than 1.0.

10.3.9 Based on your answer to questions 10.3.4–10.3.8, identify the range of values for P–B that determine whether an oxide layer is protective or nonprotective.

Concept Check 10.3.1

- **Over what range of P–B values will an oxide be protective?**

For an oxide of the general formula M_aO_b the P–B ratio is calculated as:

$$P\text{–}B = \frac{A_o \rho_m}{a A_m \rho_o} \tag{10.3.2}$$

where A_o is the molecular weight of the oxide, A_m is the molecular weight of the metal, ρ_o is the density of the oxide, ρ_m is the density of the metal, and a is the number of moles of metal atoms in the oxide (that is, the subscript on the metal in the molecular formula for the oxide). Whether or not an oxide is protective is determined by the following values for P–B:

- If P–B < 1.0: nonprotective
- If $1.0 \leq$ P–B < 2.0: protective
- If P–B ≥ 2.0: nonprotective

Guided Inquiry: Calculating the P-B Ratio

10.3.10 Determine whether a layer of SiO_2 on silicon is protective or not. The density of SiO_2 is 2.65 g/cm^3.

10.3.11 Aluminum metal is highly reactive and will burn if exposed to water. How can boat hulls be made out of aluminum?

Concept Check 10.3.2

- **What is the P–B ratio for CaO on calcium? The density of calcium oxide is 3.35 g/cm^3.**

EXAMPLE PROBLEM 10.3.1

Is vanadium (V) oxide, V_2O_5, protective for vanadium? The density of vanadium is 6.0 g/cm^3 and the density of V_2O_5 is 3.36 g/cm^3.

$$P\text{–}B = \frac{A_o \rho_m}{a A_m \rho_o}$$

$$A_m = 50.942 \text{ g/mol}$$

$$A_o = 2 \times 50.942 \text{ g/mol} + 5 \times 16.00 \text{ g/mol} = 181.884 \text{ g/mol}$$

$$\rho_m = 6.0 \text{ g/cm}^3$$

$$\rho_o = 3.36 \text{ g/cm}^3$$

$$a = \text{the number of vanadium atoms in one molecule of oxide} = 2$$

$$P\text{–}B = \frac{(181.884 \text{ g/mol})(6.0 \text{ g/cm}^3)}{(2)(50.942 \text{ g/mol})(3.36 \text{ g/cm}^3)} = 3.19$$

Since this is above 2, the oxide is not protective.

10.4 Degradation of Polymers

LEARN TO: Describe types of degradation in polymers.
Predict whether a polymer will dissolve in a solvent.

Polymers are susceptible to various environments. When exposed to heat, light, or radiation, various chemical reactions can occur that will either break the chains or cause crosslinking to occur, as shown schematically in Figure 10.4.1. Which chemical reaction occurs depends on the type of polymer. This degradation causes permanent changes in the molecules, because we are breaking or forming covalent bonds. Remember from Chapter 3 that this permanent degradation is different from melting or heating above the glass transition, because those two processes involve breaking of nonbonding interactions, which can be reformed on cooling. The permanent degradation from crosslinking or breaking the chains causes permanent changes in the properties of the polymer, as you will see in the next questions.

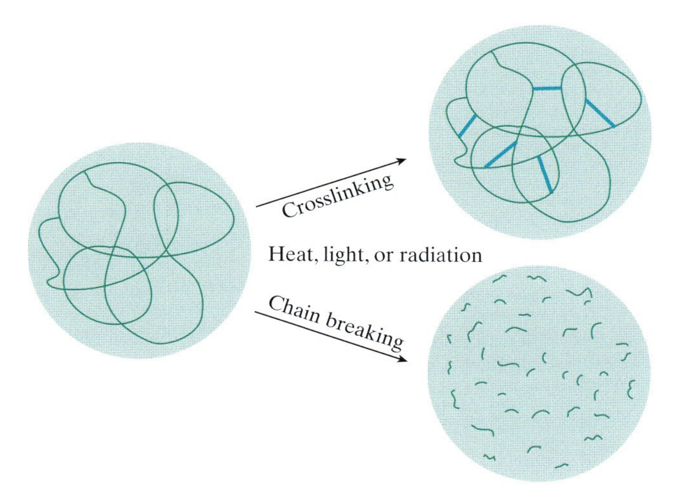

Figure 10.4.1
Illustration of degradation leading to either crosslinking or breaking the chains.

Application Spotlight **Biodegradable Polymers**

While we often think of degradation as a bad thing because it causes a loss in mechanical properties, it can also be a good thing when it is used to reduce waste. Plastic is the fourth highest component of our trash—behind paper, yard waste, and food waste—and almost all of it ends up in landfills. So how can we reduce the amount of plastic waste? There are many different ways, including recycling, but one that we will focus on here is the use of biodegradable plastics. Most plastics are very inert and will last in the environment for a very long time. But plastics can be made that degrade in the environment. There are two ways of doing this; one is to incorporate fillers, such as starch, in nonbiodegradable plastics like polyethylene. The starch breaks down in the environment and the plastic falls apart. If you buy degradable garbage bags, this is what you are getting. It is also possible to get truly degradable plastics. Some polyesters have chemical bonds that are easily hydrolyzed, or broken down by water and microbes. However, the fact that a polymer can degrade doesn't mean it will. Aerobic degradation requires oxygen, which may not be present in the center of a landfill.

(Green Stock Media/Alamy)

Guided Inquiry: Polymer Degradation

10.4.1 If the molecular weight of a polymer increases, what happens to its strength?
See Section 5.2.

10.4.2 If the molecular weight of a polymer decreases, what happens to its strength?

10.4.3 If degradation causes crosslinking in a polymer, what happens to its strength? Do you think it will become more or less brittle?

10.4.4 If degradation causes chain breaking in a polymer, what happens to its strength?

10.4.5 A polymer is to be used outside where it will be exposed to sunlight. Propose a way to reduce the amount of degradation it will undergo.
There is no single correct answer. What ideas can you come up with?

Concept Check **10.4.1**

- **If PVC is molded at too high a temperature, it turns brown and becomes brittle. Why does this happen?**

Another important effect that the environment can have on polymers is their interaction with solvents. Some polymers are more susceptible to solvents than others. For example, polyethylene is fairly solvent resistant, so it can be used for gas tanks in automobiles. On the other hand, polystyrene can be dissolved in many different solvents.

We can understand the solubility of polymers using the *solubility parameter*. First we need to define the cohesive energy density (CED):

$$CED = \frac{\text{energy of vaporization}}{\text{molar volume of liquid}} \tag{10.4.1}$$

In other words, the cohesive energy density tells us how strong the nonbonding interactions are. The stronger those interactions, the more energy it takes to boil the material. CED has units of cal/cm^3 or MPa. The solubility parameter is then defined as follows:

$$\delta = (\text{CED})^{1/2} \tag{10.4.2}$$

The units of the solubility parameter are $(cal/cm^3)^{1/2}$, which is called a hildebrand (after a famous person in thermodynamics) or $MPa^{1/2}$. The conversion between the two is that 1 hildebrand = 2.046 $MPa^{1/2}$. If two materials have similar solubility parameters, or CEDs, the nonbonding interactions between them are the same strength, and therefore of the same type (for example, hydrogen bonding). This means the two materials will "like" to interact with each other and mix together. You will use the next questions to figure out how solubility parameters can be used to predict whether a solvent will dissolve a polymer. Use Tables 10.4.1–10.4.3 to answer these questions.

TABLE 10.4.1 Solubility parameters of some common polymers

Polymer	Solubility parameter ($MPa^{1/2}$)
Polyethylene	16.2
Polystyrene	18.5
PET	20.5
Nylon 66	28.0

TABLE 10.4.2 Solubility parameters of some common solvents

Solvent	Solubility parameter ($MPa^{1/2}$)
n-hexane	14.9
Acetone	19.7
Ethanol	26.2

TABLE 10.4.3 Solubility of polymers in solvents. An X in the table means that the polymer will dissolve in that solvent

	Polyethylene	Polystyrene	PET	Nylon 66
n-hexane	X			
Acetone		X	X	
Ethanol				X

Guided Inquiry: Polymer Solubility

10.4.6 What is the solubility parameter of *n*-hexane? What are the solubility parameters of polymers that *n*-hexane will dissolve?

10.4.7 What is the solubility parameter of acetone? What are the solubility parameters of polymers that acetone will dissolve?

10.4.8 What is the solubility parameter of ethanol? What are the solubility parameters of polymers that ethanol will dissolve?

10.4.9 If the solubility parameter of the solvent is greater than that of the polymer, will this guarantee that the polymer will dissolve?

10.4.10 If the solubility parameter of the solvent is less than that of the polymer, will this guarantee that the polymer will dissolve?

10.4.11 Develop a rule for determining whether a solvent will dissolve a polymer.

10.4.12 If crosslinked polystyrene is exposed to acetone, what will happen to it? Will it dissolve?

Dissolving requires that individual molecules be separated and dispersed in the solvent.

Concept Check **10.4.2**

- **Poly(methyl methacrylate) (PMMA, or acrylic) has a solubility parameter of 19.0 MPa$^{1/2}$. Which of the solvents in Table 10.4.2 will dissolve PMMA?**

Summary

Materials are used in many kinds of environments. As engineers, we need to understand how materials behave in different environments and plan our choice of materials appropriately. From the perspective of the MSE triangle, in this chapter we focused on how the environment affects structure: changes in the oxidation state of metals, causing either corrosion through loss of mass or protection through formation of an oxide layer; changes in the structure of a polymer chain through degradation; and susceptibility of polymers to solvents.

This chapter has not linked these changes in structure to properties. A few of the end-of-chapter problems ask you to stretch your thinking a bit and figure out how properties will change. To do this you will need to connect this chapter to things you have learned in other chapters. By now you should see how the concepts you are learning throughout this text fit together through the MSE triangle.

Key Terms

Cathodic protection
Cavitation damage
Corrosion penetration rate
Crevice corrosion
Erosion corrosion
Galvanic cell
Galvanic corrosion
Galvanic series
Intergranular corrosion
Oxidation reaction
Oxidizing agent
Pilling–Bedworth ratio
Redox reaction
Reducing agent
Reduction reaction
Solubility parameter
Standard reduction potential
Stress corrosion
Uniform corrosion

Problems

Skill Problems

10.1 Write the possible balanced oxidation and reduction half-reactions that occur when magnesium is immersed in each of the following solutions:

a. Pure water
b. Water containing dissolved oxygen
c. HCl
d. An HCl solution containing dissolved oxygen

10.2 Write the possible balanced oxidation and reduction half-reactions that occur when iron is immersed in each of the following solutions:

a. Pure water
b. Water containing dissolved oxygen
c. HCl
d. An HCl solution containing dissolved oxygen

10.3 Write the possible balanced oxidation and reduction half-reactions that occur when aluminum is immersed in each of the following solutions:

a. Pure water
b. Water containing dissolved oxygen
c. HCl
d. An HCl solution containing dissolved oxygen

10.4 In a galvanic cell with tin and iron electrodes, identify which is the anode and which is the cathode. Also identify which electrode corrodes.

10.5 In a galvanic cell with silver and platinum electrodes, identify which is the anode and which is the cathode. Also identify which electrode corrodes.

10.6 In a galvanic cell with zinc and copper electrodes, identify which is the anode and which is the cathode. Also identify which electrode corrodes.

10.7 Calculate the potential of a galvanic cell with tin and iron electrodes.

10.8 Calculate the potential of a galvanic cell with silver and platinum electrodes.

10.9 Calculate the potential of a galvanic cell with zinc and copper electrodes.

10.10 The corrosion current measured on a 3.0 cm $\times$ 4.0 cm aluminum plate is 5.3×10^{-5} A. What is the corrosion penetration rate in mm/year?

10.11 The corrosion current measured on a 5.0 cm $\times$ 2.0 cm iron plate is 3.2×10^{-5} A. What is the corrosion penetration rate in mm/year?

10.12 The corrosion current measured on a 4.0 cm $\times$ 5.0 cm copper plate is 6.1×10^{-5} A. If the initial thickness of the plate is 1.0 cm, what is its thickness after 15 years?

10.13 The corrosion current measured on a 4.0 cm $\times$ 3.0 cm plate of iron is 4.1×10^{-5} A. What should the initial thickness of the plate be if the thickness after 15 years must be at least 0.75 cm? You may assume that the corrosion occurs on only one side of the plate.

10.14 For each of the metals below, compute the Pilling–Bedworth ratio. On the basis of this value, specify whether or not you would expect the oxide that forms to be protective.

Metal	Metal density (g/cm^3)	Metal oxide	Oxide density (g/cm^3)
Zr	6.51	ZrO_2	5.89
Bi	9.80	Bi_2O_3	8.90
Al	2.71	Al_2O_3	3.95
Fe	7.87	FeO	5.74

10.15 Polypropylene has a solubility parameter of 16.6 MPa-m$^{1/2}$. Which of the solvents in Table 10.4.2 will dissolve polypropylene?

10.16 PVC has a solubility parameter of 19.5 MPa-m$^{1/2}$. Which of the solvents in Table 10.4.2 will dissolve polypropylene?

Conceptual Problems

10.17 An iron bar with the dimensions shown below is being used as a structural support member, underwater in the presence of HCl. Corrosion occurs only on the top surface (marked T in the figure below), and the corrosion current measured on this surface is 1.58×10^{-2} amperes. This structural member is inspected periodically using ultrasound, and the smallest internal crack that can cause failure is 6 mm (you may assume there are no surface cracks). The critical stress intensity parameter of the iron is 41 MPa-m$^{1/2}$. The force applied to this member is 1,750 kN in the direction shown in the figure. How long can this member remain in service before it needs to be replaced?

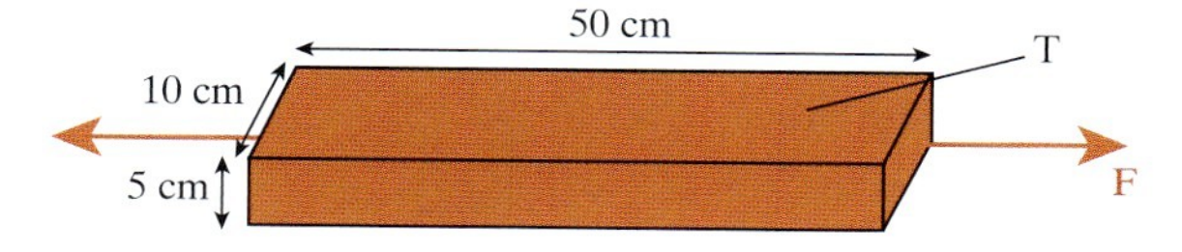

10.18 For each of the following situations give (i) the type of corrosion that occurs, and (ii) one way to prevent that type of corrosion.

a. A piece of nickel is coated with copper. A scratch develops, which quickly rusts.

b. Two pieces of metal are joined by a bolt. After some time the part fails because the bolt has corroded.

c. An underwater steel bridge support shows signs of corrosion over its entire length.

10.19 From the galvanic series, identify three metals or alloys that may be used to galvanically protect brass.

10.20 Sometimes galvanic corrosion of two metals in contact is prevented by making an electrical contact between both metals in the couple and a third metal that is anodic to the other two. Using the galvanic series, name one metal that could be used to protect a copper–aluminum couple.

10.21 The phrase "like dissolves like" is often used to describe whether a solvent will dissolve something. Explain how this phrase has a basis in the concept of the solubility parameter.

10.22 Water has a solubility parameter of 47.9 and so it will not dissolve Nylon 66. However, Nylon 66 can still absorb a substantial amount of water.

a. On the basis of structure, explain how this is possible.

b. How do you expect the properties of Nylon 66 to change when it absorbs water?

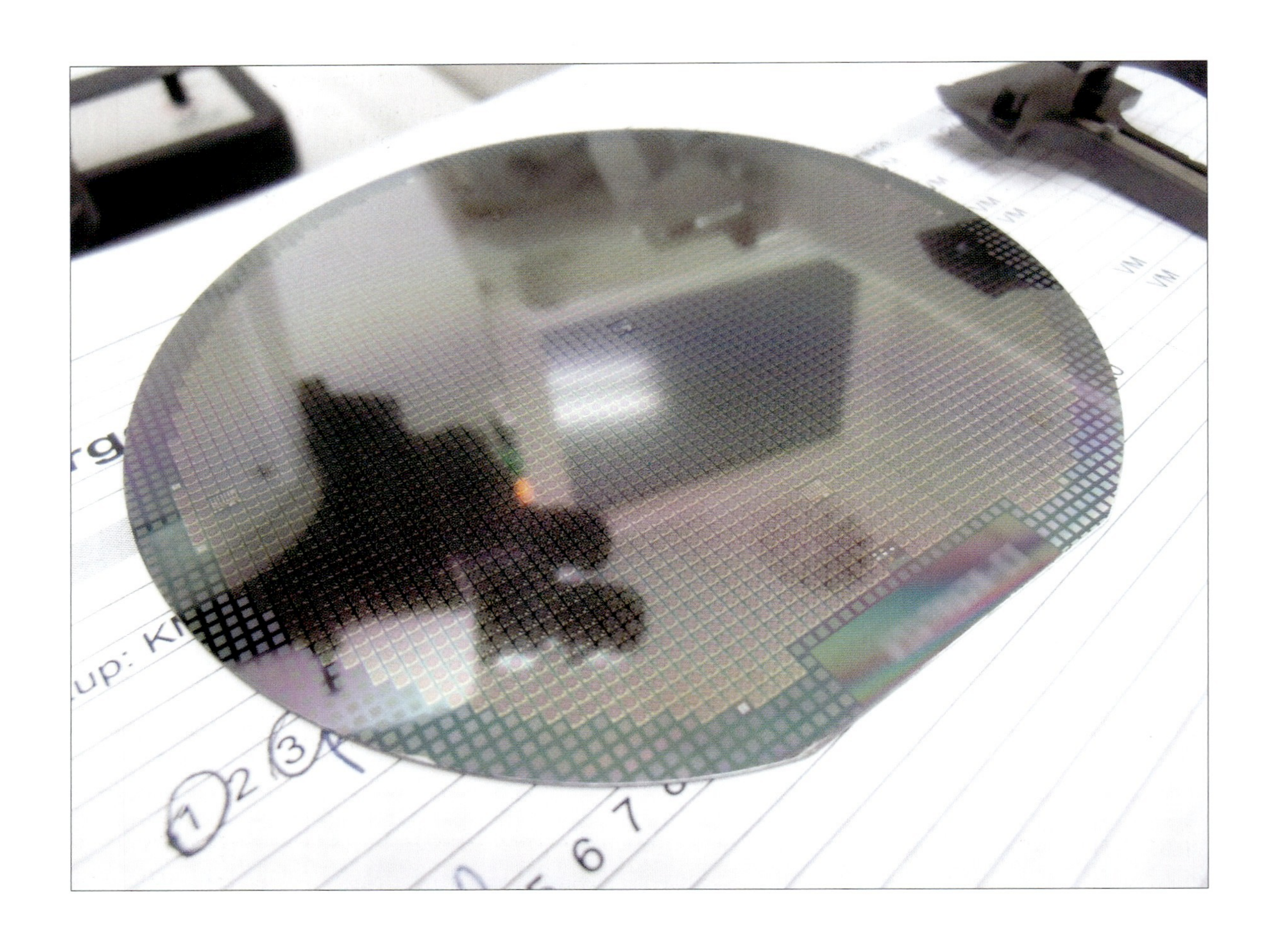

Computer chips are manufactured on slices of silicon called wafers. Each of the small squares on this wafer is a computer chip, which in turn has millions of MOSFETs. (A. Penkov/Shutterstock)

11 Electronic Behavior

Electronic properties have become a critical aspect of our current economy. From computers to smartphones to advanced lighting, much of what we rely on for our everyday activities depends on the electronic properties of the materials in the devices that we buy. While we usually think of conduction, a wide range of other electronic properties are needed to make these devices work. Figure 11.1.1 shows the conductivity values for a variety of different materials. Notice in this figure the enormous range over which conductivity varies, covering approximately 20 orders of magnitude. This is easily the widest range for any particular property. To give you some idea of what this means, if we put distances on the same scale the shortest distance would be the radius of an atom, while the largest distance would be three times the distance from Earth to the nearest star! And this range doesn't even include superconductors, which have essentially infinite conductivity. By the end of this chapter you will:

> Understand how structure of different materials affects conductivity.
>
> Be able to describe the structure and operation of a semiconductor device.

11.1 Band Structure of Materials

> LEARN TO: Describe band structure of materials.
> Describe mechanisms of conductivity.

To understand how electrons behave in solids, we will start by going back to general chemistry. In that class you learned about the energy levels of electrons in atoms (1s, 2s, 2p, etc.). For isolated atoms these are discrete energy levels, but as atoms get closer together the energy levels split and create energy bands, which the electrons from all the atoms

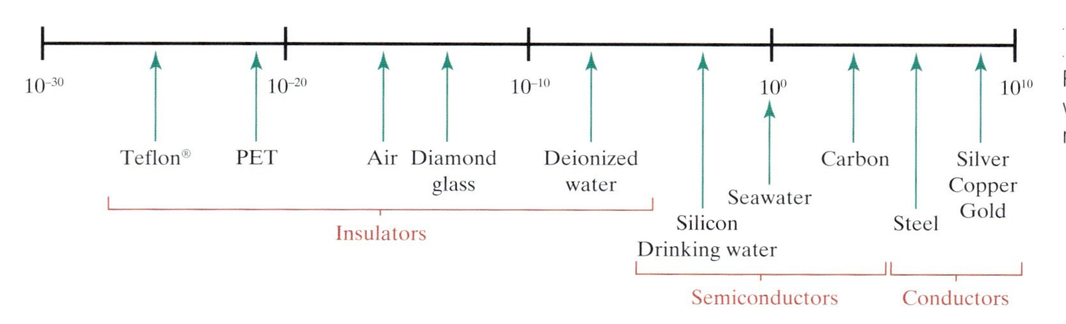

Figure 11.1.1 Range of conductivity values for different materials.

occupy, as illustrated in Figure 11.1.2. This splitting into energy bands occurs because of the Pauli exclusion principle, which requires that no two electrons have exactly the same energy. Usually we draw this band structure for the equilibrium spacing between the atoms as shown in Figure 11.1.3. When we draw the band diagram the y-axis is energy and the x-axis has no real meaning.

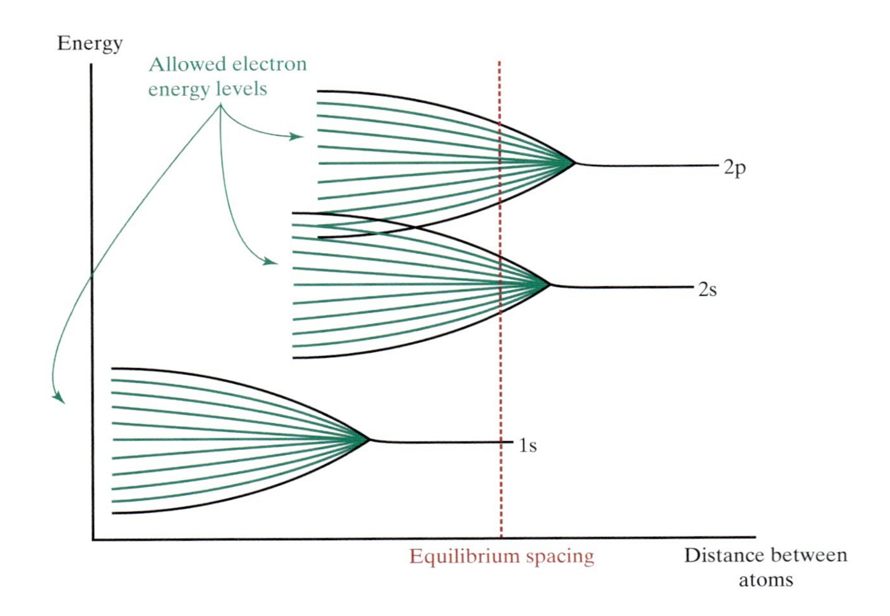

Figure 11.1.2
Illustration of how the discrete energy levels of atoms split into energy bands as atoms come closer together.

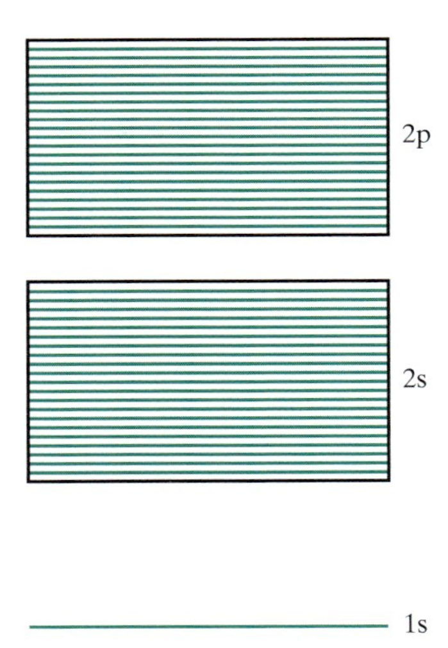

Figure 11.1.3
Energy-band diagram resulting from Figure 11.1.1.

Different kinds of materials have different kinds of band structures. These are illustrated in Figure 11.1.4. Some definitions related to this band structure are as follows:

- *Valence band*: The energy band containing the valence electrons. The difference between energy levels inside the valence band is so small that it is effectively zero.
- *Conduction band*: The next energy band above the valence band. The difference between energy levels inside the conduction band is so small that it is effectively zero.
- *Fermi energy*: An electron must have energy greater than the Fermi energy to conduct electricity.
- *Bandgap*: For some materials there is a large gap in allowed energy levels between the valence band and conduction band, called the energy gap (E_g). Bandgaps typically have energies on the order of a few eV.

The different band structures in Figure 11.1.4 result in different kinds of electronic behavior. You may already know that copper and magnesium are electrical conductors, silicon is a semiconductor, and polyethylene is an insulator. In the next set of questions you will see how the different band structures result in these different electronic properties.

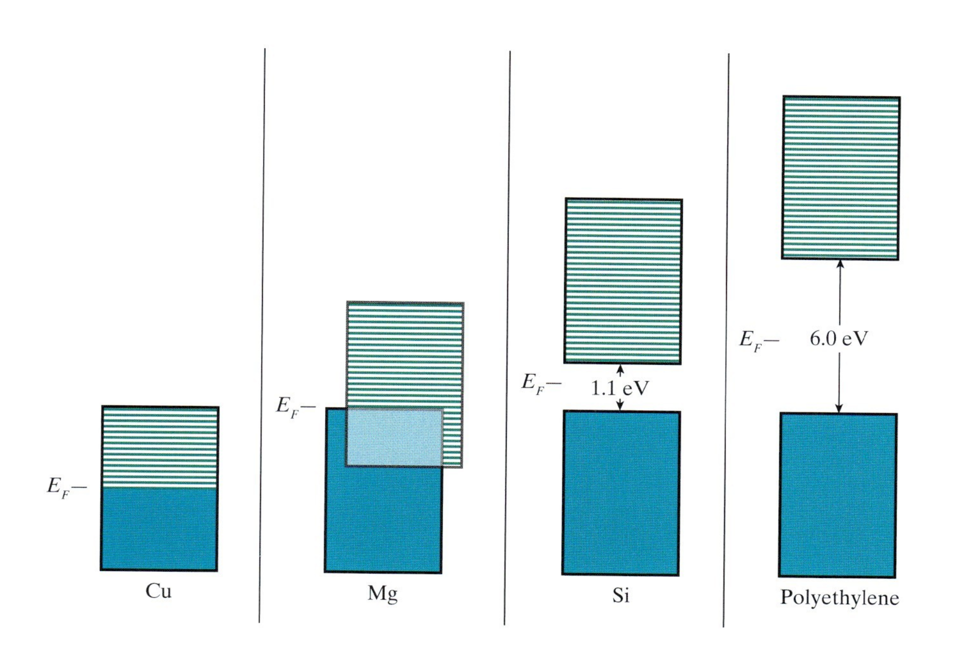

Figure 11.1.4

Band structures for different types of materials. Copper has a partially filled valence band. Magnesium, silicon, and polyethylene all have completely filled valence bands.

Guided Inquiry: Band Structures

11.1.1 Compare the band structures for copper, silicon, and polyethylene. Which one requires the least energy for an electron to be able to jump over the Fermi energy? Which requires the most?

The electron may not need to jump far. It just needs to get to an allowed energy level above the Fermi energy.

11.1.2 Which will be the best conductor of electricity: copper, silicon, or polyethylene? Which will be the worst?

How easy is it for an electron to get above the Fermi energy?

11.1.3 For copper, what is the difference in energy between the highest-energy valence electron and the next available energy level?

11.1.4 If an electron in copper is given energy of 0.5 eV, what will happen to it? Will it be able to act as a conductor?

11.1.5 What is the minimum energy required to make an electron act as a conductor in copper?

11.1.6 For silicon, what is the difference in energy between the highest-energy valence electron and the next available energy level?

11.1.7 If an electron in silicon is given energy of 0.5 eV, what will happen to it? Will it be able to act as a conductor?

11.1.8 What is the minimum energy required to make an electron act as a conductor in silicon?

11.1.9 Based on Figure 11.1.4, describe the differences between a conductor, a semiconductor, and an insulator.

Concept Check **11.1.1**

At 25° C, the thermal energy available is approximately 0.03 eV.

- **Would you expect there to be free electrons in copper at 25° C?**
- **Would you expect there to be free electrons in silicon at 25° C?**

11.2 Electronic Properties

LEARN TO: **Define electrical quantities.**

You should be familiar with Ohm's law ($V = IR$), but when we consider electronic behavior from a materials standpoint we want to look at the fundamental materials properties. At a macroscopic level the important property is the *resistivity*, or *conductivity*. *Resistance* of a wire is an extrinsic property that depends on the type of material and the shape of the wire. Resistivity is an intrinsic property that depends only on the type of material. The relationship between resistance and resistivity is shown here:

$$R = \frac{l\rho}{A} \quad (11.2.1)$$

where l is the length of the wire, A is the cross-sectional area of the wire, and ρ is the resistivity. Because the same symbol ρ is also used for density, you have to look at the context of the problem to know which meaning is intended. We can also define the conductivity of a material as $1/\rho$. The symbol σ used for conductivity is also used for stress. Again, you will have to pay attention to the context of the problem.

Guided Inquiry: Calculating Resistance

11.2.1 What are the units for resistance?

11.2.2 What are the units for length?

11.2.3 What are the units for cross-sectional area?

11.2.4 What are the units for resistivity?

11.2.5 You would like to create a 5,000 ohm resistor of boron carbide fibers that are 0.1 mm in diameter. How long should the fibers be? The conductivity of boron carbide is 1.5 $(\text{ohm-cm})^{-1}$.

Make sure to work together in your group. Does everyone agree on the answer?

Concept Check 11.2.1

- **What is the answer to question 11.2.5?**

EXAMPLE PROBLEM 11.2.1

You would like to create 100 ohm resistors from carbon fibers with a length of 10 mm. What should the diameter of the carbon fibers be? The conductivity of carbon is 0.15 (ohm-cm)$^{-1}$.

We start with the following equation:

$$R = \frac{l\rho}{A}$$

Since we want to know the diameter, we will need to solve this equation for the area:

$$A = \frac{l\rho}{R}$$

The variables are as follows:

$l = 10$ mm $= 1$ cm

$\rho = 1/\sigma = 1/(0.15\ (\text{ohm-cm})^{-1}) = 6.67$ ohm-cm

$R = 100$ ohm

$$A = \frac{(1\ \text{cm})(6.67\ \text{ohm} - \text{cm})}{100\ \text{ohm}} = 0.0667\ \text{cm}^2$$

$$A = \pi r^2$$

$$r = \sqrt{\frac{A}{\pi}} = 0.145\ \text{cm}$$

Diameter = 0.29 cm = 2.9 mm.

At the atomic level, conductivity is the result of electrons moving through a material. As the electron moves, it is scattered, as shown in Figure 11.2.1.

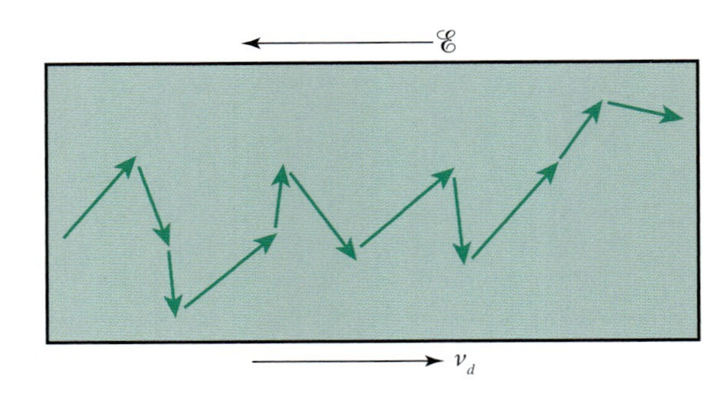

Figure 11.2.1

The motion of an electron in a solid. Although its overall motion is in the direction defined by the electric field, the electron is scattered in different directions as it moves.

The average velocity of the electron is called the drift velocity and is given by the equation

$$v_d = \mu_e \mathscr{E} \tag{11.2.2}$$

where μ_e is the mobility of the electron and $\mathscr{E}$ is the strength of the applied electric field. The mobility is also directly related to the conductivity according to

$$\sigma = n|e|\mu_e \tag{11.2.3}$$

where n is the number of free electrons—that is, electrons with an energy above E_F—and $|e|$ is the absolute value of the charge on an electron. In the case of a semiconductor, we also need to consider the "*hole*" left in the valence band when an electron jumps into the conduction band. Figures 11.2.2 and 11.2.3 show the classical view of the electrons in the bonds and the band diagram for a semiconductor. Note that the hole in the valence band is the same as the hole shown in the structure of Figure 11.2.2; these are just two different ways of representing the same thing. Holes have a positive charge.

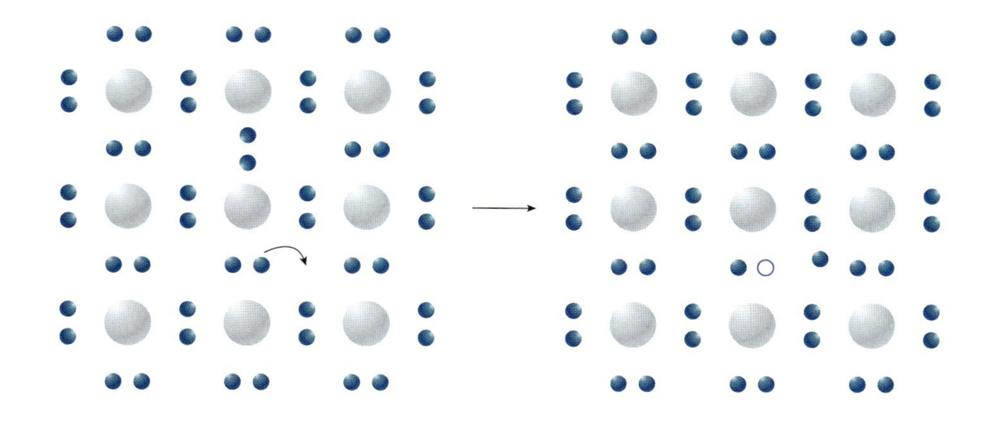

Figure 11.2.2
Classic view of what happens to an electron in a semiconductor when it gets enough energy to leave the valence band. The electron becomes free and leaves a hole in the bond where it used to be.

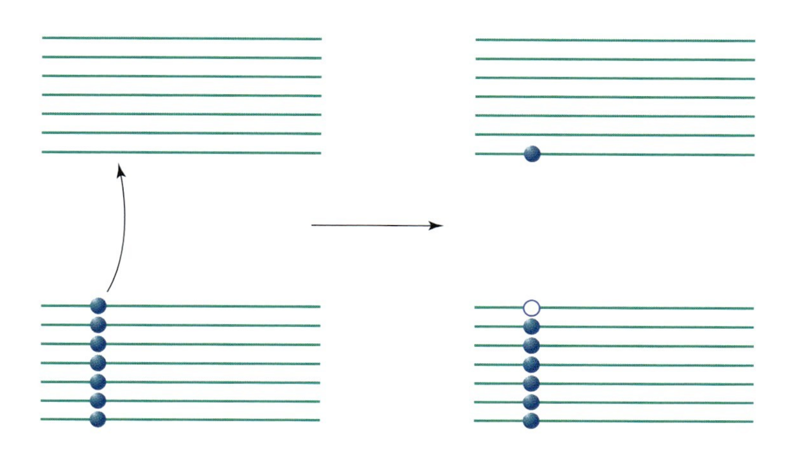

Figure 11.2.3
Band diagram of a semiconductor. When the electron jumps into the conduction band, it leaves behind a hole in the valence band.

Guided Inquiry: Semiconductors

11.2.6 In a semiconductor, if an electron jumps to the conduction band, what is left behind in the valence band?

11.2.7 Using Figure 11.2.2, imagine that an electron from a nearby bond jumps into the hole. Where is the hole now?

Draw a picture to help you see what is going on.

11.2.8 If the process described in question 11.2.7 continues, what happens to the hole?

11.2.9 If an electric field is applied, in what direction does the hole move compared to the direction in which the electron moves?

Concept Check 11.2.2

- **In a semiconductor, what charge carriers are present?**

Since a hole can act as a conductor, we need to modify Equation (11.2.3) for semiconductors. There are equal numbers of holes and electrons and they have the same charge, but the mobilities are slightly different; holes have a lower mobility than electrons. As a result, the conductivity equation becomes

$$\sigma = n|e|(\mu_e + \mu_p) \qquad (11.2.4)$$

where μ_p is the mobility of the holes. It turns out that holes are critically important for the operation of computers. By controlling the number of holes and electrons in a semiconductor we can create the devices that make up a computer chip as we will see later in this chapter.

11.3 Conductors

LEARN TO: Predict how structure affects conductivity for conductors.

For a conductor, there is enough thermal energy at room temperature for electrons to easily jump over the Fermi energy and be able to conduct electricity. This means that there are always plenty of electrons above the Fermi energy, and the number of electrons in Equation (11.2.3) doesn't change. However, the mobility of the electrons can change depending on the structure of the material. In particular, the presence of defects causes greater electron scattering, thus reducing the mobility. This is illustrated in Figure 11.3.1.

Why do defects cause greater scattering? Figure 11.3.1 shows the scattering as if the electrons were particles bouncing off the defect, but this doesn't really explain why they scatter, especially if you consider vacancies. Why would an electron bounce off an empty space? The real answer is related to the wave properties of electrons. Quantum mechanics shows us that if the crystal is perfect, the electron wave can propagate through the crystal without any interference. When there is a defect, the wave is disrupted because the crystal is no longer perfect. This disruption of the wave is really what electron scattering is.

Regardless of the explanation, the important aspect is that a defect in a crystal will cause electron scattering, resulting in a drop in electron mobility and a drop in conductivity.

In the next questions, by considering the factors that affect the perfection of a crystal, you will be able to predict which of a pair of materials will have the higher conductivity. As you work on these questions, think back to section 4.5 on crystalline defects to help you.

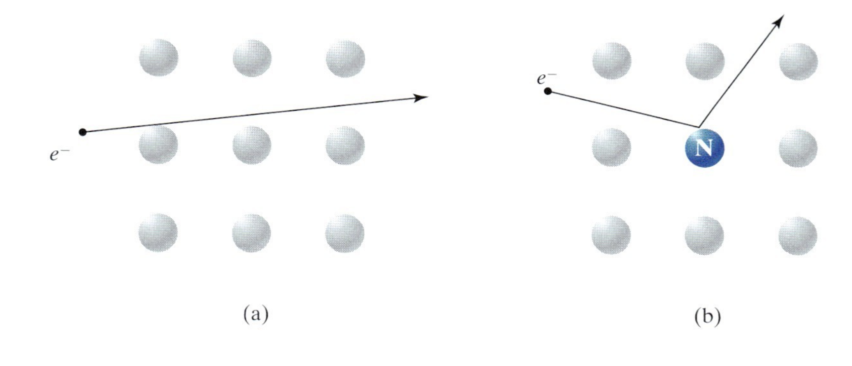

Figure 11.3.1

Illustration of the effect of defects on electron mobility: (a) a material with no defects, (b) a material with an impurity atom.

Guided Inquiry: Electron Mobility

11.3.1 Does a piece of copper held at 200° C have more or fewer vacancies than a piece of copper held at 100° C?
Go to Section 4.5 to review how temperature affects vacancies.

11.3.2 Are there more defects in a piece of copper held at 200° C than in a piece of copper held at 100° C?

11.3.3 Will there be more scattering of electrons in a piece of copper held at 200° C than in a piece of copper held at 100° C?

11.3.4 Which will have the higher conductivity; a piece of copper held at 200° C or a piece of copper held at 100° C?

11.3.5 Will conductivity of a metal increase or decrease as temperature is increased?

11.3.6 Are there more defects in pure copper or in a copper/aluminum alloy?

11.3.7 Will there be more scattering of electrons in pure copper or in a copper/aluminum alloy?

11.3.8 Which will have the higher conductivity; pure copper or a copper/aluminum alloy?

11.3.9 Will conductivity of a metal increase or decrease as impurity atoms are added?

11.3.10 Are there more defects in annealed copper or in copper that has been plastically deformed?
Go to Sections 4.5 and 9.4 to review how annealing and plastic deformation affect defects.

11.3.11 Will there be more scattering of electrons in annealed copper or in copper that has been plastically deformed?

11.3.12 Which will have the higher conductivity; annealed copper or copper that has been plastically deformed?

11.3.13 Will conductivity of a metal increase or decrease when it is plastically deformed?

11.3.14 Are there more defects in copper with 10 μm grains or with 100 μm grains?

11.3.15 Will there be more scattering of electrons in copper with 10 μm grains or with 100 μm grains?

11.3.16 Which will have the higher conductivity; copper with 10 μm grains or with 100 μm grains?

11.3.17 Will conductivity of a metal increase or decrease when the grain size increases?

11.3.18 Which will have a higher conductivity; copper solidified at 800° C or copper solidified at 600° C?

Which has the larger grains? See Section 8.1 if you need to review.

11.3.19 List the four factors that affect the conductivity of metals and describe how the conductivity changes with each of those factors.

Concept Check 11.3.1

- **Which has the higher conductivity; 24k gold or 12k gold?**
- **Which has the higher conductivity; copper solidified at 800° C or copper solidified at 600° C?**

11.4 Semiconductors

LEARN TO: Predict how structure affects conductivity for semiconductors.

In some ways semiconductors are like conductors, but in some ways they are very different. The effect of defects on mobility is the same in both materials. However, because of the bandgap in semiconductors, it is much harder for electrons to jump over the Fermi level into the conduction band and become available to conduct electricity. This means that the number of electrons in Equation (11.2.4) is not constant. In fact, the conductivity of semiconductors is generally dominated by the change in the number of conductors, rather than the mobility.

Guided Inquiry: Number of Charge Carriers

11.4.1 At higher temperatures, will electrons have more energy or less energy?

11.4.2 At higher temperatures, will it be more likely or less likely for an electron to have enough energy to jump over the bandgap?

11.4.3 At higher temperatures, will there be more or fewer electrons in the conduction band?

11.4.4 At higher temperatures, will the conductivity of a semiconductor be lower or higher?

The conductivity of a semiconductor follows the equation below:

$$\sigma = Ae^{-E_g/2kT} \tag{11.4.1}$$

where A is a constant, E_g is the bandgap, k is Boltzmann's constant, and T is temperature. This equation is the mathematical expression that describes the effect of temperature on conductivity that you determined qualitatively in the previous questions. This equation is also important because you can measure conductivity as a function of temperature and then use the equation to determine the bandgap of a semiconductor. The next questions will guide you on how to do this.

Guided Inquiry: Determining Bandgaps

11.4.5 Take the natural logarithm of both sides of Equation (11.4.1) and write the resulting equation.

Remember the rules for logarithms: ln (AB) = ln A + ln B.

11.4.6 If you plotted ln σ versus $1/T$, what shape would the graph be? What would be the slope?

To help you see this, rewrite your equation by substituting y for ln σ and x for $1/T$.

11.4.7 Propose a method to determine the bandgap of a semiconductor.

Concept Check 11.4.1

- **Draw a graph of ln σ versus $1/T$. What is the slope of this graph?**

EXAMPLE PROBLEM 11.4.1

The table below gives the intrinsic electrical conductivity of a semiconductor at two temperatures. Determine the bandgap energy in eV for this material.

T (°C)	σ ((Ω-m)$^{-1}$)
177	0.12
277	2.25

From Equation (11.4.1), a plot of ln σ versus $1/T$ should be linear with a slope of $-E_g/2k$. So we can expand the table to calculate the values needed:

T (°C)	σ ((Ω-m)$^{-1}$)	T (K)	$1/T$ (1/K)	ln σ
177	0.12	450	0.00222	−2.12
277	2.25	550	0.00182	0.81

Now we can calculate the slope:

$$\text{slope} = \frac{\ln\sigma_1 - \ln\sigma_2}{\dfrac{1}{T_1} - \dfrac{1}{T_2}} = \frac{-2.12 - 0.81}{0.00222\ \text{K}^{-1} - 0.00182\ \text{K}^{-1}} = -7325\ \text{K}$$

$$-\frac{E_g}{2k} = -7325$$

$$E_g = 2(7325\ \text{K})k = 2(7325\ \text{K})(8.62 \times 10^{-5}\ \text{eV/K}) = 1.26\ \text{eV}$$

Impurities also have a different effect in semiconductors than they do in metals. Imagine that you chose an impurity having more valence electrons than the base material—for example, a phosphorus impurity in silicon. The situation is shown in Figure 11.4.1. Again, this is a classical picture of something that is really described by quantum mechanics, but for us this classical picture is good enough. The extra electron from the phosphorus atom is not part of any bonds, so it can move and therefore conduct electricity much more easily than any of the other electrons. We can do the same thing with an impurity that has fewer electrons than the base material—for example, boron in silicon. Figure 11.4.2 shows how the missing electron—that is, a hole—can easily conduct electricity by providing a location for other electrons to hop into.

The result of these types of impurities on the band structures is shown in Figure 11.4.3. In semiconductors these impurities are called *dopants*. A dopant with an extra valence electron is called *n*-type (because it has an extra negative charge), and a dopant with fewer

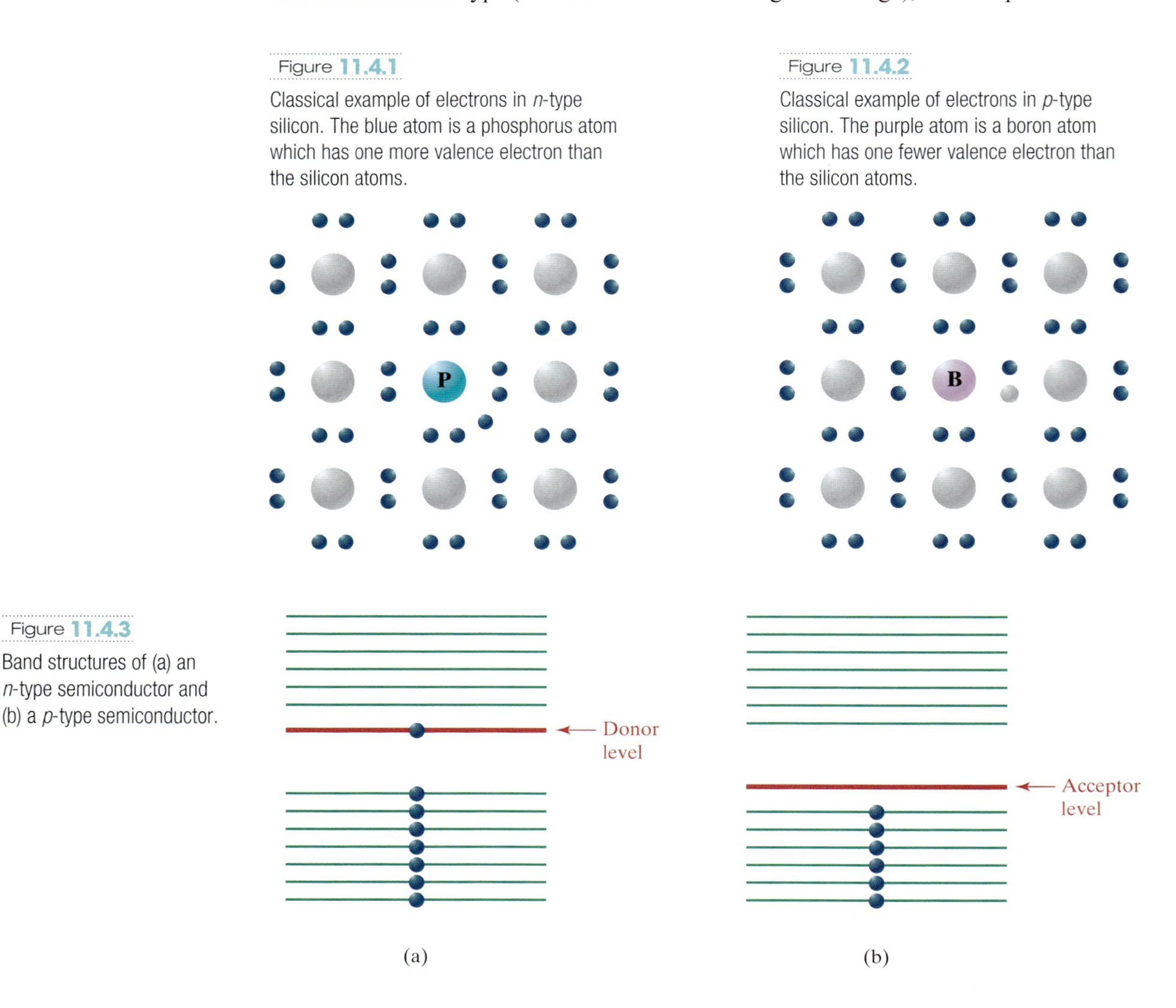

Figure 11.4.1
Classical example of electrons in *n*-type silicon. The blue atom is a phosphorus atom which has one more valence electron than the silicon atoms.

Figure 11.4.2
Classical example of electrons in *p*-type silicon. The purple atom is a boron atom which has one fewer valence electron than the silicon atoms.

Figure 11.4.3
Band structures of (a) an *n*-type semiconductor and (b) a *p*-type semiconductor.

valence electrons is called *p*-type (because it has an extra positive charge). In an *n-type semiconductor*, the energy level of the extra electron is called a donor level. In a *p-type semiconductor*, the energy level of the hole is called an acceptor level.

Guided Inquiry: Doped Semiconductors

11.4.8 In an *n*-type semiconductor, which takes less energy; an electron jumping from the donor level to the conduction band or an electron jumping from the valence band to the conduction band?

11.4.9 Which will have more electrons in the conduction band; an undoped semiconductor or an *n*-type semiconductor?

11.4.10 Which will have a higher conductivity; an undoped semiconductor or an *n*-type semiconductor?

11.4.11 In a *p*-type semiconductor, which takes less energy; an electron jumping from the valence band to the acceptor level or an electron jumping from the valence band to the conduction band?

11.4.12 Which will have more holes in the valence band; an undoped semiconductor or a *p*-type semiconductor?

11.4.13 Which will have a higher conductivity; an undoped semiconductor or a *p*-type semiconductor?

11.4.14 List the two factors that affect the conductivity of semiconductors and how the conductivity changes with each of them. Compare the effects to how those same factors affect the conductivity of conductors.

Concept Check **11.4.2**

- **Will the conductivity of silicon be higher at 25° C or 80° C?**
- **Which will have a higher conductivity; pure silicon or silicon with boron impurity atoms?**

For a doped semiconductor, the temperature dependence of conductivity is given by the following:

$$\sigma = Ae^{-\Delta E/kT} \quad (11.4.2)$$

where ΔE is the gap from the donor level to the conduction band for an n-type semiconductor or the gap from the valence band to the acceptor level for a p-type semiconductor.

Guided Inquiry: Determining Dopant Energies

11.4.15 Based on your answers to questions 11.4.5–11.4.7, propose a method for determining the gap between the donor level and conduction band for an n-type semiconductor.

Concept Check 11.4.3

- **Draw a graph of ln σ versus $1/T$ for an n-type semiconductor. What is the slope of this graph?**

11.5 Solid-State Devices

LEARN TO: Describe the operation of electronic devices.

The ability to dope semiconductors to make them n-type or p-type is what allows the creation of solid-state devices to make computer chips. Let's consider a transistor: a transistor is a device that controls the flow of electricity with a valve, just like a valve for a water pipe. When the valve is in one state the electricity flows; when you change the valve's state, the flow of electricity stops. This allows a transistor to serve as the binary code in computers; when the electricity is flowing, the code is a 1; when the electricity is not flowing, the code is a 0.

Transistors have evolved over the years. Originally vacuum tubes were used to do the job that transistors do now. The original transistors were separate devices with three wires that were attached to circuits. Now transistors are made as part of the entire computer chip, hence the name "integrated circuit."

One common type of transistor used today is the metal-oxide-semiconductor field-effect transistor, or *MOSFET*. MOSFETs come in four basic types; they can be either n-type or p-type, and either *depletion mode* or *enhancement mode*. To understand the operation of a MOSFET, we will look at a depletion-mode p-type MOSFET, drawn schematically in Figure 11.5.1. In this figure, there is a block of n-type silicon with a channel of p-type silicon at the surface. There are also various layers of metal and silicon oxide at the surface. Keep in mind that this MOSFET is very small. The channel length can be less than 100 nm!

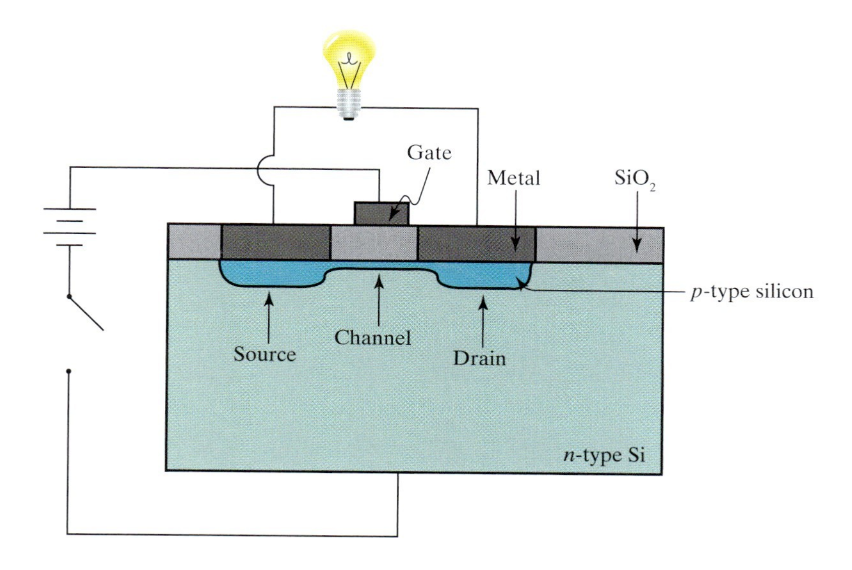

Figure 11.5.1

Schematic of a depletion-mode *p*-type MOSFET.

The key components of the MOSFET are:

- *Source:* Where electricity enters the MOSFET when the valve is set so electricity can flow.
- *Drain:* Where electricity leaves the MOSFET when the valve is set so electricity can flow.
- *Channel:* The region between the source and the drain.
- *Gate:* This is the valve. It works by creating an electric field across the channel.

The key point about how the MOSFET works is that electricity will flow only if the type of charge carrier is consistent across from the source, through the channel, to the gate. This is because electrons and holes will annihilate each other when they meet, disrupting the flow of current. For example, imagine that the source and drain are *p*-type, while the channel is *n*-type. Figure 11.5.2 shows what will happen. The dominant charge carriers in the source and the drain are holes, which move from the drain to the source (opposite the motion of electrons). However, the dominant carriers in the channel are electrons. When holes from the drain enter the channel, they meet these electrons, are annihilated, and never reach the source.

Figure 11.5.2

Example of holes and electrons annihilating each other and preventing current flow.

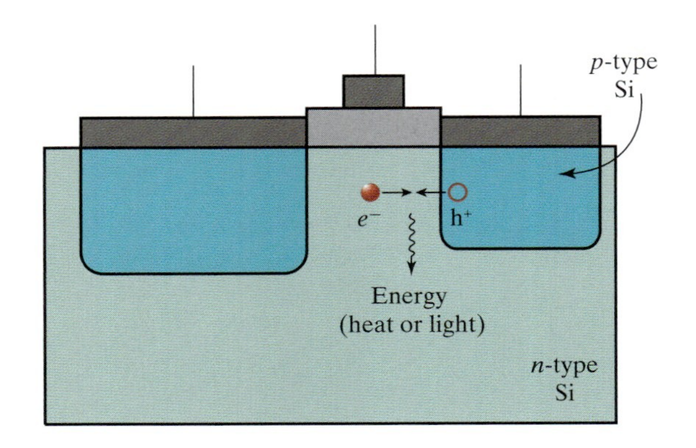

Application Spotlight **Organic LEDs and Solar Cells**

(Photo courtesy of Lifeprints Photography)

Although we usually think of polymers as insulators, some polymers are semiconductors. The Nobel Prize in Chemistry for 2000 was given for the discovery of semiconducting polymers. These polymers can be used for light-emitting diodes (LEDs) and solar cells, and are sustainable energy solutions for both electricity usage (efficient LEDs) and generation (solar cells). An LED takes advantage of the bandgap in a semiconductor to generate light. When electrons and holes enter the semiconductor from an electrical circuit, they recombine by the electron's dropping back down to the valence band. In certain materials the energy is given off as light, with a wavelength corresponding to the bandgap. A solar cell works in the same way, but in reverse; light is absorbed and excites an electron to the conduction band, which then results in a current. Most LEDs that you see are based on inorganic semiconductors. For example, LED traffic signals consist of multiple little dots, each of which is essentially a computer chip. Organic LEDs are less common because there are still some technical problems, the two primary problems being that organic LEDs have lower efficiencies and shorter lifetimes than inorganic LEDs. But organic LEDs have some advantages. One is that processing is easier because you don't need to use high temperatures. Organic LEDs can be made with low-temperature processes similar to inkjet printing and silk-screening. Another advantage is that organic LEDs can be made as large sheets. So some day, instead of having a light fixture, you may just have wallpaper that lights your room!

Guided Inquiry: MOSFETs

11.5.1 In Figure 11.5.1, what type of silicon is at the source? What kind of charge carriers are present at the source; holes or electrons?

11.5.2 In Figure 11.5.1, what type of silicon is at the drain? What kind of charge carriers are present in the drain; holes or electrons?

11.5.3 In Figure 11.5.1, what type of silicon is at the channel? What kind of charge carriers are present in the channel; holes or electrons?

11.5.4 Are the same types of charge carriers present at the source, channel, and drain?

11.5.5 Will charge carriers from the drain be annihilated when they enter the channel?
Electrons and holes annihilate each other when they meet.

11.5.6 Will electricity flow from the source to the drain? Will the lightbulb light up?
If the charge carriers are annihilated, they can't cross the channel.

11.5.7 The switch in Figure 11.5.1 is now closed so that a positive charge is created at the gate. Will holes in the channel be attracted to or repelled from the gate?

11.5.8 With the switch closed, will electrons in the *n*-type silicon be attracted to or repelled from the gate and the channel?

11.5.9 With the switch closed, what type of charge carrier will be in the channel?

11.5.10 With the switch closed, are the same types of charge carriers present at the source, channel, and gate?

11.5.11 Will charge carriers from the drain be annihilated when they enter the channel?

11.5.12 With the switch closed, will electricity flow from the source to the gate? Will the lightbulb light up?

Concept Check **11.5.1**

- **In a depletion-mode *p*-type MOSFET, when there is no voltage at the gate, will current flow from the source to the drain?**
- **In a depletion-mode *p*-type MOSFET, when a voltage is applied at the gate, will current flow from the source to the drain?**

Summary

All materials have electronic properties of some kind, whether they are insulators, semiconductors, or conductors. As you have seen, the conductivity of conductors is governed by the way structure affects scattering of electrons. The aspects of structure that are important are the same ones that influence strength; defects, grain size, plastic deformation, and annealing, as well as temperature. For conductivity of semiconductors what matters is not scattering of electrons, but rather the number of electrons that are present, so the important factors are temperature and the presence of any dopant atoms. It is important to note that for semiconductors the mobility changes in the same way as for conductors. The difference is that for semiconductors mobility is not the dominant factor affecting conductivity.

Section 11.5 was intended to give you an example of how semiconductors are used in electronic devices. The MOSFET is just one type of electronic device present in computer chips. Other devices include the junction field-effect transistor; the metal semiconductor field effect transistor; diodes, which allow current to flow in only one direction; and capacitors for storing charge. Through the MSE triangle, materials engineers continue to develop new, more efficient devices and shrink the size of existing ones. As a result, the smartphones of today have much more computing power than the desktop computers of 10 years ago, and that trend is expected to continue. We can only imagine what capabilities devices will have 10 or 20 years from now.

Key Terms

Bandgap
Conduction band
Conductivity
Depletion mode
Dopant
Enhancement mode
Fermi energy
Hole
MOSFET
***n*-type semiconductor**
***p*-type semiconductor**
Resistance
Resistivity
Valence band

Problems

Skill Problems

11.1 You would like to create a 200 ohm resistor of carbon fibers that are 0.3 mm in diameter. How long should the fibers be? The conductivity of carbon is 15 (ohm-cm)$^{-1}$.

11.2 You would like to create a 3,000 ohm resistor of boron carbide fibers that are 7 mm in length. What should the diameter of the fibers be? The conductivity of boron carbide is 1.5 (ohm-cm)$^{-1}$.

11.3 Draw the band structure of silicon doped with aluminum. What are the primary charge carriers in this material?

11.4 Draw the band structure of silicon doped with arsenic. What are the primary charge carriers in this material?

11.5 For each of the following, identify which has the higher conductivity and explain why:

a. Silver at a temperature of 20° C or silver at a temperature of 60° C.

b. Silicon at a temperature of 20° C or silicon at a temperature of 60° C.

c. Single-crystal silicon or polycrystal silicon.

d. Copper or brass.

e. Aluminum solidified at 500° C or aluminum solidified at 700° C.

f. Copper that has been plastically deformed or copper that has been annealed.

11.6 The table at top right shows the conductivity of a semiconductor measured at two different temperatures. What is the bandgap for this material?

T (°C)	σ ((Ω-m)$^{-1}$)
150	0.13
200	3.17

11.7 The table at lower right shows the conductivity of a semiconductor measured at two different temperatures. What is the bandgap for this material?

T (°C)	σ ((Ω-m)$^{-1}$)
80	0.38
120	1.12

11.8 Germanium has a bandgap of 0.7 eV, while silicon has a bandgap of 1.1 eV. Which would you expect to have the higher conductivity at room temperature, and why?

Conceptual Problems

11.9 The bandgap of silicon is 1.1 eV, and the thermal energy at room temperature is 0.3 eV. How is it possible for there to be free electrons in silicon at room temperature?

11.10 What should be the bandgap of a semiconductor for it to serve as an LED that emits blue light?

11.11 What should be the bandgap of a semiconductor for it to serve as a solar cell that absorbs the maximum intensity of sunlight?

11.12 The diagram at right shows the structure of an *n*-type enhancement-mode MOSFET. Describe the operation of this device, including the following: What happens to both the holes and electrons when no voltage is applied at the gate, and when a voltage is applied at the gate? Under what conditions will current flow from the source to the drain?

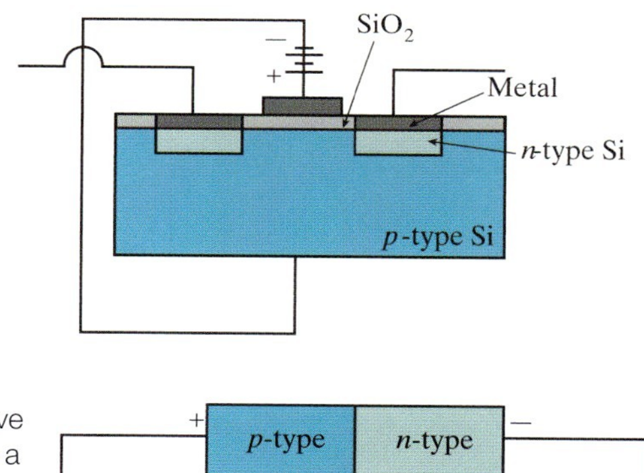

11.13 The diagram at lower right shows a *p*-*n* diode, which allows current to flow only when the voltage is applied such that the positive terminal is on the *p*-type side of the diode, as shown below (forward bias). No current can flow when the voltage is reversed from what is shown below—applied such that the positive terminal is on the *n*-type side of the diode (reverse bias). Propose a mechanism for how the electrons and holes flow under forward and reverse bias to cause the diode to work in this way.

+ *p*-type *n*-type –

Snow is actually one of the best thermal insulating materials. This is because it contains a substantial amount of air. Snow has only 8–30% of the density of water depending on how tightly it is packed. (Rita Januskeviciute/Shutterstock)

12 Thermal Behavior

As you saw in Chapter 10, the environment can affect the behavior of materials, changing their properties. One aspect of the environment that we did not address is temperature. Changes in temperature need to be considered when selecting a material or designing a new system. Obviously, extreme temperatures could lead to melting or degradation of a material. Likewise, a temperature change could cause a microstructural change with a resulting change in properties, as discussed in Chapter 8. However, we also need to consider other thermal properties that are in some ways more subtle but just as important. By the end of this chapter you will:

- Understand the mechanisms involved in heat capacity, thermal expansion, and thermal conductivity.
- Be able to calculate the effects of temperature on a material.

12.1 Heat Capacity

LEARN TO: Describe the mechanism of heat capacity.
Conduct heat capacity calculations.

Heat capacity is a measure of how much energy can be stored in a material. Energy in materials is mostly stored as vibrational energy. A greater amplitude of vibrations corresponds to a greater amount of energy in the solid. Figure 12.1.1 illustrates the vibration of the atoms in a material at two different temperatures. You will now examine how changing temperature affects the amount of energy stored.

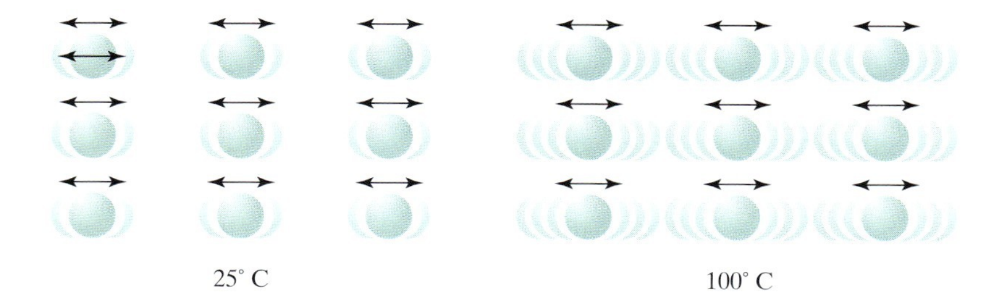

Figure 12.1.1
Atomic vibrations in a solid at two different temperatures.

Guided Inquiry: Thermal Energy

12.1.1 For the situation shown in Figure 12.1.1, at which temperature do the atomic vibrations have greater amplitude?

12.1.2 At which temperature do the atoms have greater energy?

12.1.3 At which temperature is more energy stored in the material?

Concept Check 12.1.1

- **Compare iron at 50° C and iron at 100° C. Which has more stored energy?**

For a given material, the relationship between the amount of energy it absorbs and the rise in temperature is determined by its heat capacity. Very simply, we can say that

$$Q = C\Delta T \tag{12.1.1}$$

where Q is the energy absorbed by the material, C is the heat capacity, and ΔT is the increase in temperature resulting from the energy absorbed. We can distinguish between the heat capacity measured at constant volume, C_v, and the heat capacity measured at constant pressure, C_p. However, for solids the difference between these two is very small, and we will generally just consider the constant-pressure situation.

As defined above, C_p is an extensive property that depends on the amount of material present. It is more convenient to turn this into an intensive property. At constant pressure, we can either have the *molar heat capacity*, c_m, with units of J/mol-K, or the *specific heat*, c_p, with units of J/g-K. Note that the extrinsic properties use a capital letter C, while the intrinsic properties use a lowercase letter c. Table 12.1.1. lists the molar heat capacity and the specific heat for various materials. From an engineering standpoint c_p is more convenient, but c_m has some interesting features that you will now consider.

Table 12.1.1 Heat capacities for various materials

Material	c_p (J/g-K)	c_m (J/mol-K)
Chromium	0.449	23.35
Copper	0.385	24.47
Gold	0.129	25.42
Lithium	3.58	24.8
Magnesium	1.02	24.9
Silver	0.233	24.9
Zinc	0.387	25.2

Guided Inquiry: Heat Capacity

12.1.4 What are the maximum and minimum values of c_p for the materials listed in Table 12.1.1?

12.1.5 What is the variation in c_p for the materials listed in Table 12.1.1?
Don't worry about a specific definition of variation. I just want you to get an idea of how much variation there is.

12.1.6 What are the maximum and minimum values of c_m for the materials listed in Table 12.1.1?

12.1.7 What is the variation in c_m for the materials listed in Table 12.1.1?

12.1.8 Is there an approximate universal value for c_p that applies to all materials? If so, what is that value?

12.1.9 Is there an approximate universal value for c_m that applies to all materials? If so, what is that value?

12.1.10 If you were asked to give an approximate value for the heat capacity of aluminum without looking it up, what would your answer be?

Concept Check 12.1.2

- **What is the approximate molar heat capacity of iron?**

The variation in heat capacity with temperature is shown in Figure 12.1.2. Note that the theory on which this graph is based strictly requires that we consider the constant-volume heat capacity, but since there is not much difference between constant-pressure and constant-volume heat capacities for solids, it doesn't really matter. Above the Debye temperature (θ_D) the molar heat capacity reaches a limiting value of $3R$ (R is the ideal gas constant). This is called the Dulong–Petit law, after two chemists who discovered it in 1819. For most materials the Debye temperature is below or close to room temperature, so if we are working at or above room temperature we can consider the heat capacity to be a constant value.

Figure 12.1.2

Change in heat capacity as a function of temperature.

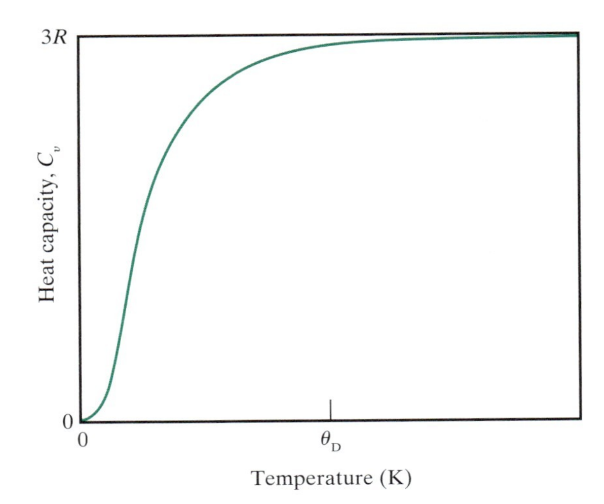

We can modify Equation (12.1.1) to use the specific heat, resulting in

$$q = mc_p \Delta T \qquad (12.1.2)$$

where m is the mass of the sample and q is the heat absorbed per gram of sample.

The next questions are based on the following scenario: A copper pipe weighing 0.5 kg is at room temperature (25° C). As a result of hot water flowing through the pipe, it absorbs 3.5 kJ of energy.

Guided Inquiry: Heat Capacity Calculations

12.1.11 List the values of the known quantities, with appropriate units, for the problem above.

12.1.12 Calculate the temperature of the pipe after it has absorbed the energy. *Try doing this problem without looking at the example problem.*

Concept Check 12.1.3

- What is the final temperature of the pipe in question 12.1.12?

EXAMPLE PROBLEM 12.1.1

A piece of zinc weighing 100 g is at room temperature (25° C). What is its temperature after absorbing 1.2 kJ of energy?

We begin with the following equation:

$$q = mc_p\Delta T$$

The values for the variables are as follows:

$q = 1200$ J

$m = 100$ g

$c_p = 0.387$ J/g-K

$$\Delta T = \frac{q}{mc_p} = \frac{1200\text{ J}}{(100\text{ g})\left(0.387\frac{\text{J}}{\text{g}-\text{K}}\right)} = 31\text{ K}$$

The temperature increases by 31 K, which is also an increase of 31° C, so the final temperature is

$$T = 25°\text{ C} + \Delta T = 25°\text{ C} + 31°\text{ C} = 56°\text{ C}.$$

12.2 Thermal Expansion

LEARN TO: **Describe the mechanism of thermal expansion.**
Conduct thermal expansion calculations.

We all know that materials expand when they are heated. Our goal in materials science and engineering is to relate this property to the structure of materials so we can predict and control their thermal expansion behavior. To understand thermal expansion we go back to the bond-energy curve in Chapter 9. Just as the bond-force curve can explain the modulus of materials based on bond strength, we can use the bond-energy curve to explain thermal expansion based on bond strength.

The bond-force curve is shown in Figure 12.2.1, but some things have been changed from the way it looked in Chapter 9. In Chapter 9 we defined the equilibrium bond distance as r_0, however, this is true only at a temperature of 0 Kelvin; above absolute zero we need to consider the energy of the system. Figure 12.2.1(a) illustrates the vibration of the atoms at a temperature above absolute zero. As the atom vibrates, it can follow the energy curve up to the maximum energy level for that temperature. The average bond distance is actually the midpoint of the possible distances it can have, which is the same as the midpoint of the line shown for the energy level. This means that at a temperature corresponding to an energy level of E_1, the average bond distance is actually r_1. Figure 12.2.1(b) shows several energy levels corresponding to different temperatures. The next set of questions will show you how the bond-energy curve is related to thermal expansion.

Figure 12.2.1

Illustration of atomic vibrations at different temperatures. (a) At a temperature above absolute zero the atom vibrates along the bond-energy curve up to a maximum energy of E_1. The average spacing between atoms is given by r_1. (b) Several possible energy levels at different temperatures. The temperature increases from E_1 to E_2 to E_3.

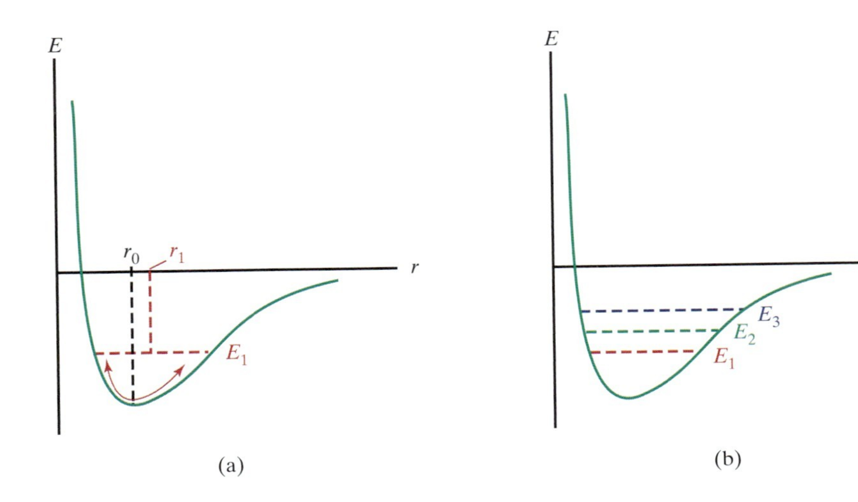

Guided Inquiry: Thermal Expansion Mechanism

12.2.1 In Figure 12.2.1(b), what is the equilibrium bond distance at absolute zero?

12.2.2 In Figure 12.2.1(b), mark the midpoint of the lines for each energy level.

12.2.3 In Figure 12.2.1(b), mark the equilibrium bond distance for each energy level.
The bond distance is not the same at the different energy levels.

12.2.4 How does the bond distance change as the temperature increases?

12.2.5 Why do materials expand as the temperature increases?

Concept Check 12.2.1

- **How does the length of the bonds in sodium chloride change as the temperature increases?**
- **Figure 12.2.2 shows a symmetrical bond-energy curve. If a material were to have this kind of bond-energy curve, how much would it expand as the temperature increased?**

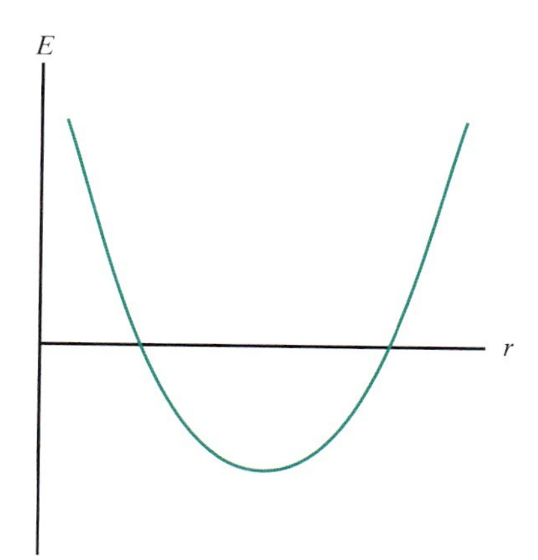

Figure 12.2.2

Symmetrical bond-energy curve. Note that no materials actually have a curve like this.

Figure 12.2.3 shows the difference in bond-energy curves as the strength of the bonds changes. As you can see in this figure, when bonds are weaker the curve is more asymmetric. In the next questions you will work out the relationship between bond strength and thermal expansion.

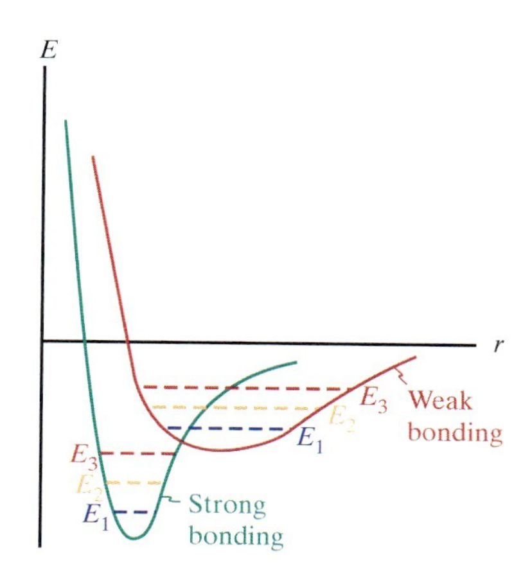

Figure 12.2.3

Comparison of bond-energy curves for materials with different bond strengths.

Guided Inquiry: Bond Strength and Thermal Expansion

12.2.6 For each of the energy levels shown in Figure 12.2.3, identify the average bond distance.

12.2.7 Which material shows a greater increase in bond distance as the temperature increases?

12.2.8 Which material will expand more as the temperature increases?

12.2.9 What kinds of bonds are present in copper? What kinds of bonds are present in silicon?
Go back to Chapter 3 if you need to review the types of bonds in materials.

12.2.10 Which material has stronger bonds; copper or silicon?

12.2.11 Which material will expand more when heated; copper or silicon?

Concept Check 12.2.2

- **Rank the following materials in order from least expansion when heated to the most expansion: zinc, polyethylene, sodium chloride.**

We can quantify the change in length when a material is heated using

$$\frac{l - l_0}{l_0} = \frac{\Delta I}{l_0} \alpha \Delta T \qquad (12.2.1)$$

where l_0 is the initial length, l is the length after heating, ΔT is the change in temperature, and α is the *linear coefficient of thermal expansion*. For anisotropic materials like single crystals and composites α will be different in different directions. We can also calculate the change in volume with

$$\frac{V - V_0}{V_0} = \frac{\Delta V}{V_0} = \beta \Delta T \qquad (12.2.2)$$

where V_0 is the initial volume, V is the the volume after heating, ΔT is the change in temperature, and β is the *volumetric coefficient of thermal expansion*.

Table 12.2.1 lists thermal expansion coefficients for various materials. Note that the units for thermal expansion are usually given as $10^{-6}/°C$ or ppm/°C. This means that the length change is 10^{-6} length units per °C. For example, a thermal expansion coefficient of 1 ppm/°C means that a piece of the material that is 1 cm long will expand by 10^{-6} cm for every 1° C increase in temperature. You will now practice doing calculations related to thermal expansion.

TABLE 12.2.1 Thermal expansion coefficients at 20° C

Material	α (ppm/°C)	β (ppm/°C)
Aluminum	23	69
Copper	17	51
Iron	12	36
PVC	52	156
Polypropylene	200	600
Quartz (crystalline SiO_2)	0.33	1
Silicon	3	9

Guided Inquiry: Thermal Expansion Calculations

12.2.12 What is the linear thermal expansion coefficient of iron?

12.2.13 If a piece of iron 1 cm long is heated so its temperature increases 1° C, how much will the length increase?

12.2.14 If a piece of iron 1 cm long is heated so its temperature increases 10° C, how much will the length increase?

12.2.15 A piece of aluminum is 10 cm long at room temperature (25° C). What will the length be if it is heated to 500° C?

Make sure you work together in your group and that everyone agrees on the answer.

Concept Check 12.2.3

- **If a material has a linear thermal expansion coefficient of 123 ppm/°C, how much will a 1 cm piece expand if the temperature is increased by 10° C?**
- **Why does polypropylene have a larger linear thermal expansion coefficient than PVC?**

EXAMPLE PROBLEM 12.2.1

A copper pipe that is 0.5 m long at 25° C has a hot fluid flowing through it, which causes the pipe's temperature to increase to 250° C. How much will the pipe expand due to this heating?

We begin with the following equation:

$$\frac{\Delta l}{l_0} = \alpha \Delta T$$

Values for the variables are as follows:

$l_0 = 0.5$ m

$\alpha = 17$ ppm/° C $= 17 \times 10^{-6°}$ C^{-1}

$\Delta T = 225°$ C

Rearrange the equation to solve for the change in length:

$$\Delta l = l_0 \alpha \Delta T = (0.5 \text{ m})(17 \times 10^{-6°} \text{C}^{-1})(225° \text{ C}) = 1.9 \times 10^{-3} \text{m} = 1.9 \text{ mm}$$

12.3 Thermal Conductivity

LEARN TO: Describe the mechanism of thermal conductivity.
Conduct thermal conductivity calculations.

As we saw in Section 12.1, heat capacity relates to the amount of energy that a substance can absorb. Thermal conductivity relates to how quickly energy travels through a material. A good way to think about it is in terms of how time affects each of the properties. Thermal conductivity has to do with how long it takes a material to transmit energy. Heat capacity has to with how much energy is absorbed if we wait long enough for it to absorb the maximum amount possible.

There are two ways energy gets transmitted through a material. If free electrons are present (that is, electrons above the Fermi energy), they can transport energy. As mentioned in Chapter 11, to get a full understanding of how this energy transport occurs we would need to consider the wave nature of electrons, which is beyond what we want to do in this book. The other way energy gets transmitted is through atomic vibrations. Again, for a full understanding we would need to consider these atomic vibrations as waves, which are called *phonons*. Electrons turn out to be better thermal conductors than phonons. By considering the relative contributions of electrons and phonons for a given material, you will be able to compare thermal conductivities for different materials.

Guided Inquiry: Thermal Conductivity Mechanism

12.3.1 Copper has a partially filled valence band. Is copper a conductor, a semiconductor, or an insulator?

Go back to Chapter 11 if you need to review electronic properties.

12.3.2 Does copper have free electrons at room temperature?

Will there be electrons above the Fermi energy?

12.3.3 What will be the primary mechanism for thermal conductivity in copper; electrons or phonons?

12.3.4 Quartz has a bandgap of 8.9 eV. Is quartz a conductor, a semiconductor, or an insulator?

12.3.5 Does quartz have free electrons at room temperature?

12.3.6 What will be the primary mechanism for thermal conductivity in quartz; electrons or phonons?

12.3.7 Which are better thermal conductors; electrons or phonons?

12.3.8 Which will have a higher thermal conductivity at room temperature; copper or quartz?

Concept Check 12.3.1

- **What is the primary mechanism for thermal conductivity in polyethylene?**
- **Which will have a higher thermal conductivity; polystyrene or steel?**

Phonons and electrons behave the same way in materials. This means that the same things that scatter electrons will also scatter phonons. If you need to remind yourself of how electron scattering works, go back and review Section 11.3. For conductors, in which electrons are the primary mechanism for thermal conduction, this means that scattering reduces thermal conductivity in the same way it reduces electrical conductivity. And since phonons are also scattered by defects, thermal conductivity in insulators will also be reduced by defects. The next questions will show you how structure affects thermal conductivity.

Guided Inquiry: Electron and Phonon Mobility

12.3.9 Does a piece of quartz held at 200° C have more or fewer vacancies than a piece of quartz held at 100° C?

These questions are analogous to the questions in Section 11.3 on scattering of electrons.

12.3.10 Are there more defects in a piece of quartz held at 200° C than in one held at 100° C?

12.3.11 Will there be more scattering of phonons in a piece of quartz held at 200° C than in one held at 100° C?

12.3.12 Which will have the higher thermal conductivity; a piece of quartz held at 200° C or one held at 100° C?

If your instructor has given you a time limit for these questions, make sure you are keeping track of time so you don't get behind.

12.3.13 Will thermal conductivity of a material increase or decrease as temperature is increased?

12.3.14 Are there more defects in pure copper or in a copper/aluminum alloy?

12.3.15 Will there be more scattering of electrons in pure copper or in a copper/aluminum alloy?

12.3.16 Which will have the higher thermal conductivity; pure copper or a copper/aluminum alloy?

12.3.17 Will thermal conductivity of a material increase or decrease as impurity atoms are added?

12.3.18 Are there more defects in annealed copper or in copper that has been plastically deformed?

12.3.19 Will there be more scattering of electrons in annealed copper or copper that has been plastically deformed?

12.3.20 Which will have the higher thermal conductivity; annealed copper or copper that has been plastically deformed?

12.3.21 Will thermal conductivity of a material increase or decrease when it is plastically deformed?

12.3.22 Are there more defects in quartz with 10 μm grains or with 100 μm grains?

12.3.23 Will there be more scattering of phonons in quartz with 10 μm grains or with 100 μm grains?

12.3.24 Which will have the higher thermal conductivity; quartz with 10 μm grains or with 100 μm grains?

12.3.25 Will thermal conductivity of a material increase or decrease when the grain size increases?

12.3.26 Which will have a higher thermal conductivity; copper solidified at 800° C or copper solidified at 600° C?

12.3.27 List the four factors that affect the thermal conductivity of materials and describe how the thermal conductivity changes with each of them.

Concept Check **12.3.2**

- **Which has the higher thermal conductivity; 24k gold or 12k gold?**
- **Which has the higher thermal conductivity; copper solidified at 800° C or copper solidified at 600° C?**

The mathematics for thermal conductivity is very much like the mathematics for diffusion. *Fourier's law* is

$$q = -k\frac{dT}{dx} \tag{12.3.1}$$

where q is the heat flux (J/m^2-s or W/m^2), k is the *thermal conductivity* of the material (W/m-K), and dT/dx is the temperature gradient. Compare Fourier's law to Fick's first law in Section 7.2 and you will see they are basically the same. Table 12.3.1 provides thermal conductivity values for some materials.

TABLE 12.3.1 Thermal conductivity values at 25° C

Material	k (W/m-K)
Air	0.024
Aluminum	237
Brick	0.69
Concrete	1.28
Copper	385
Iron	80
Polystyrene (foam)	0.033
Polystyrene (solid)	0.15
Silica glass	0.93

Application Spotlight **Thermal Insulation**

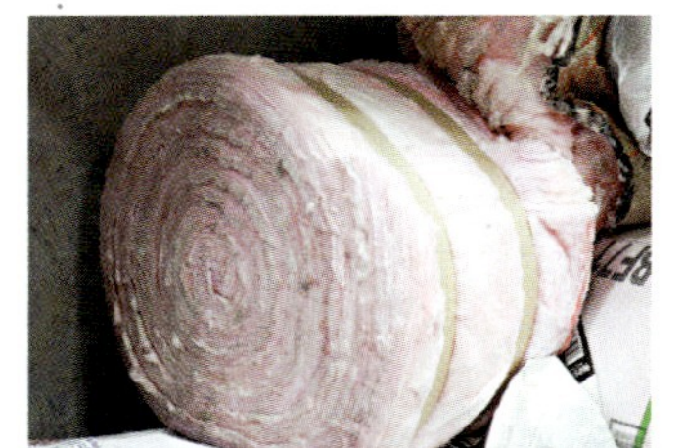

(Photo courtesy of Lifeprints Photography)

Thermal insulation is an important way of conserving energy. Reducing the transfer of heat either into or out of a building means less energy is consumed in heat or air conditioning. You may be familiar with the R-value of thermal insulation, but what exactly is it? Very simply, the R-value is the thickness of the insulation divided by its thermal conductivity. To increase the effectiveness of insulation you can either increase its thickness or decrease its thermal conductivity. Looking at Table 12.3.1, you can see that air has the lowest thermal conductivity by at least a factor of 10 over most other materials. This is why the best insulators, such as polystyrene foam or fiberglass, are porous: the air in the pores reduces the thermal conductivity.

The next questions are based on the following scenario: Bricks that are 10 cm thick are used to build an oven that operates at 500° C. The heat flux through the oven wall is 3 kW/m^2.

Guided Inquiry: Thermal Conductivity Calculations

12.3.28 In the above problem, what are the units and value for q?

12.3.29 What are the units and value for k?

12.3.30 What are the units for T?

Make sure you are consistent with the units for other quantities in the problem.

12.3.31 What are the units for x?

12.3.32 What are the units and value for dx?

12.3.33 What is the outside temperature of the oven?

Concept Check 12.3.3

- **What is the outside temperature of the oven in question 12.3.33?**

EXAMPLE PROBLEM 12.3.1

You measure the temperature of the outside of a 10 cm thick iron container and find that it is 30° C. If the heat flux through the container wall is 12 kW/m², what is the temperature inside the container?

We start by drawing a diagram showing the situation:

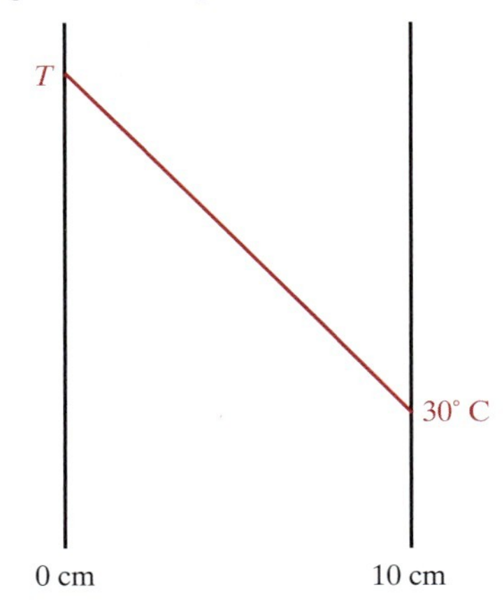

Now we use Fourier's law

$$q = -k\frac{dT}{dx} = -k\frac{T_h - T_l}{x_h - x_l}$$

The values of the variables are

$q = 12{,}000 \text{ W/m}^2$

$k = 80 \text{ W/m-K}$

$T_h = ?$

$T_l = 303 \text{ K}$

$x_h = 0 \text{ m}$

$x_l = 10 \times 10^{-2} \text{ m}$

Note that the high-temperature side has to be defined as $x = 0$ in order to account for the negative sign in Fourier's law. Rearranging the equation and solving gives

$$T_h = \frac{-q(x_h - x_l)}{k} + T_l = \frac{-(12000 \text{ W/m}^2)(0 \text{ m} - 0.10 \text{ m})}{80 \text{ W/m} - \text{K}} + 303 \text{ K} = 318 \text{ K} = 45° \text{ C}$$

Summary

The three properties discussed in this chapter, heat capacity, thermal expansion, and thermal conductivity, all need to be considered when selecting materials. A good example of how these properties affect performance is a computer chip. The electronic circuit generates a lot of heat, which needs to be conducted away. Materials with high thermal conductivity are needed to help dissipate that heat. The elevated temperatures also cause thermal expansion of the materials. Many different materials are used in a computer chip, so there can be a significant difference in the thermal expansion coefficients. This differential thermal expansion causes stresses, which can lead to failure of the materials and loss of performance of the computer chip. As you can see, careful matching of thermal properties is an important issue in design.

Key Terms

Fourier's law
Heat capacity
Linear coefficient of thermal expansion
Molar heat capacity
Phonon
Specific heat
Thermal conductivity
Volumetric coefficient of thermal expansion

Problems

Skill Problems

12.1 A piece of copper weighing 300 g is initially at a temperature of 25° C. What is its temperature after absorbing 3.2 kJ of energy?

12.2 A piece of gold weighing 300 g is initially at a temperature of 25° C. What is its temperature after absorbing 3.2 kJ of energy?

12.3 How much energy does it take to heat a 100 g piece of lithium from 25° C to 100° C?

12.4 How much energy does it take to heat a 100 g piece of gold from 25° C to 100° C?

12.5 Rank the following materials from lowest to highest thermal expansion: copper, diamond, polystyrene. Explain your answer.

12.6 A piece of copper 1 m long is heated from 20° C to 500° C. How much does the length increase?

12.7 A piece of polypropylene 2 m long is heated from 20° C to 150° C. How much does the length increase?

12.8 To what temperature would you have to heat a 2 m long piece of aluminum in order to increase its length by 2 mm? Its initial temperature is 20° C.

12.9 To what temperature would you have to heat a 2 m long piece of iron in order to increase its length by 1.5 mm? Its initial temperature is 20° C.

12.10 For each of the following, identify which has the higher thermal conductivity and explain why:

- **a.** Silver at a temperature of 20° C or silver at a temperature of 60° C.
- **b.** Silicon at a temperature of 20° C or silicon at a temperature of 60° C.
- **c.** Single-crystal silicon or polycrystal silicon.
- **d.** Copper or brass.
- **e.** Aluminum solidified at 500° C or aluminum solidified at 700° C.
- **f.** Copper that has been plastically deformed or copper that has been annealed.

12.11 A container made out of 5 cm thick concrete has an outside temperature of 25° C. If the heat flux through the container wall is 3.5 kW/m^2, what is the inside temperature of the container?

12.12 A container made out of 20 cm thick iron has an inside temperature of 300° C. If the heat flux through the container wall is 23 kW/m^2, what is the outside temperature of the container?

Conceptual Problems

12.13 When you step out of a shower onto a metal surface, it feels colder than if you stepped onto a wood surface, even when the metal and wood are at the same temperature. Why does this occur?

12.14 Polystyrene has a specific heat of 1.3 J/g-K. If a 200 g piece of polystyrene is initially at room temperature, how much heat can it absorb and still stay solid?

12.15 Would crosslinked polystyrene and linear polystyrene have the same linear coefficient of thermal expansion? If not, which would be higher? Explain your answer.

12.16 Why are disposable coffee cups made out of Styrofoam® instead of clear polystyrene?

12.17 You decide to move to Antarctica to study penguins and want a house that will keep you warm through the winter. You decide to build the house out of concrete shaped like a box with a flat roof with walls, roof, and foundation that are 3 feet thick. The overall dimensions of the house are 20 feet by 30 feet by 10 feet high. You want to maintain the inside temperature at 70° F, and you assume that the outside temperature will be a constant −28° C (the average temperature for the month of August). You have a tank of heating oil which you can use to heat your house. Your heating system is 75% efficient, meaning that 75% of the energy content of the heating oil actually goes toward heating the house. How big a tank do you need to store enough oil to heat your home for a week? The energy content of the oil is 46.2 MJ/kg and its density is 0.85 g/cm^3.

The design of a car is complex and needs to account for many factors, including materials choices. The two most important factors in selecting materials for a car are cost and weight. (Rob Wilson/Shutterstock)

13 Materials Selection and Design

Design is an important aspect—some would say the central activity—of engineering. Whether designing large structures like bridges and buildings, or small devices like computer chips, engineers take fundamental engineering and scientific knowledge and apply it to the creation of useful things. You may have already been introduced to design. Perhaps your school has a freshman engineering program that includes design, or you are a senior taking this course as an elective and have been part of a design project in your major. But you have probably not seen design from the materials engineering perspective. In materials engineering, design is primarily about selecting appropriate materials. Sometimes a design is substantially completed before the materials are chosen. Other times materials selection is an integral part of the design process, with the design changing to match the materials' capabilities. Either way, materials selection generally involves trade-offs in properties. For example, stiff materials tend to have low fracture toughness, so if you need both high stiffness and high fracture toughness you'll have to compromise in some way. This chapter will show you how we do materials selection. By the end of this chapter you will:

Understand the procedures used for materials selection.

Be able to select a material, given design criteria.

13.1 Ranking Procedures

LEARN TO: Select a material for an application using ranking procedures.

When we conduct design, we often have to decide among materials with competing properties. How do we make that decision? The first set of questions will start by showing the difficulty that can be involved.

TABLE 13.1.1 Properties for three different materials

Material	Modulus (GPa)	Linear coefficient of thermal expansion (ppm/°C)	Thermal conductivity (W/m-K)
1040 steel	207	11.3	51.9
Polystyrene	2.50	100	0.13
Soda–lime glass	69	9.0	1.7

Guided Inquiry: Materials Selection

13.1.1 In Table 13.1.1, which material has the highest modulus? Which has the lowest modulus?

13.1.2 In Table 13.1.1, which material has the highest coefficient of thermal expansion? Which has the lowest coefficient of thermal expansion?

13.1.3 In Table 13.1.1, which material has the highest thermal conductivity? Which has the lowest thermal conductivity?

13.1.4 You are selecting a material for a design which requires a high modulus, low thermal expansion, and low thermal conductivity. Which of the materials in Table 13.1.1 would you choose? Justify your answer.

Concept Check 13.1.1

- **You are selecting a material for a design which requires a high modulus, low thermal expansion, and high thermal conductivity. Which of the materials in Table 13.1.1 would you choose?**

You probably had difficulty in selecting a material because the information in Table 13.1.1 is not sufficient to discriminate among them. You needed to rely on some kind of subjective interpretation to make a decision. When different materials each have a different property that is optimal for the application, we need some objective way of deciding. One way is through a formal *ranking procedure*. To do this we rank the materials in order of each of the properties, and assign a *weighting factor* based on the importance of each property to the design. By calculating a total score from the rankings and weighting factors we can determine which material has the best combination of properties. The next questions will show you how this procedure works.

TABLE 13.1.2 Property ranking for the three materials in Table 13.1.1

	Modulus Weighting = 5		Thermal expansion Weighting = 2		Thermal conductivity Weighting = 10		
Material	**Ranking**	**Score**	**Ranking**	**Score**	**Ranking**	**Score**	**Total score**
1040 steel	3	15	2	4			29
Polystyrene	1			2		30	
Soda–lime glass	2		3				

Guided Inquiry: Ranking Procedures

13.1.5 Compare the modulus values in Table 13.1.1 with the modulus rankings in Table 13.1.2. Which gets the highest ranking; lowest modulus or highest modulus?

13.1.6 What is the requirement for modulus for this design?
See question 13.1.4.

13.1.7 Compare the thermal expansion values in Table 13.1.1 with the thermal expansion rankings in Table 13.1.2. Which gets the highest ranking; lowest thermal expansion or highest thermal expansion?

13.1.8 What is the requirement for thermal expansion for this design?

13.1.9 Explain how to assign the rankings in Table 13.1.2.

13.1.10 Fill in the empty cells in Table 13.1.2.

13.1.11 Is the best material the one with the highest total score or the lowest total score?
Consider your answer to question 13.1.9.

13.1.12 Based on Table 13.1.2, which material would you choose for this design? Is this the same one you chose in question 13.1.4?

Concept Check **13.1.2**

- **A different design uses the same properties listed in Table 13.1.2, but different weighting factors. For this new design the weighting factors are; modulus = 10, thermal expansion = 5, thermal conductivity = 2. Which material would you choose for this new design?**

One advantage of using ranking procedures is that you can include properties that are not easily quantified. For example, you can rank materials based on their availability, recyclability, or aesthetics. However, there are some disadvantages, the main one being the choice of the weighting factors. There are ways to determine weighting factors, although that is beyond the scope of this book. Another disadvantage is that you can't use ranking procedures to see the relationship between two different properties—for example, modulus and density. To deal with these issues we need to turn to another approach, described in the next section.

13.2 Ashby Plots

LEARN TO: **Select a material for an application using Ashby plots.**

The use of ranking procedures in the previous section depends on the choice of weighting factors. You can arbitrarily change the result of the ranking process just by changing the weighting factors. A more objective approach is to use what are called *Ashby plots*, named after Michael F. Ashby, a materials engineer from England. These plots allow you to compare two different properties across a wide range of materials. Figure 13.2.1 is an example of an Ashby plot which shows modulus versus strength, both on logarithmic scales. Materials are placed on this plot based on their property values and then grouped into categories. You will start with a simple selection process based on a minimum property requirement.

(Photo courtesy of Lifeprints Photography)

Application Spotlight **Life-Cycle Analysis**

There are many ways to look at the environmental impacts of technologies. One way is through a life-cycle analysis, in which you examine the materials and energy impacts of a technology, calculating the amounts of energy and material used during the life of a product. There are several different types of analysis: cradle-to-grave, which goes from manufacture to use to disposal; cradle-to-gate, which covers only manufacturing; cradle-to-cradle, which goes from manufacture to use to recycling; and gate-to-gate, which examines only one process in the production chain. The materials aspect of life-cycle analysis is straightforward; you just determine how much material is consumed or generated at each step. The energy aspect is harder. Theoretically you could use thermodynamics, but that becomes very difficult as you try to account for inefficiencies in processes, energy losses, etc. Instead, the approach used is to consider the embodied energy of a material, which is the energy needed to create 1 kg of the material. This energy is carried along from step to step of the process, and as material is generated or consumed its embodied energy is added to or subtracted from the total. Various software packages for life-cycle analysis are available which also include databases of materials and processing methods. Using life-cycle analysis you can determine the impact of your materials choices on the environment.

Figure 13.2.1

Ashby plot for modulus versus strength.

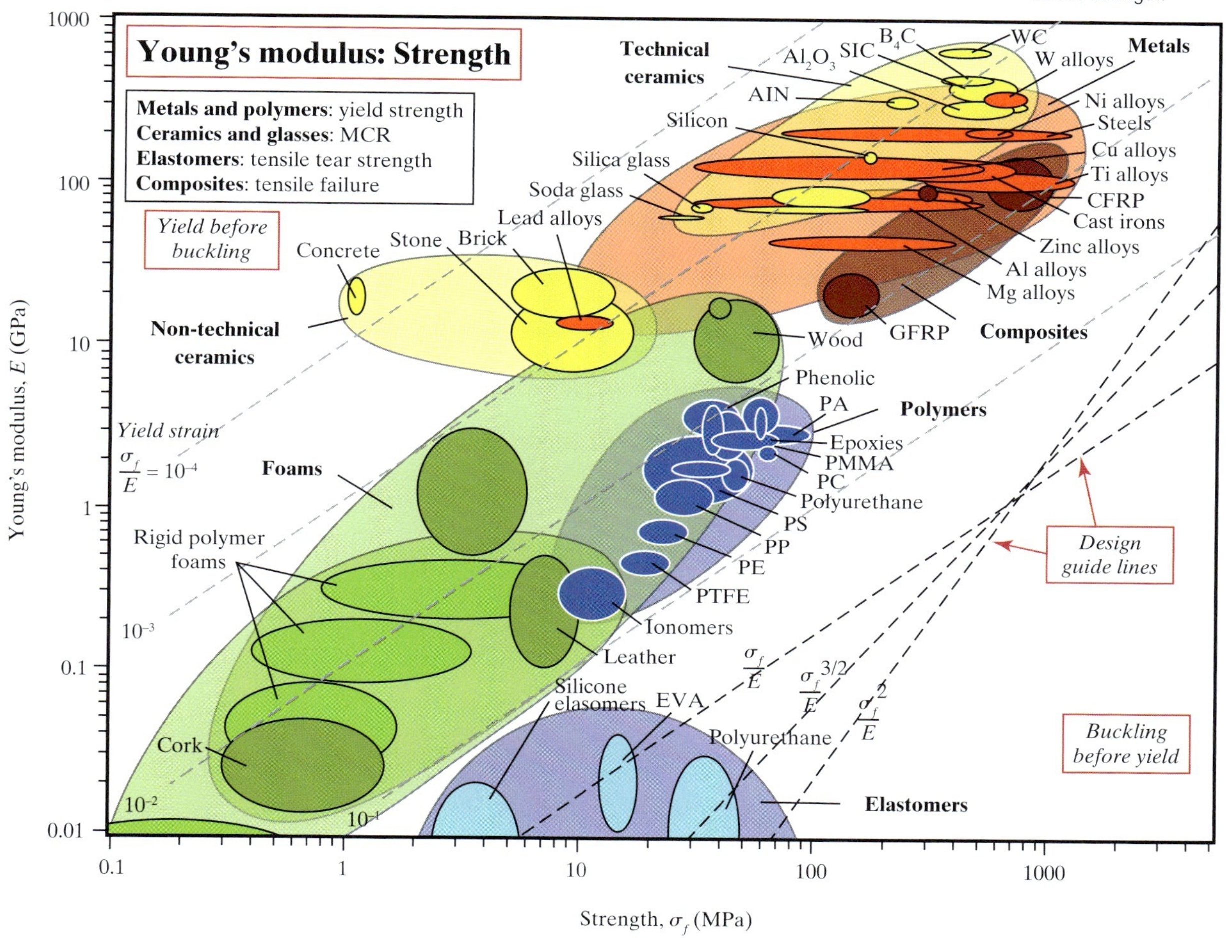

Granta CES EduPack. Printed with permission of Granta Design Ltd.

Guided Inquiry: Property Selection

13.2.1 According to Figure 13.2.1, which tend to have a higher modulus; polymers or metals?

13.2.2 According to Figure 13.2.1, which tend to have a higher strength; polymers or metals?

13.2.3 According to Figure 13.2.1, which has a higher modulus; polyethylene (PE) or polystyrene (PS)?

13.2.4 According to Figure 13.2.1, which has a higher strength; polyethylene (PE) or polystyrene (PS)?

13.2.5 If you wanted a material with a high modulus, which would you choose; wood or brick?

13.2.6 If you wanted a material with a high strength, which would you choose; wood or brick?

13.2.7 You have a design that requires a material with a minimum modulus of 5 GPa. Draw a line on Figure 13.2.1 corresponding to 5 GPa.
Remember that the axes are in a log scale.

13.2.8 List three materials that meet the modulus requirement.

13.2.9 Your design also requires a minimum strength of 100 MPa. Draw a line on Figure 13.2.1 corresponding to 100 MPa.

13.2.10 Does wood meet the modulus requirement for your design? Does it meet the strength requirement?

13.2.11 List three materials that meet both the modulus and strength requirements.

Concept Check 13.2.1

- **You are told to select a material that has a modulus no greater than 1 GPa and a minimum strength of 100 MPa. What materials can you select to meet these requirements?**

The previous questions dealt with the simple case of selecting a material based on specific property values. But we can also use Ashby plots for more complicated selection procedures using something called a *materials index*. A materials index allows us to optimize the choice between two different properties. Let's say, for example, that we want a material for a beam that is both stiff in bending (high modulus) and light (low density). In this case the materials index is as follows:

$$\frac{E^{1/2}}{\rho} \tag{13.2.1}$$

The corresponding Ashby plot for modulus and density is shown in Figure 13.2.2. On this plot there are dotted lines labeled as "Guide lines for minimum mass design." If you draw a line with the slope corresponding to the guide line anywhere on the plot, materials that fall on that line all have an equivalent balance between stiffness and weight. Materials that fall above the line are better (higher stiffness/lower weight), while materials falling below the line are worse (lower stiffness/higher weight). The next questions will show you how this approach works, using the red dotted guide line as an example.

Figure 13.2.2

Ashby plot for modulus vs. density.

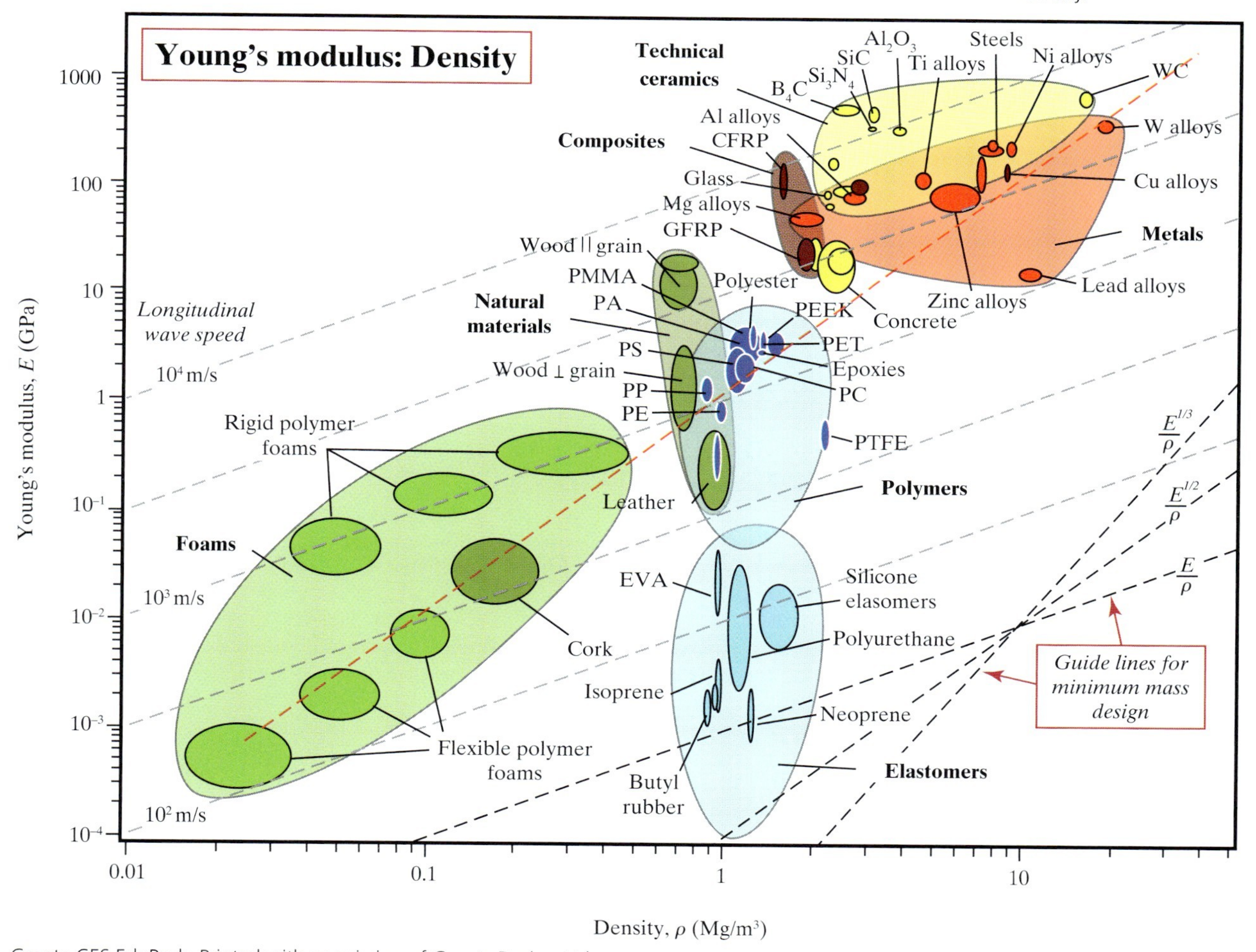

Guided Inquiry: Performance Selection

13.2.12 List two materials that fall on the red dotted line in Figure 13.2.2.

13.2.13 List two materials that fall below the red dotted line in Figure 13.2.2.

13.2.14 List two materials that fall above the red dotted line in Figure 13.2.2.

13.2.15 List a metal alloy that behaves equivalently to cork for a light, stiff beam in bending.
Make sure everyone in your group understands how to do this question before moving on.

13.2.16 List a polymer that behaves better than cork for a light, stiff beam in bending.

13.2.17 List a polymer that behaves worse than cork for a light, stiff beam in bending.

Concept Check 13.2.2

The materials index for a light, stiff plate in bending is

$$\frac{E^{1/3}}{\rho}$$

- **Identify a material that behaves equivalently to cork for a light, stiff plate in bending.**

In reality, we need to combine the two approaches you have just seen. The initial screening is used to identify materials that meet minimum requirements, and then the materials index can be used to identify materials that optimize different properties. In the next questions you will practice doing that.

Guided Inquiry: Property and Performance Selection

13.2.18 You are selecting a material to be used as a light, stiff plate in bending. Your design requires a minimum modulus of 100 GPa. List three materials that meet these requirements.

13.2.19 The design in question 13.2.18 also requires that the plate perform as well as or better than a plate made out of cork. Identify a material that will meet all of the requirements.

13.2.20 You are selecting a material to be used as a light, stiff tie in tension. Your design requires a minimum modulus of 10 GPa and a minimum strength of 400 MPa. List three materials that meet these requirements.

13.2.21 The design in question 13.2.20 also requires that the tie perform as well as or better than a tie made out of GFRP (glass fiber reinforced composite). Identify a material that will meet all of the requirements.

The materials index for a light, stiff tie in tension is $\frac{E}{\rho}$

Concept Check 13.2.3

- **Identify a material that has a maximum modulus of 100 GPa, a maximum strength of 40 MPa, and behaves equivalently to lead alloys for a light, stiff tie in tension.**

The two Ashby plots you have seen in this chapter are just examples. You can create Ashby plots for any combination of properties you want, including cost. Table 13.2.1 lists several materials indices. And although I have given you the materials indices to use, it is possible to derive them, so don't feel you are limited to the specific properties you have seen here. It is possible to optimize materials selection for any set of properties needed in a design.

TABLE 13.2.1 Materials indices

Design requirement	Materials index
Tie in tension, high stiffness and light weight	$\frac{E}{\rho}$
Beam in bending, high stiffness and light weight	$\frac{E^{1/2}}{\rho}$
Beam in bending, high strength and light weight	$\frac{\sigma_f^{2/3}}{\rho}$
Beam in bending, high stiffness and low cost	$\frac{E^{1/2}}{C\rho}$
Beam in bending, high strength and low cost	$\frac{\sigma_f^{2/3}}{C\rho}$
Column in compression, high stiffness and low cost	$\frac{E^{1/2}}{C\rho}$
Plate in bending, high stiffness and light weight	$\frac{E^{1/3}}{\rho}$
Thermal insulation, low heat flux and low cost	$\frac{1}{kC\rho}$

Note: E = modulus; ρ = density; σ_f = strength; C = cost per unit mass; k = thermal conductivity.

Summary

In this chapter you saw two methods for materials selection; ranking procedures with weighting factors and Ashby plots. Both methods allow you to compare competing properties and make decisions that balance the needs of the design. What these procedures don't do, however, is consider other aspects of materials selection that are unrelated to materials properties. You will need additional supporting information to make a final decision. For example, you need to consider particular strengths and weaknesses of the material. Perhaps recycling or sustainability is important. Have any concerns been noted in previous applications or any issues with processing? Availability can also be an important consideration. You may have identified the perfect material, but if there is only one supplier, you could have a serious problem if that supplier ran into difficulties and couldn't provide the material. The customer may have aesthetic requirements that need to be taken into consideration. While the information in this chapter forms the basis of your technical decision for materials selection, you will need to consider many other factors before making a final decision.

Key Terms

Ashby plot
Materials index
Ranking procedure
Weighting factor

Problems

Skill Problems

13.1 A material is needed that has a high modulus, low thermal expansion, and low thermal conductivity. The weighting factors are: modulus = 8; thermal expansion = 5; thermal conductivity = 2. Which of the materials in Table 13.1.1 would you select?

13.2 A material is needed that has a low modulus, low thermal expansion, and low thermal conductivity. The weighting factors are: modulus = 10; thermal expansion = 5; thermal conductivity = 8. Which of the materials in Table 13.1.1 would you select?

13.3 A material is needed that has a low modulus, low thermal expansion, and high thermal conductivity. The weighting factors are: modulus = 3; thermal expansion = 8; thermal conductivity = 5. Which of the materials in Table 13.1.1 would you select?

13.4 You must select a material with a minimum modulus of 70 GPa and a minimum strength of 200 MPa. What material would you select?

13.5 You must select a material with a minimum modulus of 70 GPa and a maximum density of 2 g/cm^3. What material would you select?

13.6 You must select a material for a light, stiff tie in tension with a maximum density of 4 g/cm^3 that performs as well as or better than steel. What material would you select?

13.7 You must select a material for a light, stiff beam in bending with a minimum modulus of 100 GPa that performs as well as or better than wood perpendicular to the grain. What material would you select?

13.8 You must select a material for a light, stiff beam in bending with a minimum strength of 80 MPa and a minimum modulus of 5 GPa that performs as well as or better than concrete. What material would you select?

Conceptual Problems

13.9 After you select the material for problem 13.1, the client tells you that the material must be recyclable. How will you change the selection procedure to account for this new requirement?

13.10 You are tasked with selecting a material for the casing of a cellphone. Propose three properties that you would use to rank candidate materials.

13.11 You are tasked with selecting a material for a jet turbine blade. Propose three properties that you would use to rank candidate materials.

13.12 You are tasked with selecting a material for a stiff electrical conductor. What properties would you plot on an Ashby plot to help you make your selection?

13.13 You are tasked with selecting a low-cost material that is fracture resistant. What properties would you plot on an Ashby plot to help you make your selection?

Appendix: Physical Properties of Materials

APPENDIX 1A Tensile and bend test data for selected engineering materials

Material	E [GPa (psi)]	E_{flex} [MPa (ksi)]	E_{Dyn} [MPa (ksi)]	Y.S. [MPa (ksi)]	T.S. [MPa (ksi)]	Flexural strength [MPa (ksi)]	Compressive strength [MPa (ksi)]	Percent elongation at failure
Metal alloys								
1040 carbon steel	$200(29 \times 10^6)$			600(87)	750(109)			17
8630 low-alloy steel				680(99)	800(116)			22
304 stainless steel	$193(28 \times 10^6)$			205(30)	515(75)			40
410 stainless steel	$200(29 \times 10^6)$			700(102)	800(116)			22
L2 tool steel				1,380(200)	1,550(225)			12
Ferrous superalloy (410)	$200(29 \times 10^6)$			700(102)	800(116)			22
Ductile iron, quench	$165(24 \times 10^6)$			580(84)	750(108)			9.4
Ductile iron, 60–40–18	$169(24.5 \times 10^6)$			329(48)	461(67)			15
3003-H14 aluminum	$70(10.2 \times 10^6)$			145(21)	150(22)			8–16
2048, plate aluminum	$70.3(10.2 \times 10^6)$			416(60)	457(66)			8
AZ31B magnesium	$45(6.5 \times 10^6)$			220(32)	290(42)			15
AM100A casting magnesium	$45(6.5 \times 10^6)$			83(12)	150(22)			2
Ti–5Al–2.5Sn	$107–110(15.5–16 \times 10^6)$			827(120)	862(125)			15
Ti–6Al–4V	$110(16 \times 10^6)$			825(120)	895(130)			10
Aluminum bronze, 9% (copper alloy)	$110(16.1 \times 10^6)$			320(46.4)	652(94.5)			34
Monel 400 (nickel alloy)	$179(26 \times 10^6)$			283(41)	579(84)			39.5
AC41A zinc					328(47.6)			7
50:50 solder (lead alloy)				33(4.8)	42(6.0)			60
Nb–1 Zr (refractory metal)	$68.9(10 \times 10^6)$			138(20)	241(35)			20
Dental gold alloy (precious metal)					310–380(45–55)			20–35
Ceramics and glasses								
Mullite (aluminosilicate) porcelain	$69(10 \times 10^6)$					69(10)		
Steatite (magnesia aluminosilicate) porcelain	$60(10 \times 10^6)$					140(20)		
Superduty fireclay (aluminosilicate) brick	$97(14 \times 10^6)$					5.2(0.75)		

APPENDIX 1A Tensile and bend test data for selected engineering materials

Material	E [GPa (psi)]	E_{flex} [MPa (ksi)]	E_{Dyn} [MPa (ksi)]	Y.S. [MPa (ksi)]	T.S. [MPa (ksi)]	Flexural strength [MPa (ksi)]	Compressive strength [MPa (ksi)]	Percent elongation at failure
Alumina (Al_2O_3) crystals	380(55 × 10^6)					340–1,000 (49–145)		
Sintered alumina (~5% porosity)	370(54 × 10^6)					210–340(30–49)		
Alumina porcelain (90–95% alumina)	370(54 × 10^6)					340(49)		
Sintered magnesia (~5% porosity)	210(30 × 10^6)					100(15)		
Magnesite (magnesia) brick	170(25 × 10^6)					28(4)		
Sintered spinel (magnesia aluminate) (~5% porosity)	238(35 × 10^6)					90(13)		
Sintered stabilized zirconia (~5% porosity)	150(22 × 10^6)					83(12)		
Sintered beryllia (~5% porosity)	310(45 × 10^6)					140–280(20–41)		
Dense silicon carbide (~5% porosity)	470(68 × 10^6)					170(25)		
Bonded silicon carbide (~20% porosity)	340(49 × 10^6)					14(2)		
Hot-pressed boron carbide (~5% porosity)	290(42 × 10^6)					340(49)		
Hot-pressed boron nitride (~5% porosity)	83(12 × 10^6)					48–100(7–15)		
Silica glass	72.4(10.5 × 10^6)					107(16)		
Borosilicate glass	69(10 × 10^6)					69(10)		
Polymers								
Polyethylene								
High-density	0.830(0.12 × 10^6)				28(4)			15–100
Low-density	0.170(0.025 × 10^6)				14(2)			90–800
Polyvinylchloride	2.80(0.40 × 10^6)				41(6)			2–30
Polypropylene	1.40(0.20 × 10^6)				34(5)			10–700
Polystyrene	3.10(.045 × 10^6)				48(7)			1–2
Polyesters		8,960(1,230)			158(22.9)			2.7
Acrylics (Lucite)	2.90(0.42 × 10^6)				55(8)			5

(Continued)

APPENDIX 1A Tensile and bend test data for selected engineering materials *(con't.)*

Material	E [GPa (psi)]	E_{flex} [MPa (ksi)]	E_{Dyn} [MPa (ksi)]	Y.S. [MPa (ksi)]	T.S. [MPa (ksi)]	Flexural strength [MPa (ksi)]	Compressive strength [MPa (ksi)]	Percent elongation at failure
Polyamides (nylon 66)	$2.80(0.41 \times 10^6)$	2,830(410)			82.7(12.0)			60
Cellulosics	$3.40–28.0(0.50–4.0 \times 10^6)$				14–55(2–8)			5–40
ABS	$2.10(0.30 \times 10^6)$				28–48(4–7)			20–80
Polycarbonates	$2.40(0.35 \times 10^6)$				62(9)			110
Acetals	$3.10(0.45 \times 10^6)$	2,830(410)			69(10)			50
Polytetrafluoroethylene (Teflon)	$0.41(0.060 \times 10^6)$				17(2.5)			100–350
Polyester-type thermoplastic elastomers		585(85)			46(6.7)			400
Phenolics (phenolformaldehyde)	$6.90(1.0 \times 10^6)$				52(7.5)			0
Urethanes					34(5)			—
Urea-melamine	$10.0(1.5 \times 10^6)$				48(7)			0
Polyesters	$6.90(1.0 \times 10^6)$				28(4)			0
Epoxies	$6.90(1.0 \times 10^6)$				69(10)			0
Polybutadiene/polystyrene copolymer								
Vulcanized	$0.0016(0.23 \times 10^3)$		0.8(0.12)		1.4–3.0(0.20–0.44)			440–600
Vulcanized with 33% carbon black	$0.003–0.006(0.4–0.9 \times 10^3)$		8.7(1.3)		17–28(2.5–4.1)			400–600
Polyisoprene								
Vulcanized	$0.0013(0.19 \times 10^3)$		0.4(0.06)		17–25(2.5–3.6)			750–850
Vulcanized with 33% carbon black	$0.003–0.008(0.44–1.2 \times 10^3)$		6.2(0.90)		25–35(3.6–5.1)			550–650
Polychloroprene								
Vulcanized	$0.0016(0.23 \times 10^3)$		0.7(0.10)		25–38(3.6–5.5)			800–1,000
Vulcanized with 33% carbon black	$0.003–0.005(0.4–0.7 \times 10^3)$		2.8(0.41)		21–30(3.0–4.4)			500–600
Polyisobutene/polyisoprene copolymer								
Vulcanized	$0.0010(0.15 \times 10^3)$		0.4(0.06)		18–21(2.6–3.0)			750–950
Vulcanized with 33% carbon black	$0.003–0.004(0.4–0.6 \times 10^3)$		3.6(0.52)		18–21(2.6–3.0)			650–850

APPENDIX 1A Tensile and bend test data for selected engineering materials

Material	E [GPa (psi)]	E_{flex} [MPa (ksi)]	E_{Dyn} [MPa (ksi)]	Y.S. [MPa (ksi)]	T.S. [MPa (ksi)]	Flexural strength [MPa (ksi)]	Compressive strength [MPa (ksi)]	Percent elongation at failure
Silicones					7(1)			4,000
Vinylidene fluoride/ hexafluoropropylene					12.4(1.8)			
Composites								
E-glass (73.3 vol %) in epoxy (parallel loading of continuous fibers)	56(8.1×10^6)				1,640(238)	—	—	2.9
Al_2O_3 whiskers (14 vol %) in epoxy	41(6×10^6)				779(113)	—	—	—
C (67 vol %) in epoxy (parallel loading)	221(32×10^6)				1,206(175)	—	—	—
Kevlar (82 vol %) in epoxy (parallel loading)	86(12×10^6)				1,517(220)	—	—	—
B (70 vol %) in epoxy (parallel loading of continuous filaments)	210–280($30–40 \times 10^6$)				1,400–2,100(200–300)	—	—	—
Al_2O_3 (10 vol %) dispersion-strengthened aluminum	—				330(48)	—	—	—
W (50 vol %) in copper (parallel loading of continuous filaments)	260(38×10^6)				1,100(160)	—	—	—
W particles (50 vol %) in copper	190(27×10^6)				380(55)	—	—	—
SiC whiskers in Al_2O_3	—				—	800(116)	—	—
SiC fibers in SiC	—				—	750(109)	—	—
SiC whiskers in reaction-bonded Si_3N_4	—				—	900(131)	—	—

APPENDIX 1B Miscellaneous mechanical properties data for selected engineering materials

	Poisson's ratio ν	Brinell hardness number	Rockwell hardness R scale	Charpy impact energy [J(ft·lb)]	Izod impact energy [J(ft·lb)]	K_{Ic} (MPa $\sqrt{m}$)	Fatigue limit [MPa (ksi)]
Metals and alloys							
1040 carbon steel	0.30	235		180(133)			280(41)
Mild steel						140	
Medium-carbon steel						51	
8630 low-alloy steel	0.30	220		51(41)			400(58)
304 stainless steel	0.29						170(25)
410 stainless steel		250		34(25)			
L2 tool steel				26(19)			
Rotor steels (A533; Discalloy)						204–214	
Pressure-vessel steels (HY130)						170	
High-strength steels (HSS)						50–154	
Ductile iron	0.29	167		9(7)			
Cast iron						6–20	
Pure ductile metals (e.g., Cu, Ni, Ag, Al)						100–350	
Be (brittle hcp metal)						4	
3003-H14 aluminum	0.33	40					62(9)
2048, plate aluminum				10.3(7.6)			
Aluminum alloys (high strength-low strength)						23–45	
AZ31B magnesium	0.35	73		4.3(3.2)			
AM100A casting magnesium	0.35	53		0.8(0.6)			69(10)
Ti–5Al–2.5Sn	0.35	335		23(17)			410(59)
Ti–6Al–4V	0.33						
Titanium alloys						55–115	
Aluminum bronze, 9% (copper alloy)	0.33	165		48(35)			200(29)
Monel 400 (nickel alloy)	0.32	110–150		298(220)			290(42)
AC41A zinc		91					56(8)
50:50 solder (lead alloy)		14.5		21.6(15.9)			
Nb–1 Zr (refractory metal)				174(128)			
Dental gold alloy (precious metal)		80–90					

APPENDIX 1B Miscellaneous mechanical properties data for selected engineering materials

	Poisson's ratio ν	Brinell hardness number	Rockwell hardness R scale	Charpy impact energy [J(ft·lb)]	Izod impact energy [J(ft·lb)]	K_{Ic} (MPa $\sqrt{m}$)	Fatigue limit [MPa (ksi)]
Ceramics and glasses							
Al_2O_3	0.26					3–5	
BeO	0.26						
CeO_2	0.27–0.31						
MgO						3	
Cordierite ($2MgO \cdot 2Al_2O_3 \cdot 5SiO_2$)	0.31						
Mullite ($3Al_2O_3 \cdot 2SiO_2$)	0.25						
SiC	0.19					3	
Si_3N_4	0.24					4–5	
TaC	0.24						
TiC	0.19						
TiO_2	0.28						
Partially stabilized ZrO_2	0.23					9	
Fully stabilized ZrO_2	0.23–0.32						
Glass-ceramic ($MgO–Al_2O_3–SiO_2$)	0.24						
Electrical porcelain						1	
Cement/concrete, unreinforced						0.2	
Soda glass ($Na_2O–iO_2$)						0.7–0.8	
Borosilicate glass	0.20						
Glass from cordierite	0.26						
Polymers							
Polyethylene							
High-density			40		1.4–16(1–12)	2	
Low-density			10		22(16)	1	
Polyvinylchloride			110		1.4(1)	—	
Polypropylene			90		1.4–15(1–11)	3	
Polystyrene			75		0.4(0.3)	2	
Polyesters			120		1.4(1)	0.5	40.7(5.9)
Acrylics (Lucite)			130		0.7(0.5)	—	

(Continued)

APPENDIX 1B Miscellaneous mechanical properties data for selected engineering materials *(con't.)*

	Poisson's ratio ν	Brinell hardness number	Rockwell hardness R scale	Charpy impact energy [J(ft·lb)]	Izod impact energy [J(ft·lb)]	K_{Ic} (MPa $\sqrt{m}$)	Fatigue limit [MPa (ksi)]
Polyamides (nylon 66)	0.41		121		1.4(1)	3	
Cellulosics			50 to 115		3–11(2–8)	—	
ABS			95		1.4–14(1–10)	4	
Polycarbonates			118		19(14)	1.0–2.6	
Acetals	0.35		120		3(2)	—	31(4.5)
Polytetrafluoroethylene (Teflon)			70		5(4)	—	
Polyester-type thermoplastic elastomers							
Phenolics (phenolformaldehyde)			125		0.4(0.3)	—	
Urethanes			—		—	—	
Urea-melamine			115		0.4(0.3)	—	
Polyesters			100		0.5(0.4)	—	
Epoxies			90		1.1(0.8)	0.3–0.5	
Composites							
E-glass (73.3 vol %) in epoxy (parallel loading of continuous fibers)						42–60	
B (70 vol %) in epoxy (parallel loading of continuous filaments)						46	
SiC whiskers in Al_2O_3						8.7	
SiC fibers in SiC						25.0	
SiC whiskers in reaction-bonded Si_3N_4						20.0	

APPENDIX 1C Thermal properties data for selected materials

	Specific heat c_p	Linear coefficient of thermal expansion α [mm/(mm · °C) × 10^6]			Thermal conductivity k [J/(s · m · K)]			
	[J/kg · K]	27° C	527° C	0–1000° C	27° C	100° C	527° C	1000° C
Metals								
Aluminum	900	23.2	33.8		237		220	
Copper	385	16.8	20.0		398		371	
Gold	129	14.1	16.5		315		292	
Iron (α)	444				80		43	
Lead	159							
Nickel	444	12.7	16.8		91		67	
Silver	237	19.2	23.4		427		389	
Titanium	523				22		20	
Tungsten	133	4.5	4.8		178		128	
Ceramics and glasses								
Mullite ($3Al_2O_3 \cdot 2SiO_2$)				5.3		5.9		3.8
Porcelain				6.0		1.7		1.9
Fireclay refractory				5.5		1.1		1.5
Al_2O_3	160			8.8		30		6.3
Spinel ($MgO \cdot Al_2O_3$)				7.6		15		5.9
MgO	457			13.5		38		7.1
UO_2				10.0				
ZrO_2 (stabilized)				10.0		2.0		2.3
SiC	344			4.7				
TiC						25		5.9
Carbon (diamond)	519							
Carbon (graphite)	711							
Silica glass				0.5		2.0		2.5
Soda–lime–silica glass				9.0		1.7		
Polymers								
Nylon 66	1,260–2,090	30–31			2.9			
Phenolic	1,460–1,670	30–45			0.17–0.52			
Polyethylene (high-density)	1,920–2,300	149–301			0.33			
Polypropylene	1,880	68–104			2.1–2.4			
Polytetrafluoroethylene (PTFE)	1,050	99			0.24			

Appendix: Electronic Properties of Materials

2

APPENDIX 2A Electrical conductivities of selected materials at room temperature

Material	Conductivity, σ ($\Omega^{-1} \cdot m^{-1}$)
Metals and alloys	
Aluminum (annealed)	35.36×10^6
Copper (annealed standard)	58.00×10^6
Gold	41.0×10^6
Iron (99.9+%)	10.3×10^6
Lead (99.73+%)	4.84×10^6
Magnesium (99.80%)	22.4×10^6
Mercury	1.04×10^6
Nickel (99.95% + Co)	14.6×10^6
Nichrome (66% Ni + Cr and Fe)	1.00×10^6
Platinum (99.99%)	9.43×10^6
Silver (99.78%)	62.9×10^6
Steel (wire)	$5.71 - 9.35 \times 10^6$
Tungsten	18.1×10^6
Zinc	16.90×10^6
Semiconductors	
Silicon (high purity)	0.40×10^{-3}
Germanium (high purity)	2.0
Gallium arsenide (high purity)	0.17×10^{-6}
Indium antimonide (high purity)	17×10^3
Lead sulfide (high purity)	38.4
Ceramics, glasses, and polymers	
Aluminum oxide	$10^{-10} - 10^{-12}$
Borosilicate glass	10^{-13}
Polyethylene	$10^{-13} - 10^{-15}$
Nylon 66	$10^{-12} - 10^{-13}$

APPENDIX 2B Properties of some common semiconductors at room temperature

Material	Energy gap, E_g (eV)	Electron mobility, μ_e [m²/(V·s)]	Hole mobility, μ_h [m²/(V·s)]	Carrier density, $n_e (= n_h)$ (m^{-3})
Elements				
Si	1.107	0.140	0.038	14×10^{15}
Ge	0.66	0.364	0.190	23×10^{18}
III–V Compounds				
AlSb	1.60	0.090	0.040	—
GaP	2.25	0.030	0.015	—
GaAs	1.47	0.720	0.020	1.4×10^{12}
GaSb	0.68	0.500	0.100	—
InP	1.27	0.460	0.010	—
InAs	0.36	3.300	0.045	—
InSb	0.17	8.000	0.045	13.5×10^{21}
II–VI Compounds				
ZnSe	2.67	0.053	0.002	—
ZnTe	2.26	0.053	0.090	—
CdS	2.59	0.034	0.002	—
CdTe	1.50	0.070	0.007	—
HgTe	0.025	2.200	0.016	—

Glossary

Activation energy The energy barrier that must be overcome for a thermally activated process to occur. Activation energies are present in diffusion, nucleation, and electronic conduction in semiconductors, as well as many other processes.

Annealing The process of heating a material to a temperature below its melting temperature and holding it there for some period of time to change its microstructure, and consequently its properties. In metals, the process of annealing results in the material becoming more ductile.

Ashby plot A plot which has two different properties on its axes, for example strength and density. Materials are placed on the plot based on those properties. These plots are used for materials selection.

Atactic A structure of polymers in which the position of the side groups relative to the chain backbone is random.

Atomic packing factor The fraction of space in a unit cell that is occupied by atoms. It is calculated by dividing the volume of atoms in a unit cell by the volume of the unit cell.

Austenite A phase of plain carbon steel that is present at temperatures above the eutectoid temperature of 727° C.

Bainite A microstructure of steel consisting of thin needles of cementite in ferrite. Bainite forms during isothermal transformations at low temperatures.

Bandgap The gap in energy in semiconductors and insulators between the valence band and conduction band. The bandgap represents "forbidden" energies, and electrons cannot have energies that fall in the bandgap.

Binary phase diagram A phase diagram that has two components.

Body-centered cubic A cubic unit cell consisting of one atom on each corner and one atom in the center.

Bragg's law The equation relating diffraction angle, interplanar spacing in a crystal, and x-ray wavelength.

Branch In a polymer, a string of repeat units that comes off the main polymer chain.

Branched polymer A polymer that contains branches.

Burgers vector A vector that defines a dislocation. It is obtained by drawing a circuit of equal-sized steps around a dislocation. The Burgers vector goes from the end of the circuit to the beginning of the circuit.

Cathodic protection The use of a more easily oxidized metal to protect another metal from corrosion by placing the two metals in contact.

Cavitation damage Corrosion caused by bubbles damaging the protective layer on a metal. This occurs, for example, on boat propellers.

Cementite A phase of plain carbon steel. It is a line compound with the formula Fe_3C and a composition of 6.70 wt% carbon.

Close-packed direction The direction in a crystal in which the atoms are touching. It has a density of 1.0.

Close-packed plane The plane in a crystal which has the maximum possible packing for a set of spheres. Among cubic unit cells only FCC has close-packed planes, which are the {111} planes.

Component A pure substance. A component may be an element or it may be a compound.

Composite A physical mixture of two materials which has properties superior to either material alone. Typically a composite consists of a reinforcement material embedded in a matrix material.

Composition The relative amounts of the components in a mixture.

Concentration gradient The change in concentration of a species with respect to distance in a material, dc/dx. The concentration gradient is the driving force for diffusion.

Conduction band The next energy band above the valence band. At zero Kelvin the conduction band is empty. In semiconductors, electrons in the conduction band serve as charge carriers.

Conductivity An intrinsic property of a material that indicates how well it conducts electricity. The conductivity is the inverse of the resistivity.

Continuous cooling transformation diagram A diagram that shows the microstructures that form when a material is cooled at a constant rate.

Continuous fiber composite A composite in which the fibers span the length of the material.

Coordination number For a particular atom in a crystal, the number of other atoms that touch that atom.

Copolymer A polymer that contains more than one type of repeat unit. The repeat units may be arranged in the polymer chain in different ways, with the most common being random and block.

Corrosion penetration rate The rate at which uniform corrosion occurs. Typical units are mm/yr.

Covalent bond A bond formed when electrons are shared between two atoms. It forms between atoms that have a small difference in electronegativity and a large average electronegativity.

Crevice corrosion Corrosion that preferentially occurs in cracks or crevices between two pieces of metal.

Critical resolved shear stress The minimum shear stress in the slip direction to cause dislocation motion.

Critical stress intensity parameter An intrinsic material property that describes resistance to fracture.

Crosslinked polymer A polymer in which all of the polymer chains are connected by covalent bonds.

Crystal A material with a regular repeating arrangement of atoms. The arrangement is described by the unit cell.

Degree of polymerization The number of repeat units in a polymer molecule.

Depletion mode A MOSFET in which the channel is the same type of semiconductor (*n*-type or *p*-type) as the source and the drain. Current flows from the source to the drain when no voltage is applied at the gate.

Diffraction The scattering of waves by a crystal. The waves can only be scattered at specific angles, which are given by Bragg's law.

Diffusion The process of atoms moving through a material.

Diffusion coefficient A constant relating the diffusion flux to the concentration gradient. It depends on the diffusing species, the host material, and the temperature.

Diffusionless transformation A transformation that occurs by local shifts of atoms rather than through diffusion. Formation of martensite is an example.

Discontinuous fiber composite A composite in which the fibers do not span the length of the material.

Dislocation Line defects in crystals, in which a row of atoms is displaced from its normal position in the crystal lattice.

Dopant An impurity added to a semiconductor to introduce additional energy levels and charge carriers. *n*-type dopants add donor levels and electrons, *p*-type dopants add acceptor levels and holes.

Ductility The strain at which a materials fails. It is the maximum possible elongation for that material.

Edge dislocation A dislocation that can be pictured as formed by an extra plane of atoms inserted partway into a crystal. The dislocation is the line of atoms at the end of this extra plane.

Elastic There are two definitions. One is the range of stresses over which a material can be deformed and still completely recover. On a stress–strain curve the elastic region is the initial linear part of the curve. A second definition is a material that exhibits elastic behavior for all stress levels, in the same way as a spring. When a constant stress is applied to an elastic material its length will remain constant.

Electronegativity An atomic property that describes how tightly an atom holds on to its electrons.

Enhancement mode A MOSFET in which the channel is the other type of semiconductor (*n*-type or *p*-type) compared to the source and the drain. Current flows from the source to the drain when a voltage is applied at the gate.

Equilibrium phase diagram A more complete name for what is commonly called a phase diagram. The word equilibrium indicates that phase diagrams show the thermodynamically stable states.

Erosion corrosion Corrosion that is caused by fluid flow damaging the protective layer on a metal.

Eutectic composition The composition corresponding to the point at which a eutectic reaction occurs.

Eutectic phase diagram A phase diagram that contains a eutectic reaction.

Eutectic reaction A microstructural transformation on cooling in which a liquid phase transforms to two solid phases.

Eutectic solid The solid formed as a result of the eutectic reaction. It consists of alternating layers, or lamellae, of the two phases.

Eutectic temperature The temperature corresponding to the point where a eutectic reaction occurs.

Eutectoid reaction A reaction in which one solid phase transforms to two different solid phases.

Face-centered cubic A cubic unit cell consisting of one atom on each corner and one atom in the center of each face.

Family of directions A group of directions that have the same linear density. They are designated with the notation <hkl>.

Family of planes A group of planes that have the same planar density. They are designated with the notation {hkl}.

Fatigue life The number of cycles to failure at a given stress amplitude when a material is subjected to fatigue.

Fatigue strength The stress amplitude that causes failure for a given number of cycles when a material is subjected to fatigue.

Fermi energy The energy required for an electron to act as a charge carrier and conduct electricity. In metals, the Fermi energy occurs in the valence band, or in the region where the valence band and conduction band overlap. In this case it takes effectively no energy for an electron to jump above the Fermi energy. In semiconductors the Fermi energy occurs in the bandgap, and thus sufficient energy to jump over the bandgap is needed for an electron to become a conductor.

Ferrite A phase of plain carbon steel. At the eutectoid temperature of 727° C ferrite has a composition of 0.022 wt% carbon.

Fick's first law An equation that describes diffusion under steady-state conditions.

Fick's second law An equation that describes diffusion under nonsteady-state conditions.

Fourier's law An equation that describes thermal conduction.

Fracture toughness Generically, it is the resistance of a material to fracture. There are three specific ways it can be defined: the area under the stress–strain curve; the strain energy release rate, G_c; and the critical stress intensity parameter, K_{Ic}.

Frenkel defect A defect in an ionic crystal consisting of a vacancy and an interstitial impurity of the same type as was removed to create the vacancy.

Galvanic cell An electrochemical cell consisting of a metal anode and metal cathode immersed in ionic solutions, connected through an electrical circuit and a salt bridge. The resulting redox reaction generates a voltage and an electrical current in the circuit.

Galvanic corrosion Corrosion that occurs as a result of two different metals being in contact with each other.

Galvanic series A list of metals in order from most easily reduced to most easily oxidized when placed into seawater.

Glass A material in which the atoms are arranged randomly with no long-range order.

Glass transition The temperature at which a glass transforms from a solid to a liquid upon heating.

Grain A small region of a crystal that has a uniform orientation of its lattice.

Grain boundary The boundary between crystalline grains in a material. At the grain boundary, the crystalline order is disrupted.

Grain boundary diffusion Diffusion that occurs through the grain boundaries of a material.

Growth The stage of microstructural transformations, such as solidification, in which the previously formed phase grows larger in size.

Hardness A measure of the resistance of a material to having an indentation made in its surface. There are several different ways to measure hardness and various hardness scales: Brinell, Vickers, Knoop, and Rockwell. Hardness can be correlated to strength.

Heat capacity A constant that relates the amount of energy absorbed by a material and its rise in temperature. It is an extensive property that depends on the mass of the material.

Hole A charge carrier that is represented as an electron removed from the valence band. Holes move opposite to electrons when an electric field is applied.

Hooke's law The linear relationship between stress and strain in the elastic region.

Hydrogen bond A nonbonding interaction that involves interaction between a highly electronegative atom and a hydrogen atom that is bonded to a highly electronegative atom.

Induced dipole A nonbonding interaction that results from random fluctuations in the electron distribution in atoms, resulting in partial negative and positive charges.

Intergranular corrosion Corrosion that occurs preferentially at grain boundaries.

Interstitial diffusion Diffusion that occurs through motion of atoms through interstitial sites in a crystal.

Interstitial impurity An impurity atom that occupies an interstitial site in a crystal.

Interstitial site A space in a unit cell that is not occupied by the atoms defining the unit cell, and that can be occupied by another atom.

Ionic bond A bond formed when one or more electrons are transferred from one atom to another atom, resulting in a negative charge on one atom and a positive charge on the other. It forms between atoms that have a large difference in electronegativity and a large average electronegativity.

Isotactic A structure of polymers in which the side groups are all on the same side of the chain backbone.

Isothermal transformation diagram A diagram that shows the microstructures that form when a material is held at a constant temperature.

Kinetics The rate at which a process occurs.

Lamellae Alternating layers of material. Eutetic solid consists of lamellae.

Lattice parameter The length of the edge of a unit cell. It is given the symbol ***a***.

Lever rule The technique used to determine the amount of each phase in a mixture. Mathematically it is given in Equation (6.2.1).

Linear coefficient of thermal expansion A constant that relates change in temperature to how much a material will expand in a single direction. It may have different values in different directions for the same material.

Linear density The fraction of a direction in a crystal that is occupied by atoms.

Linear polymer A polymer that has no branches or crosslinks.

Line compound A phase that has a specific composition. It appears as a vertical line on a phase diagram.

Longitudinal modulus The modulus of a fiber-reinforced composite in the direction parallel to the fibers.

Martensite A phase of plain carbon steel that has a body-centered tetragonal unit cell. It is formed by a diffusionless transformation from the FCC structure of austenite.

Materials index A ratio of materials properties that defines performance for a given application. It can be used in combination with an Ashby plot for materials selection.

Matrix The continuous component of a composite. The matrix can be thought of as the glue that holds the reinforcement together.

Melting temperature The temperature at which a crystal transforms from a solid to a liquid upon heating.

Metallic bond The bond formed when atoms are shared equally among atoms, forming a sea of electrons. It forms between atoms that have a small difference in electronegativity and a small average electronegativity.

Microconstituent A specific part of a microstructure.

Microstructure The size and arrangement of phases in a material.

Miller indices A set of three integers that define directions and planes in crystals. Individual directions have the notation [*hkl*]; families of directions the notation <*hkl*>; individual planes the notation (*khl*); and families of planes the notation {*hkl*}.

Mixed dislocation A dislocation with partial edge and partial screw character.

Modulus The constant that defines the relationship between stress and strain in the elastic region. It represents the stiffness of the material.

Molar heat capacity A constant that relates the amount of energy absorbed per mole of a material and its rise in temperature.

MOSFET Metal-oxide-semiconductor field effect-transistor. It acts as a valve, controlling whether or not electricity flows.

Network former An additive for glass that forms tetrahedra at the atomic level, and is added to change the properties.

Network modifier An additive for glass that breaks up the silica tetrahedra and lowers the softening point.

Network polymer Another name for a crosslinked polymer.

Nonbonding interaction Interactions that occur due to attraction between partial charges on different molecules.

Nonsteady-state diffusion Diffusion that occurs with a flux that changes with time. The concentration gradient and the concentration at any given point in the material also change with time.

***n*-type semiconductor** A semiconductor to which atoms with extra electrons have been added.

Nucleation The stage of microstructural transformations, such as solidification, in which a small particle of the new phase spontaneously forms.

Nuclei The initial small particles of a phase that spontaneously form during a microstructural transformation.

Number average molecular weight An average molecular weight of a polymer which is weighted by the number of moles of polymer molecules of each molecular weight that are present.

Oxidation reaction A reaction in which an atom loses electrons.

Oxidizing agent A material which is more easily reduced, and therefore causes oxidation in another material.

Pearlite The eutectic microstructure of plain carbon steel, consisting of lamellae of ferrite and cementite.

Peritectic reaction A microstructural transformation on cooling in which a two-phase mixture of liquid plus solid transforms to a different solid phase.

Permanent dipole A nonbonding interaction that results from the dipole of a polar bond interacting with another dipole. The permanent partial and negative charges of the dipole are attracted to each other.

Phase A region of a material with uniform composition and structure.

Phase amount The amounts of a phase present in a mixture.

Phase composition The amounts of the components in a particular phase.

Phase diagram A map showing the phases present as a function of composition and temperature for a given set of components.

Phase transformations The process of one phase changing into one or more different phases.

Phonon Atomic vibrations that transmit thermal energy. In an insulator, phonons are the dominant mechanism of thermal conductivity.

Pilling–Bedworth ratio The ratio of the volume of an oxide to the volume of the metal used to create that oxide. The value of this ratio determines whether or not the oxide layer on a metal surface will protect it from corrosion.

Planar density The fraction of a plane in a crystal that is occupied by atoms.

Plane of highest density The plane that has the highest planar density for a particular crystal structure. The planes of highest density are for SC, {100}; for BCC, {110}; and for FCC {111}.

Plastic There are two definitions. One is a common term for a rigid or semirigid polymer. A second definition is the range of stresses over which a material can be deformed and not completely recover. On a stress-strain curve the plastic region is the nonlinear portion of the curve at higher values of strain.

Polycrystal A crystalline material made up of many grains, each of which has a different orientation.

Polydispersity index The ratio of weight average molecular to number average molecular weight of a polymer. It represents how disperse the range of molecular weights in the polymer is.

Primary bond The sharing of electrons between atoms to satisfy the octet rule.

Primary solid In a eutectic phase diagram, the solid formed above the eutectic temperature upon cooling a liquid.

***p*-type semiconductor** A semiconductor to which atoms with fewer electrons have been added.

Ranking procedure A procedure for selecting materials based upon ranking the relevant properties, multiplying those ranks by weighting factors, and adding those products to give a final score for each material.

Redox reaction The combination of an oxidation and a reduction reaction which represents the overall reaction in a system. It does not contain any electrons.

Reducing agent A material which is more easily oxidized, and therefore causes reduction in another material.

Reduction reaction A reaction in which an atom loses electrons.

Reinforcement The discontinuous component of a composite. Typical reinforcements are particles or fibers.

Repeat unit The unit of a polymer that repeats to make up the entire structure.

Resistance The extent to which a piece of a material resists conducting electricity. It is an extrinsic property that depends on the resistivity and the dimensions.

Resistivity An intrinsic property of a material that indicates how much it resists conducting electricity. The resistivity is the inverse of the conductivity.

Scattering angle The angle at which diffraction occurs for a given plane of atoms in a crystal. It is generally presented as 2θ, where θ is the angle used in Bragg's law.

Schottky defect A defect in an ionic crystal consisting of a cation vacancy and an anion vacancy.

Screw dislocation A dislocation that can be pictured as formed by twisting one half of a crystal relative to the other half. The dislocation is the line of atoms at the axis of the twist.

Semicrystalline A material that contains both amorphous and crystalline phases. Many polymers are semicrystalline.

Short-circuit diffusion Another name for grain boundary diffusion.

Side-group A chemical group in a polymer that is part of the repeat unit.

Simple cubic A cubic unit cell consisting of one atom on each corner.

Single crystal A crystalline material with a perfect lattice. It does not have any grain boundaries.

Slip Motion of a dislocation to cause plastic deformation.

Solubility limit The maximum amount of a component that can be incorporated into another to form a single phase. If an additional component is added, the system will become two-phase.

Solubility parameter A number that can be used to determine whether a solvent will dissolve a solute. It is calculated as the square root of the energy of vaporization per molar volume.

Specific heat A constant that relates the amount of energy absorbed per gram of a material and its rise in temperature.

Spheroidite A microstructure of steel consisting of spheres of cementite in ferrite. Spheroidite can be formed by heating pearlite or bainite to 700° C and holding for 24 hours.

Stable crack growth Growth of a crack that occurs only when the stress is increased. In the stable region, if the stress is constant the crack will not get longer.

Standard reduction potential The voltage generated by a metal when it is one of the electrodes in a standard electrochemical cell, defined as operating at 25 C with all gases at 1 atm of pressure and all solutions at a 1 M concentration. It is given as the voltage for the metal as the cathode.

Steady-state diffusion Diffusion that occurs with a flux that is constant with time. The concentration gradient and the concentration at any given point in the material are also constant with time.

Strain The change in dimensions of a material when a stress is applied, divided by the initial dimension. The specific dimensions used depend on the type of strain being calculated (tensile, shear, etc.).

Stress The force applied to a material divided by the area over which the force is applied.

Stress corrosion Corrosion that occurs in combination with stress. The presence of the stress can cause the corrosion to occur faster than it would without the stress.

Substitutional impurity An impurity atom that takes the place of the host atom in a material.

Syndiotactic A structure of polymers in which the side groups are on alternating sides of the chain backbone.

Tacticity The arrangement of the side groups of polymers relative to the chain backbone. A polymer of one type of tacticity cannot be converted another type. Conversion would require breaking covalent bonds.

Tempered martensite A microstructure of steel consisting of very small spheres of cementite in ferrite. Tempered martensite can be formed by heating martensite to 700° C and holding for 1 hour. It has the best balance of strength and ductility for steel.

Thermal conductivity A constant relating the thermal flux to the temperature gradient.

Tieline A line in a two-phase region of a phase diagram that is horizontal to the composition axis, and extends between the boundaries with the one-phase regions on either side. The tieline is used to determine the compositions and amounts of the phases in a two-phase region.

Transverse modulus The modulus of a fiber-reinforced composite in the direction perpendicular to the fibers.

Ultimate tensile strength The maximum stress that a material can withstand under tensile loading.

Uniform corrosion Corrosion that occurs uniformly over the surface of a material. The thickness of the material decreases at a steady rate, given by the corrosion penetration rate.

Unit cell A piece of a crystal that repeats throughout the entire material.

Unstable crack growth Growth of a crack that occurs spontaneously. In the unstable region the crack will get longer even if the stress is constant, usually at ultrasonic speeds.

Vacancy A location in a crystal that is missing an atom that would normally be part of the crystal structure.

Vacancy diffusion Diffusion that occurs through motion of atoms through vacancies in a crystal.

Valence band The energy band that contains the valence electrons.

Viscoelasticity Behavior of a material that is partially elastic and partially viscous. When a constant stress is applied to a viscoelastic material, it will increase in length at a decreasing rate.

Viscous A material that behaves as an ideal fluid. When a constant stress is applied to a viscous material, it will increase in length at a constant rate.

Volumetric coefficient of thermal expansion A constant that relates change in temperature to how much a material will increase in volume.

Weight average molecular weight An average molecular weight of a polymer which is weighted by the mass of polymer molecules of each molecular weight that are present.

Weighting factor A factor that is applied to the ranking of different properties for materials selection. The values of the weighting factors are chosen based on the relative importance of the properties for the design.

Yield strain There are several ways to define yield strain. It is the strain at which plastic deformation occurs, and is ideally the point where the stress–strain curve becomes nonlinear. In practice it is difficult to identify this point, so it is defined as the 0.02% offset for metals, or the first maximum for polymers.

Yield stress There are several ways to define yield stress. It is the stress at which plastic deformation occurs, and is ideally the point where the stress–strain curve becomes nonlinear. In practice it is difficult to identify this point, so it is defined as the 0.02% offset for metals, or the first maximum for polymers. It is also the minimum stress applied in an arbitrary direction required for slip to occur.

Concept Check Answers

1.1.1 5 calories.
1.1.2 Yes.
2.1.1 Ceramic.
Polymer.
2.1.2 There are multiple possible answers, such as light weight (low density), fracture resistance, corrosion resistance, high strength.
2.1.3 All materials (ceramics, metals, polymers) are electronic materials.
Metals can be used for multiple functions: structural materials, biomaterials, and electronic materials.
2.2.1 It will be the same as after experiment 2; it will break easily when bent.
3.1.1 K-Br.
Br-Br.
3.2.1 Ionic.
Covalent.
3.2.2 Covalent.
3.3.1 Covalent primary bonds; hydrogen bonds for non-bonding interactions.
3.3.2 The primary bonds do not change; the non-bonding interactions are broken.
4.1.1 It would not change.
4.2.1 1.
3.
4.2.2 0.52.
4.2.3 $a = 2R\sqrt{2}$.
$a = \frac{4R}{\sqrt{3}}$.
4.3.1 $[\bar{1}\bar{4}3]$.

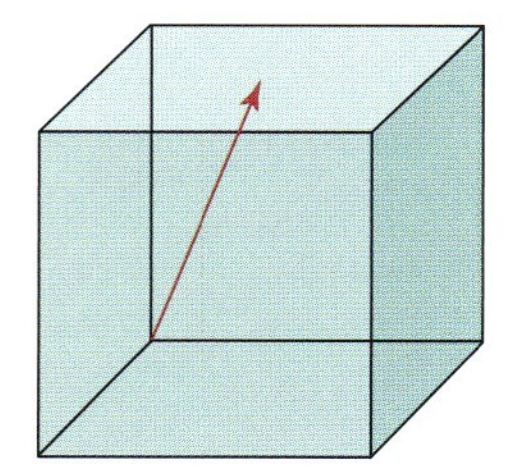

4.3.2 $(\bar{4}03)$.

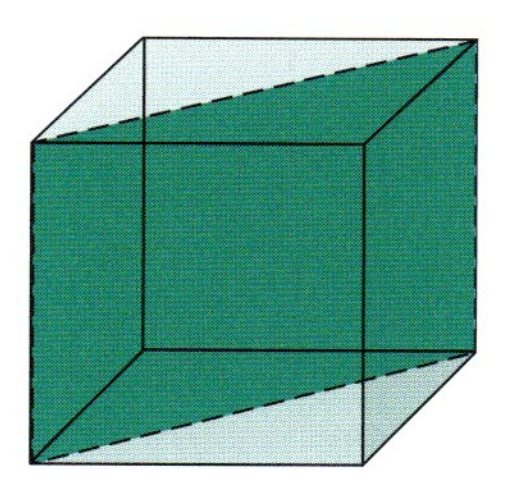

4.3.3 [hkl] is the normal to (hkl).
4.4.1 0.71.
1.0.
4.4.2 0.56.
4.4.3 The entire family consists of [110], [101], [011], $[\bar{1}10]$, $[1\bar{1}0]$, $[\bar{1}\bar{1}0]$, $[\bar{1}01]$, $[10\bar{1}]$, $[\bar{1}0\bar{1}]$, $[0\bar{1}1]$, $[01\bar{1}]$, $[0\bar{1}\bar{1}]$.
There are none.
4.5.1 Substitutional.
4.5.2 At absolute zero.
4.5.3 <111>.
0.252 nm.
4.6.1 Chlorine.
4.6.2 FCC unit cell of sulfur ions with zinc ions in half the tetrahedral sites.
4.7.1 Create two chlorine vacancies.
4.7.2 One Fe^{2+} vacancy for every two Fe^{3+} ions that form.
Ca^{2+} vacancy.
4.8.1 There is no scattering.
5.1.1 The repeat unit molecular weights are different.
5.1.2 832.
Unitless.
5.1.3 It is a network or crosslinked polymer.
5.1.4 Isotactic.
5.2.1 202,500 g/mol.
280,000 g/mol.
5.2.2 200,000 g/mol.
100,000 g/mol.

5.3.1 Polyethylene.
Polyethylene with short branches.
Isotactic polypropylene.
Polypropylene.
Polyethylene.
PET held at 150° C for 1 hour.

5.4.1 Isotactic polypropylene.
Polypropylene.
Polystyrene with Mn = 5000 g/mol.
PVC mixed with solvent.

6.1.1 It is different because in nail polish remover the two liquids are intimately mixed.

6.1.2 1.
2.
2.
2.

6.1.3 Heat it up to a temperature above approximately 40° C, or add water so the composition is less than approximately 63 wt% sugar.

6.1.4 A phase that is 100% aluminum and a phase that is 50% aluminum/50% copper.
90% pure aluminum and 10% aluminum/copper mixture.

6.2.1 Liquid that is 20 wt% Ni and 80 wt% Cu.
Cannot determine.
100% liquid.
Cannot determine.

6.2.2 Liquid that is 25 wt% Ni and 75 wt% Cu; solid that is 37 wt% Ni and 63 wt% Cu. Note that these are approximations from the phase diagram so your numbers might be slightly different.
17 wt% liquid and 83 wt% solid. Note that these are approximations from the phase diagram so your numbers might be slightly different.

6.3.1 There would be 5 grains, and they would be smaller.

6.4.1 No.

6.4.2 At I the large solid particles are α. At J the large solid particles are β.

6.5.1 73 wt%.
12 wt%.
15 wt%.

6.6.1 A eutectic reaction is one liquid transforming to two different solids. A peritectic reaction is a solid plus a liquid transforming to a different solid.

6.7.1. It is a range of compositions on the phase diagram, not a single exact composition.

7.1.1 Right to left.
No diffusion will occur.
Increase.

7.1.2 Decrease.

7.1.3 Interstitial.

7.2.1 Column A.

7.2.2 3×10^{-11} m^2/s.

7.2.3 110.4 hours.
Fick's first law.
Full Fick's second law.

7.2.4 873 K.
566° C.

8.1.1 Two spheres.

8.1.2 It will disappear.

8.1.3 Medium temperatures; not very high and not very low.
High temperatures (but below the melting temperature).

8.2.1 90°.

8.2.2 Heterogeneous nucleation.

8.3.1 Heat up the material to allow diffusion to occur.

8.4.1 50% fine pearlite, 25% bainite, 25% martensite.
100% fine pearlite.
50% bainite, 50% tempered martensite.

8.5.1 Cannot be made with continuous cooling.
Cannot be made with continuous cooling.
Cool at a rate between 35° C/s and 140° C/s.

9.1.1 690 kN.

9.2.1 The red curve.

9.2.2 PVC < copper < potassium bromide.

9.3.1 Without a dislocation.

9.3.2 [111].
(110).

9.3.3 Stronger.

9.3.4 73.5 MPa.

9.4.1 Aluminum with 10 wt% copper.

9.4.2 Copper with 10 μm grains.

9.4.3 Copper that has been plastically deformed.

9.4.4 Unannealed copper.

9.5.1 Fine pearlite > coarse pearlite > spheroidite.

9.6.1 It has a higher initial slope.
It has a higher strain at failure.

9.6.2 Opaque and flexible.
Opaque and rigid.
HDPE is more rigid than LDPE.

9.7.1 They all have the same viscosity at the softening point.
Fused silica.

9.7.2

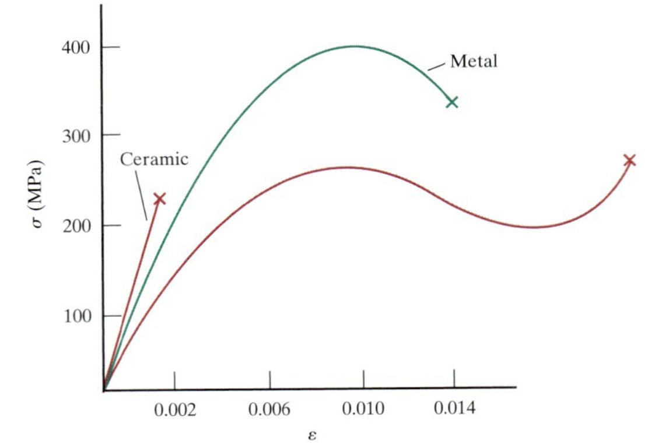

9.8.1 Increasing.
Decreasing.

9.8.2 The one with 1.0 mm cracks.
0.58 mm.

9.9.1 125 MPa at a mean stress of 100 MPa; 290 MPa at a mean stress of 0 MPa.
50 MPa.

9.10.1 Aluminum alloy 2117-T4.

9.11.1 The strain is reduced by half.

9.11.2

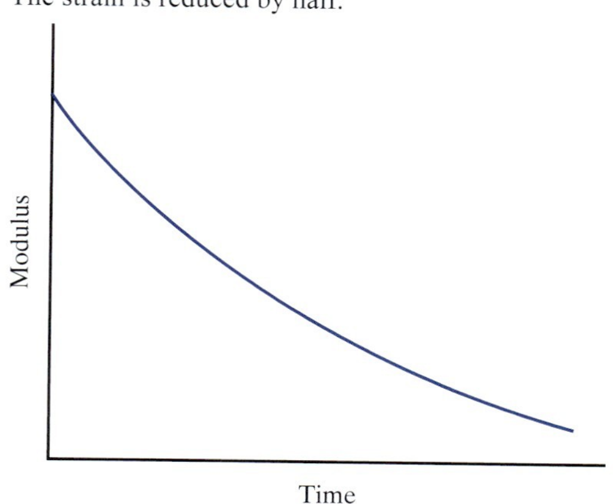

9.11.3 Below T_g.

9.12.1 Uniaxial continuous = cross-ply continuous > random in 3 dimensions discontinuous.
0.76.

10.1.1 Ni^{2+}.
Aluminum.
$Al \rightarrow Al^{3+} + 3e^-$.
$2Al + 3Ni^{2+} \rightarrow 2Al^{3+} + 3Ni$.

10.1.2 $Ni + 2Cu^+ \rightarrow Ni^{2+} + 2Cu$.

10.1.3 Titanium.
Titanium.
1.23 V.

10.1.4 Brass.

10.2.1 0.05 mm/year.

10.2.2 Corrosion rate decreases.
Aluminum.

10.2.3 It can occur anytime there is a gap between two pieces of metal.

10.2.4 Any metals lower than iron on the table of standard reduction potentials or the galvanic series (depending on which list is more applicable to the environment). Some examples are aluminum, zinc, and magnesium.

10.3.1. P–B should be close to 1.0. In actual practice, the oxide layer will be protective if P–B is between 1 and 2.

10.3.2 0.65.

10.4.1 The high molding temperature causes degradation, decreasing its strength because of the change in molecular weight.

10.4.2 Acetone.

11.1.1 Yes.
No.

11.2.1 6 mm.

11.2.2 Holes and electrons.

11.3.1 24k gold.
Copper solidified at 800° C.

11.4.1

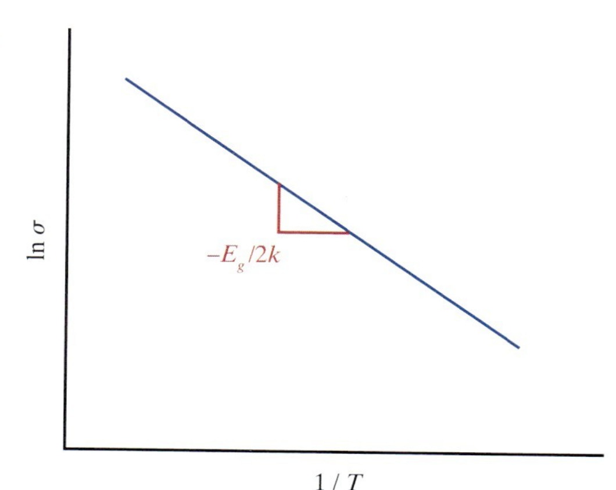

11.4.2 80° C.
Silicon with boron impurities.

11.4.3

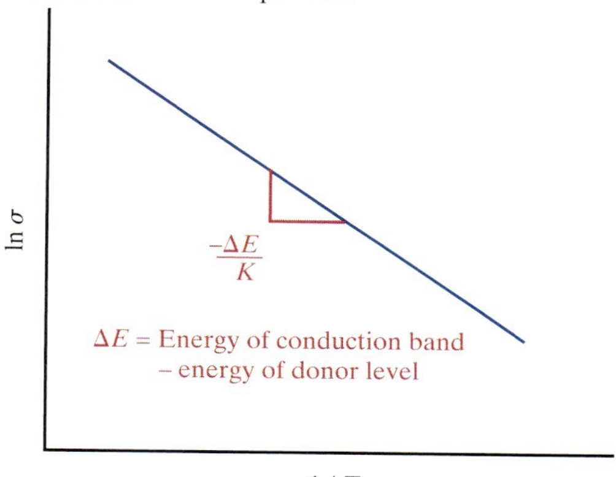

11.5.1. Yes.
No.

12.1.1 Iron at 100° C.

12.1.2 25 J/mol-K.

12.1.3 43° C.

12.2.1 Increases.
Would not expand at all.

12.2.2 Sodium chloride < zinc < polyethylene.

12.2.3 12 μm.
Polyethylene has weaker nonbonding interactions; induced dipoles for polyethylene versus permanent dipoles for PVC.

12.3.1 Phonons.
Steel.

12.3.2 24k gold.
Copper solidified at 800° C.

12.3.3 65° C.

13.1.1 There is no single, definitive answer. It depends on how you justify the relative importance of the different properties.

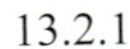

13.1.2 1040 steel.

13.2.1

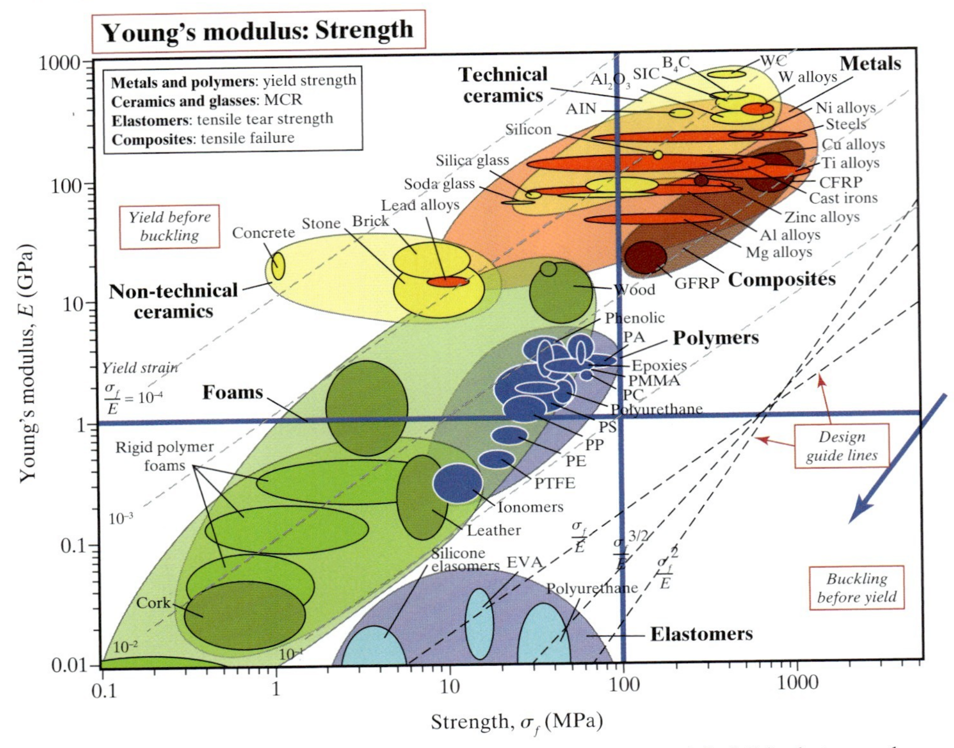

Select any material in the area indicated by the arrow. Since no materials fall in that area, there are no materials that will meet the requirements.

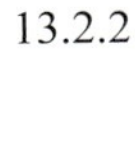

13.2.2

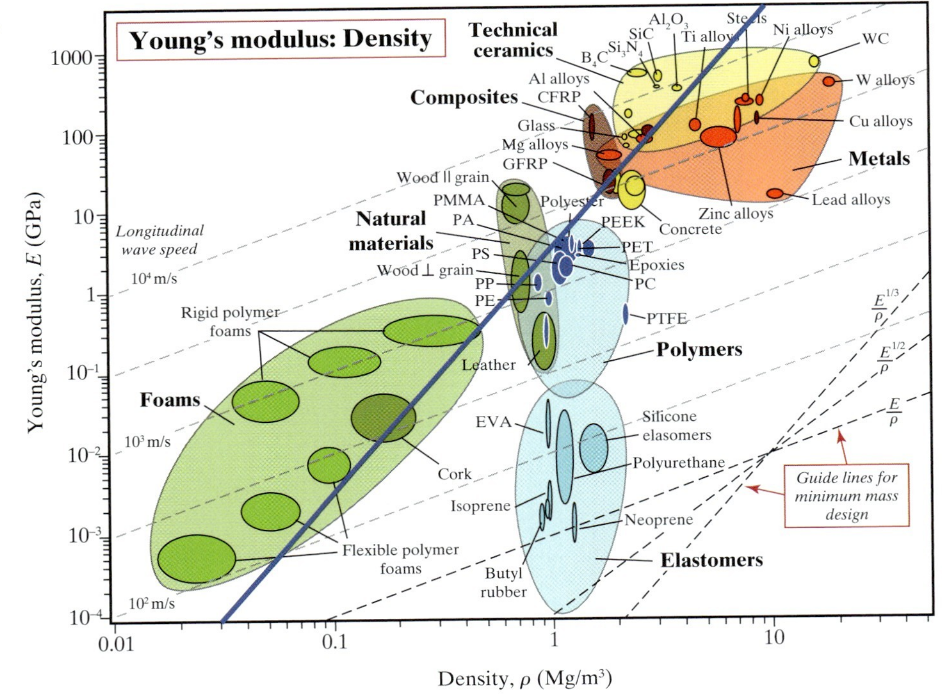

Select any material that falls on the line.

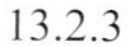

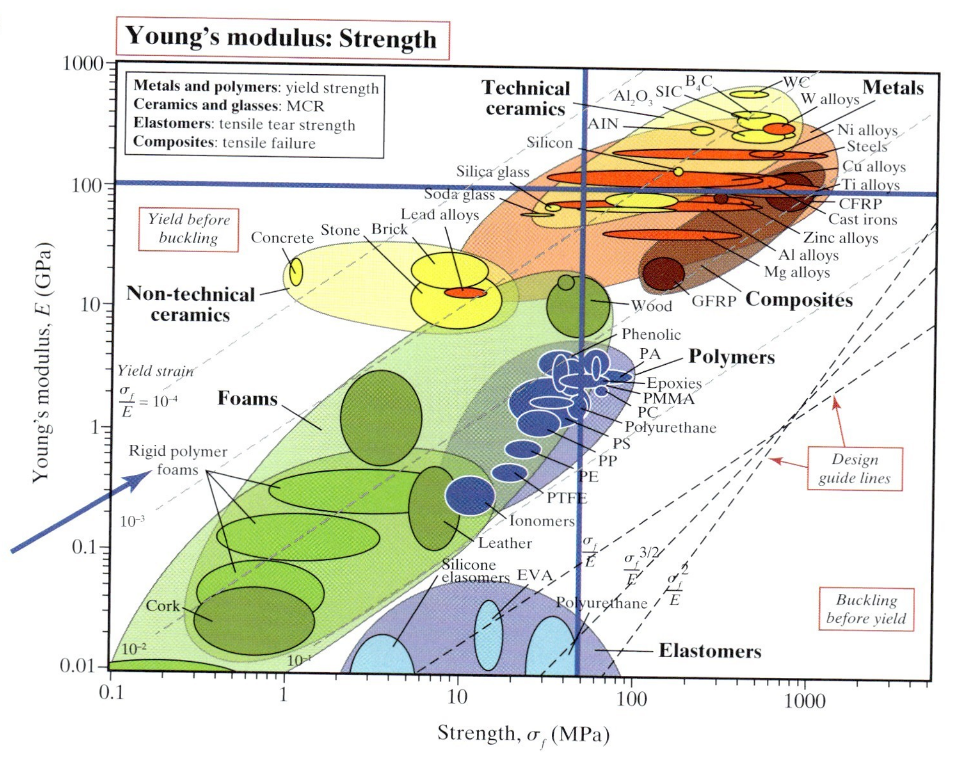

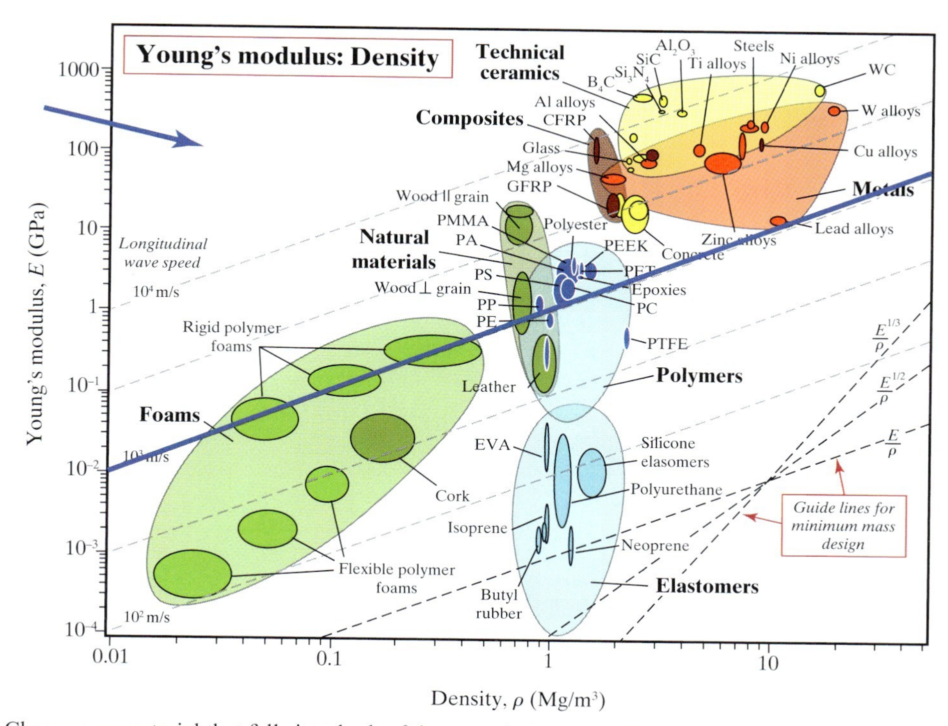

Choose any material that falls into both of the areas indicated by the arrows.

Index

C

D

E

F

G

H

I

K

L

M

N

O

P

Q

R

S

T